A Handbook of Horticultural Sciences

Volume 3

NIPA® GENX ELECTRONIC RESOURCES & SOLUTIONS P. LTD.
New Delhi-110 034

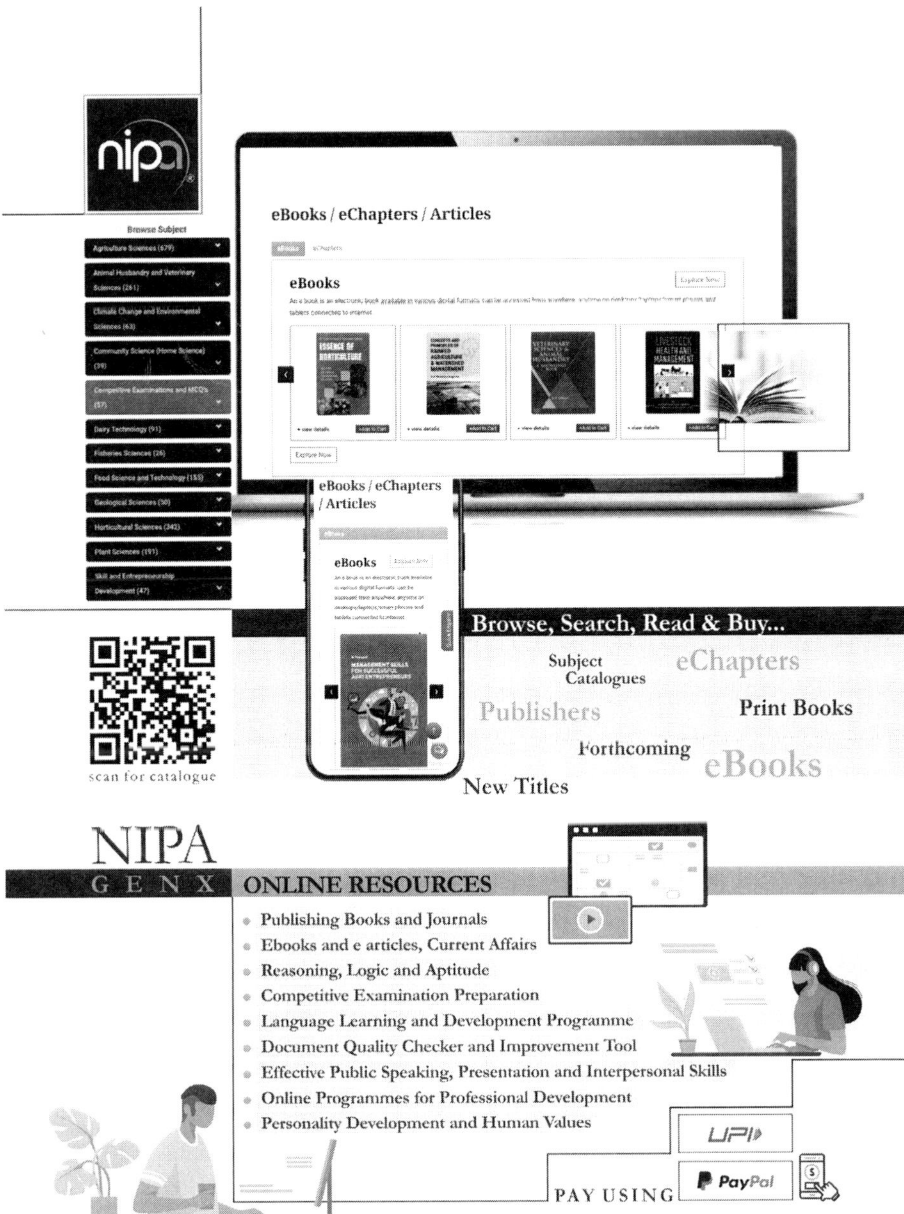
nipa
Browse Subject
Agriculture Sciences (679)
Animal Husbandry and Veterinary Sciences (261)
Climate Change and Environmental Sciences (63)
Community Science (Home Science) (39)
Competitive Examinations and MCQ's (57)
Dairy Technology (91)
Fisheries Sciences (26)
Food Science and Technology (155)
Geological Sciences (30)
Horticultural Sciences (342)
Plant Sciences (191)
Skill and Entrepreneurship Development (47)
eBooks / eChapters / Articles
eBooks
Explore Now
scan for catalogue
Browse, Search, Read & Buy...
Subject Catalogues
eChapters
Publishers
Print Books
Forthcoming
eBooks
New Titles
NIPA
GENX
ONLINE RESOURCES
Publishing Books and Journals
Ebooks and e articles, Current Affairs
Reasoning, Logic and Aptitude
Competitive Examination Preparation
Language Learning and Development Programme
Document Quality Checker and Improvement Tool
Effective Public Speaking, Presentation and Interpersonal Skills
Online Programmes for Professional Development
Personality Development and Human Values
PAY USING
UPI
PayPal

About the Authors

Dr. K. Vanangamudi, Former Dean (Agriculture), graduated B.Sc., (Ag.) in 1975 from Agricultural College and Research Institute, Tamil Nadu Agricultural University, Coimbatore; M.Sc., (Ag.) and Ph.D. in Seed Science and Technology, respectively in 1977 and 1981 from TNAU; and Post Doctoral Fellow (1988-1990) at Mississippi State University, USA.

He joined Department of Seed Science and Technology, TNAU as Assistant Professor in 1982 and served the University for 32 years as Associate Professor (1986-1988) at Agricultural Research Station, Bhavanisagar; Post-Doctoral Fellow (1988-1990), Mississippi State University, USA; Professor (1992-2001) at Forest College & Research Institute, Mettupalayam; Professor and Head (2001-2003) of Department of Seed Science and Technology, TNAU, Coimbatore; Dean, Adhiparasakthi Agricultural College (2003-2006), Kalavai; and Dean (Agriculture), Agricultural College and Research Institute, TNAU, Coimbatore (2006-2010). He retired from service on August 31, 2013.

He has undergone Post Doctoral Programme in Tree Seed Technology at Mississippi State University, USA from 1988-1990 and visited USA, Japan and Canada in connection with Forestry education and research under USAID-Winrock International program. He underwent training on e-learning and web based teaching at UC, Davis, California and Cornell University, Ithaca, New York, USA during 2007.

Dr. Vanangamudi's significant contributions in Agricultural Education are:

- Started B.Tech (Bioinformatics), B. Tech (Agricultural Information Technology) and BS (Agribusiness Management) degrees at AC&RI, Coimbatore,
- Started B.Sc (Horti.) degree at Adhiparasakthi Agricultural College, Kalavai
- He introduced e-education and online examination
- Under NAIP project on "Development of e-courses for B.Sc., (Ag) degree program as Co.PI, he developed e-learning materials for all the 50 courses and hosted in TNAU and ICAR websites
- Video streaming of lecture presentation by the teachers in the classroom (Video filming, editing and publishing).
- Prepared course materials for 90 courses for Industrial Training Institute (ITI) program in Agriculture, Horticulture, Agricultural Engineering, Animal Husbandry, Poultry and Food Processing under "Modular Employment Skills" and "Centre of Excellence" funded by National Instructional Media Institute, Ministry of Labour & Employment, GOI, Chennai, for the students and instructors of ITI all over India.

As a researcher, he was the Principal Investigator of 11 research projects funded by ICAR-NAIP, ICAR-World bank, ICFRE-World bank, Tamil Nadu Forest department, Ministry of Labour and Employment (GOI), CIDA-Gulpeh, Canada and Private agencies.

Dr. Vanangamudi has published 2 review papers, 35 research papers in International journals, 112 research papers in National journals, 29 popular articles in English and 53 in Tamil. He has written 46 books in English and 10 in Tamil. He has also written 11 technical bulletins in English and 29 in Tamil.

His recent books are:

- A Handbook of Agricultural Sciences Vol 1 & 2.
- Objective Seed Science and Technology – 2nd Edition, Revised & Enlarged
- MCQ's in Plant Breeding, Biotechnology & Seed Science
- Model test paper of Seed Science and Technology
- Seed Science and Technology: 2nd Enlarged and Fully Revised Edition
- Competitive Seed Science and Technology
- Competitive Agriculture

In recognition of his contribution in Agricultural Education, Research and Extension, he was awarded with

- Best PG teacher award in 1998 by TNAU
- Best scientist award in 1999 by TNAU
- Best PG teacher award 2000 by Madras Agricultural Students Union (MASU), Coimbatore
- Best Seed scientist award in 2003 by Hisar Agricultural University (HAU), Hisar
- Best Research paper award in 2005 by Indian Society of Seed Technology (ISST), New Delhi
- Tamil Nadu scientist award in 2006 by Tamil Nadu State Council for Science and Technology (TNSCST), Chennai
- Bharat Ratna Dr. C. Subramaniam best teacher award in 2006 by Indian Council of Agricultural Research (ICAR), New Delhi.

He is a recognized guide of Bharathiar University, Coimbatore; Bharathidasan University, Trichy and Forest Research Institute, Dehradun.

He was a Research and Technical Consultant to Little's Orientals Balm and Pharmaceuticals Ltd., Chennai for seed priming, seed film coating and setting up of Seed Research Lab; to a farmer for 200 acres of farm land development; to a EU-KKID-Peace trust project for end term evaluation; and to a Kerala Forest Department project (World Bank) for developing training course syllabi and establishment of seed processing and testing units. He served as Senior Advisor and Consultant to Dr Mohan's Health Care Products Ltd., Chennai.

Dr. N. Chezhiyan, worked as a Professor and Head, Department of Spices and Plantation Crops, Horticultural College and Research Institute, Tamil Nadu Agricultural University, Coimbatore; Professor and Head, Horticultural Reasearch Station, Thadiyankudisai and Professor and Head of Vegetable Crops, Adhiparasakthi Agricultural College, Kalavai. He graduated B.Sc., (Agri.) in 1972 and Ph.D (Floriculture) in 1991 from TNAU, Coimbatore; M.Sc., (Horticulture) from Allahabad Agricultural Institute, Allahabad.

Dr. N. Chezhiyan served in TNAU in various capacities for 30 years since 1979 to 2008. He also served as an Agricultural officer in the Department of Agriculture, Government of Tamil Nadu from 1972 to 1976.

He has rich experience in teaching, research and extension; and guided 18 M.Sc., (Horticulture) students and 16 Ph. D scholars in Horticulture. Two of his Ph.D scholars were awarded with ICAR Jawaharlal Nehru Award for Best Thesis in Pomology. He released 13 varieties in fruits, vegetables, spices and flower crops. He has published 175 research papers in International and National journals, 10 books and 150 popular articles.

Dr. Chezhiyan attended 20 trainings and 20 workshops/seminars; and organized 35 trainings and 20 workshops/seminars. He served as a member of Academic Council and Board of Studies (Horticulture), TNAU, Coimbatore.

Ms. M. Kokila is now persuing PG program in Seed Science & Technology. She worked as a Faculty, Trinity Cultural Academy, Coimbatore. She graduated B.Sc., (Agri.) in 2019 from Agricultural College and Research Institute (TNAU), Kudumiyanmalai, Pudukkottai, Tamil Nadu. She has published 4 books related to Agricultural sciences as follows:

- A Handbook of Agricultural Sciences Vol 1 & 2.
- Model test paper of Seed Science and Technology
- Competitive Agriculture

She has participated in Personality Development Workshop held on 21st & 22nd January 2019 in AC & RI, Kudumiyanmalai. She has also undergone ten days internship with NGO (Vrutti). She also served as a National Service Scheme volunteer (2015-17) and committee leader in various college functions (2018-2019).

Dr. M. Prabhu is currently working as an Associate Professor in Horticulture at Horticultural College and Research Institute for Women, Tiruchirappalli,Tamil Nadu Agricultural University. He has 17 years of experience in teaching, research and extension activities related to Horticultural Sciences.

He worked as a researcher in Fruit Science, Vegetable Science, Ornamental Horticulture, Medicinal plants, Agro-forestry and Palmyrah. He handled 10 research and development projects related to Horticultural Sciences. He is one of the contributors for the development of brinjal hybrid COBH 2, which is moderately tolerant to shoot and fruit borer infestation.

He handled 75 courses for undergraduate, 10 courses for postgraduate degree programs, and 5 courses for open and distance learning programs. He was recognized as question paper setter and External Examiner for TNAU, Annamalai University, Kerala Agricultural University, University of Horticultural Sciences (Bagalkot) and Pondicherry Central University.

He has published 9 books with ISBN, 5 text books for Open and Distance Learning Programs of TNAU, Coimbatore, 65 research articles in reputed scientific journals, 105 popular articles in popular magazines and newspapers, 45 chapters in reputed books with ISBN and 60 abstracts in the proceedings of seminars and conferences.

He was associated in the conduct of International training programme on Agro-forestry for the experts from Asia and African countries funded by World Agroforestry Regional Centre, New Delhi as co-organizer. He has participated and presented papers in 47 national and 10 international conferences and 30 webinars related to horticulture, medicinal plants and life sciences. He attended 10 training programmes related to horticulture, life sciences and education technologies. He served as organizing committee member for various conferences, workshops and training programmes organized by Tamil Nadu Agricultural University.

He is a recipient of the Junior Scientist Award (2006), Certificate of Excellence Award (2008), Best poster presentation award (2011), Best Tamil Book Award (2012), Best presentation award (2013 & 2015), Best Outstanding Scientist National Award (2020) for his research contributions.

At present, he is actively involved in the Farmers Participatory Contractual Seed Production Programme and ICAR All India Coordinated Research Project on Vegetable Crops

A Handbook of Horticultural Sciences

Volume 3

(Spices, Plantation, Medicinal & Aromatic Crops and Post Harvest Management)

Authors:

K. Vanangamudi

N. Chezhiyan

M. Kokila

M. Prabhu

NIPA® GENX ELECTRONIC RESOURCES & SOLUTIONS P. LTD.

New Delhi-110 034

NIPA® GENX ELECTRONIC
RESOURCES & SOLUTIONS P. LTD.

101,103, Vikas Surya Plaza, CU Block
L.S.C. Market, Pitam Pura, New Delhi-110 034
Ph : +91 11 27341616, 27341717, 27341718
E-mail: newindiapublishingagency@gmail.com
www: www.nipabooks.com

For customer assistance, please contact
Phone: + 91-11-27 34 17 17
Fax: + 91-11- 27 34 16 16
E-Mail: feedbacks@nipabooks.com

ISBN: 978-93-95763-21-9

Composed and Designed by NIPA®.

Preface

Spices and condiments are natural plant products or mixtures used in whole or ground form as food adjuncts, mainly for imparting flavour, aroma and pungency to foods. It is also used for seasoning of foods and flavouring of beverages, liquors, and pharmaceutical, cosmetic and perfumery products. From time immemorial, India has been known as the 'Land of Spices'.

Most of the plantation crops are export oriented (Cashew nut, beetle vine, areca nut, tea, coffee etc.). Plantation crops earn foreign exchange for the country and they occupy 75% of the total earnings from the export of the agricultural produce. They support many of the ancillary industries and rural cottage industries.

Medicinal and aromatic plants constitute a major segment of the flora, which provides raw materials for use in the pharmaceuticals, cosmetics, and drug industries. The indigenous systems of medicines (Ayurveda, Siddha, Unani), developed in India for centuries, make use of many medicinal herbs.

Despite the COVID Pandemic, spices export from India has continued its upward trend during 2020-21 and has attained an all-time high of US $ 4.0 billion mark for the first time in the history of spices export. During 2021-22, the export of spices/spice products from the country has been 1.53 million tons valued Rs.30, 576 crores. Plantation crops occupy less than 2 per cent of the total cultivated area, but they generate an income of around Rs. 16,000 million or about 12.72 per cent of the total export earnings of all commodities. Medicinal and aromatic plants have a high market potential with the world demand of herbal products growing at the rate of 7 per cent per annum.

A considerable amount of fruits and vegetables produced in India is lost due to improper post-harvest operations and as a result, there is a considerable gap between the gross production and net availability. Furthermore, only a small fraction of fruits and vegetables are utilized for processing and export. Post harvest losses in fruits and vegetables are very high (20-40%). About 10-15% fresh fruits and vegetables shrivel and decay, lowering their market value and consumer acceptability.

Considering the economic importance of Spices, Condiments & Plantation Crops, Medicinal & Aromatic Crops and Post Harvest Management, 'A Handbook of Horticultural Sciences Volume 3': Spices, Plantation, Medicinal & Aromatic

Crops and Post Harvest Management is prepared. Each chapter with respect to crops deals with origin and distribution, economic importance, species, varieties, hybrids and its characteristics, climatic and soil requirements, propagation, propagating materials and planting, cultural practices, plant protection measures, physiological disorders, harvesting, post harvest handling and storage. Apart from the above information, separate unit on post harvest technology of horticultural crops is included which covers the different techniques of post harvest handling and management, and value addition of all kinds of horticultural crops. This book has been prepared for the benefit of scientists, students, extension personnel, progressive growers and entrepreneurs and in particular the horticulture and agriculture graduates who are appearing for competitive exams at national and state level.

The authors extend their sincere thanks to Mr. Sumit Pal Jain, NIPA Genx Electronic Resoruces & Solutions P.Ltd., New Delhi for publishing this book and his continued support.

K. Vanangamudi
& Co-authors

Contents

Section III: Production Technology of Plantation Crops

Section IV: Facts and Figures of Spices, Condiments and Plantation Crops

Unit IX - Production Technology of Medicinal and Aromatic Crops

Section I: Production Technology of Medicinal Plants

Section II: Production Technology of Aromatic Crops

Section III: Facts and Figures of Medicinal and Aromatic Crops

Unit X: Post-Harvest Technology of Horticultural Crops

Section I: Post-Harvest Management of Horticultural Crops

Section II: Value Addition in Horticultural Crops

Unit VIII
Production Technology of Spices, Condiments and Plantation Crops
Section I: Production Technology of Spices

1

Black Pepper (*Piper nigrum* L.), Piperaceae, 2n = 52

1.1. Origin and distribution

- Originated in Western Ghats and Eastern Himalayas.
- Cultivated in Kerala, Karnataka, Tamil Nadu, Maharashtra, North Eastern States, and Andaman and Nicobar Islands.
- It is the king of spices.

1.2. Economic importance

- Spiciness is due to the chemical compound piperine
- In foods, used as a spice.
- High in antioxidant and anti-inflammatory properties.
- May benefit for brain.
- May improve blood sugar control and lower cholesterol levels.
- May have cancer fighting properties.

1.3. Varieties

- Panniyur 1, 2, 3, 4, 5, 6, 7 and 8, Karimunda, Sreekara, Subhakara, Panchami, Pournami, IISR Thevam, IISR Malabar Excel, IISR Girimunda, IISR Sakthi, PLD-2, Vadakkan tropical cultivar.
- Lower elevation and less shady areas: Panniyur 1.
- Higher elevation and more shady areas: Karimunda.
- Inter cropping in arecanut: Panniyur 5.

1.3.1. Varieties and their characteristics

- **Panniyur 1:** Suited to all regions. Not suited to heavily shaded areas. Average yield is 1242 kg/ha. Oleoresin is 11.8%. Piperine is 5.3%. Essential oil is 3.5%.
- **Panniyur 2:** Shade-tolerant variety of black pepper. Average yield is 2570 kg/ha. Oleoresin is 10.9%. Piperine is 6.6%. Essential oil is 3.4%.

- **Panniyur 3:** Late-maturing variety of black pepper, suited to all pepper growing regions. Average yield is 1953 kg/ha. Oleoresin is 12.7%. Piperine is 5.2%. Essential oil is 3.1%.
- **Panniyur 4:** Performs well under a variety of conditions. Stable yielder. Average yield is 1277 kg/ha. Oleoresin is 9.2%. Piperine is 4.4%. Essential oil is 2.1%.
- **Panniyur 5:** Tolerant to nursery diseases and shade. Average yield is 1098 kg/ha. Oleoresin is 12.33%. Piperine is 5.5%. Essential oil is 3.8%.
- **Subhakara:** Suited to all pepper growing regions. Average yield is 2352 kg/ha. Oleoresin is 12.4%. Piperine is 3.4%. Essential oil is 6.0%.
- **PLD-2**: Selection from Kottanadan (Coll.2559), yield 2475 kg dry pepper/ha, potential yield 4731 kg dry pepper/ha. Piperine is 3.0%. Oleoresin is 15.45%. Essential oil is 4.8%.
- **Sreekara:** Suited to all pepper growing regions. Average yield is 2677 kg/ha. Oleoresin is 13.0%. Piperine is 5.1%. Essential oil is 7.0%.
- **Panchami:** Late-maturing, suited to all pepper growing regions. Average yield is 2828 kg/ha. Oleoresin is 12.5%. Piperine is 4.7%. Essential oil is 3.4%.
- **Pournami:** Tolerant to root-knot nematode. Average yield is 2333 kg/ha. Oleoresin is 13.8%. Piperine is 4.1%. Essential oil is 3.4%.
- **IISR-Girimunda**: Hybrid of Narayakodi x Neelamundi. Yield is 6.14 kg/vine. Piperine is 2.2%. Oleoresin is 9.65%. Essential oil is 3.4%.
- **IISR-Shakthi**: Open pollinated progeny of Perambramundi. Yield is 5.216 kg/vine. Piperine is 3.3%. Oleoresin is 10.2%. Essential oil is 3.7%.
- **IISR-Thevam**: Clonal selection from Thevanmundi. Yield is 5.17 kg/vine. Piperine is 1.6%. Oleoresin is 8.5%. Essential oil is 3.1%.
- **IISR-Malabar Excel**: Hybrid of Cholamundi x Panniyur 1. Yield is 2.78 kg/vine. Piperine is 2.96%. Oleoresin is 13.5%. Essential oil is 3.2%.

1.4. Soil and climate

- Elevation: Lower altitude to 1500 m above msl.
- pH 4.5 to 6.6.
- Temperature: 10-40°C
- Ideal: 25-35°C.
- Grown mainly as a rainfed crop.

- High rainfall: 150-250 cm, high humidity, warm climate and elevations up to 1500 m.
- Thrives best on virgin red lateritic and alluvial soils rich in humus content.

1.5. Season

- June-December

1.6. Propagation

- Three types of shoots
- Vertical orthotropic flower producing
- Horizontal plagiotropic and geotropic which hang towards the ground.
- Cuttings are raised from runner shoots.
- Orthotropic terminal shoots are used as propagules currently in large scale.
- Plagiotropic branches are useful for raising bush pepper.

1.7. Production of rooted cuttings

1.7.1. Traditional method

- Runner shoots are kept coiled on wooden pegs to prevent shoots from coming in contact with soil and striking roots.
- Orthotropic shoots treated with IBA 5000 ppm recorded higher rooting
- Single node cuttings from runner shoots treated with IBA 2500 ppm gave higher rooting.
- They are separated from the vine during February-March.
- Leaf trimmed cuttings of 2-3 nodes are planted either in nursery beds or in polythene bags filled with soil, sand and FYM in 1:1:1 ratio.
- Adequate shade has to be provided and polythene bags are to be irrigated frequently.
- Ready for planting during May-June.

1.7.2. Rapid multiplication method

- Modified the Sri Lanka technique
- A trench of 45 cm depth, 30 cm width and convenient length is made; filled with rooting medium i.e., soil, sand and FYM in 1:1:1 ratio.
- Split halves of bamboo with septa or split halves of PVC pipes of 1.25-1.50 m length and 8-10 cm diameter provided with plastic septa at 30 cm intervals are fixed at 45° angle on a strong support.
- Rooted cuttings are planted in the trench at one cutting for each bamboo split.

- Lower portions of bamboo splits are filled with weathered coir dust and FYM at 1:1 ratio.
- Growing vine is tied to bamboo split using dried banana leaf sheath fibres or coir rope and nodes are pressed to the rooting medium.
- Irrigate regularly.
- As the cuttings grow, bamboo splits are filled with rooting medium and each node is pressed in rooting medium and tied.
- Apply nutrient solution (1 kg urea + 0.75 kg super phosphate + 0.5 kg muriate of potash + 0.25 kg magnesium sulphate dissolved in 250 litres of water) at 0.25 litre/vine at monthly intervals.
- When the vine reaches the top (3-4 months after planting), terminal bud is nipped off and vine is crushed at about three nodes above the base to activate the axillary buds.
- After about 10 days, vine is cut at crushed point, removed from rooting medium and cut between each node.
- Each cutting with bunch of roots is planted in polythene bags filled with fumigated potting mixture.
- *Trichoderma asperellum* at 1 g and AM at 100 cc/kg of soil can be added to the potting mixture.
- Polythene bags should be kept in a cool and humid place, or should be covered with thin polythene (200 gauge) sheet to retain humidity.
- Buds start developing in about 3 weeks and the polybags can then be removed and kept in shade.

1.7.3. Trench method

- A pit of 2.0 x 1.0 x 0.5 m size is dug under a cool and shaded area.
- Single nodes of 8-10 cm length with leaf intact are planted in polythene bags (25 x 15 cm, 200 gauge) filled at lower half with a mixture of sand, soil, coir dust and cow dung in equal proportion.
- Leaf axil should be above potting mixture.
- Bags are arranged in the pit and covered with a polythene sheet by placing weights on the corners.
- Watered 5 times a day with rose cane and pit should be covered with polythene sheet immediately.
- Drench the cuttings 2-3 times with copper oxychloride (2g/litre).
- After one month, new shoots start emerging from the leaf axil.

- Keep the pit open for one hour per day to hardening.
- After two months, cuttings can be taken out and kept in a shaded area and watered twice a day.
- 80-85% success can be obtained.

1.7.4. Serpentine method

- Rooted cuttings are planted in polythene bags in a nursery shed with roofing sheet or shade net; serve as mother plants.
- As plant grows and produces few nodes, small polythene bags (20 x 10 cm) filled with potting mixture may be kept under each node.
- Press node in to mixture with help of a flexible twig such as mid rib of a coconut leaflet for rooting.
- Process of keeping potting mixture filled polythene bags at every node to induce rooting is repeated.
- In three months, first 10 to 12 nodes would have rooted profusely and will be ready.
- Each node with polythene bag is cut and cut end is also buried into mixture to induce more roots.
- Rooted nodes produce new sprouts in a week time and ready for field planting in 2-3 months.
- 60 cuttings can be harvested per mother plant in a year.

1.8. Planting

- Slopes facing West and South should be avoided.
- Pits of 50 x 50 x 50 cm size are dug at a spacing of 2 to 3 m in either direction (Panniyur 1-3 x 3 m).
- Pits are filled with 5 to 10 kg of FYM mixed with top soil.
- Rooted cuttings are planted in June - July at two per standard tree like silver oak, dadap, spined drumstick, coconut and jack fruit which are planted two months before planting of pepper vines at a spacing of 7-8 m.
- *Ailanthus malabarica, Grevillea robusta, Ethrythina indica, Cananga pinnata, Ceiba pentandra, Artocarpus netroplum, Morinda tinctoria*, are few standard trees.

1.9. Manures and Fertilizers

- Apply FYM or compost at 10 kg/vine.
- Apply 100 g N, 40 g P_2O_5 and 140 g K_2O/vine in two split doses in May-June and September-October.

- Slaked lime or dolomite at 500 g/vine is applied in alternate years during May-June to maintain the optimum soil pH level.
- Apply *Azospirillum* at 100 g/vine one month after application of chemical fertilizers.
- Fertilizers are applied around vine at a distance of 30 cm from base and incorporated into the soil.

1.10. Irrigation

- 7 lit per vine per day during October-May.
- Protective irrigation in basins during December-May at 10 days interval.

1.11. After cultivation

- Two weedings are given during June-July and at October-November.
- Vines are trained to the standards.
- Prune excessive foliage of standards and limit height of standards to about 6 m.
- Spray NAA at 40 ppm to increase the berry size.

1.12. Mulching

- Mulching can be done with green and dry leaves.
- Plant legume crops in field edges to control soil erosion.

1.13. Pruning

- Pepper vine tips can be pinched to improve the lateral branch growth.
- Hanging vines after reaching the height of supporting trees are removed.

1.14. Fruit drop

- Spray diammonium phosphate at 1.0 % four times *viz.,* before flower initiation (May), during new leaves and flower emergence (June), before spike initiation (July) and pinhead stage of berries (August).

1.15. Harvest

- Harvesting commences from third year onwards.
- Harvest berries in the spike start turning red colour in November to March.
- Berries are separated and dipped in hot water (80°C) for one minute and sun dried for 7 to 10 days.

1.16. Yield

- About 2 to 3 kg/vine/year.

- Panchami: 2828 kg of dried pepper/ha
- Cirikara: 2677 kg of dried pepper/ha
- Malabar excel: 2.78 kg of dried pepper/vine
- Thevam and Kirimunta: 6.14 kg of dried pepper/vine
- PLD-2: 4.97 kg green pepper/vine
- Subhakara: 4.2 kg green pepper/vine
- Pournami: 4.7 kg green pepper/vine

1.17. White pepper

- Harvest well matured fruits and keep for ripening in 50 kg gunny bag up to 1-2 days.
- They are soaked in water for 7-10 days.
- Then, extract the pieces, wash and dry by spreading in cement or bamboo mats until 11% moisture.
- Pack in clean poly or gunny bag.
- Sumatra is the major source of this pepper which is called muntok white pepper.

1.18. Black pepper

- The berries are harvested at younger stage heaped for a day or two and then dried in the sun.
- The flesh blackens and wrinkles over the seed.
- In Sri Lanka, the berries are dipped in hot water for ten minutes and then, dried in the sun.
- There are different forms of black pepper. They may be 1. Malabar pepper, 2. Tellicherry black pepper, 3. Lampong black pepper, 4. Sarawak black pepper, 5. Brazilian black pepper, 6. Ceylon black pepper etc.

1.19. Bush pepper

- Vertical grown branches of the plagiotropics by rooting is known as bush pepper, and it derives name from bushy nature of growth.
- It needs no support for trailing and can be harvested without climbing.
- The bush pepper is grown in pots in the households in the urban areas.
- Since bush pepper tolerates shade, it can be grown as an under storey crop in coconut and home steads.
- It may also be raised in kitchen garden and other agroforestry system.

- The growth and yield of bush pepper is superior under 50% shade under field conditions
- The crop can be raised by 2 x 2 m spacing
- Only drought resistant lines are recommended.
- From 100 kg berries, 28 kg of white pepper is obtained.
- The moisture may be 11%.

2

Cardamom (*Elettaria cardamomum* Maton), Zingiberaceae, 2n = 48, 52; Mysore type 2n=50; Malabar type 2n=48

2.1. Origin and distribution

- Queen of spices, native to East Indies/ India, originating in the forests of the Western Ghats in Southern India.
- Also grows in Sri Lanka, Guatemala, Indo China and Tanzania.
- Cultivated in Kerala, Karnataka and Tamil Nadu.

2.2. Economic importance

- Used for flavouring various preparations of food, confectionery, beverages and liquors.
- Used as whole pod, seeds or ground.
- Used in garam masala
- Spice mixture seasons meat and vegetable dishes, and masala chai and Turkish coffee.
- Antioxidant and diuretic properties
- May lower blood pressure.
- May contain cancer-fighting compounds.
- May protect from chronic diseases due to anti-inflammatory effects.
- May help with digestive problems including ulcers.
- May treat bad breath and prevent cavities.

2.3. Varieties

- Three cultivars are commonly recognized
  1. Malabar
  2. Mysore
  3. Vazhukka (Natural hybrid of Malabar x Mysore)

- Malabar type: Mudigree 1, Mudigree 2, PV 1, ICRI 1, ICRI 3, TKD 4, IISR Suvarna, IISR Vijetha, IISR Avinash.
- Mysore type: ICRI 2, Vazhukka, PV2, Njallani (Green gold).

Large cardamom: *Amomum subulatum*

- It is a perennial herb with subterranean rhizomes and 50-140 aerial leafy shoots.
- Each shoot has height of 1.7 to 2.6 m and possess 9 to 13 leaves in each tiller.
- Leaves are glabrous on both sides with a prominent mid rib.
- Inflorescence is a condensed spike with yellowish perianth.
- Each spike has 10-15 fruits.
- Fruit is round or oval shape capsule with reddish brown colour.
- Each capsule is trilocular with many seeds.
- Large cardamom is cultivated in the Sub-Himalayan region of North Eastern India, Nepal and Bhutan.
- It is grown in cold humid conditions under shade of trees at an altitude between 800-2000 m above MSL, with an average precipitation of 3000-3500 mm spread over about 200 days and with temperature ranging from 6-30°C.

2.3.1. Varieties and their characteristics

- **IISR Vijetha (Small Cardamom):** Malabar type. Selection from field resistant plants for Katte. Yield is 643 kg dry capsules/ha. Potential yield is 979 kg dry capsules/ha.
- **Kodagu Cardamom-Suvasini (CCS-1) (Small Cardamom):** Malabar type. Selection from open pollinated progeny of CL-37. Yield is 745 kg dry capsules/ha. Potential yield is 1322 kg dry capsules/ha.
- **IISR-Avinash (RR1) (Small Cardamom):** Malabar type. Selection from open pollinated progeny of CCS-1. Yield is 847 kg dry capsules/ha. Potential yield is 1483 kg dry capsules/ha.
- **Appangala 2:** Malabar type. Hybrid of Appangala-1 (Moderate yielder) x NKE-19 (Mosaic resistant). Yield is 927 kg dry capsules/ha. Potential yield is 1393 kg dry capsules/ha.

Varietal name	Type	Yield potential (Dry capsules/ ha in kg)	Other attributes
Coog Cardamom Malabar	Malabar	408	Released by IISR, Cardamom Research Centre. Seedling of clone 37. Dry recovery: 22%, Essential oil content: 8.7%.
Selection-1 (CCS-1)			
ICRI-1	Malabar	325-rainfed; 650-irrigated	Released by Indian Cardamom Research Institute, Myladumpara. Dry recovery: 22.9%, Essential oil: 8.3%.
ICRI-2	Mysore	375-rainfed;	Released by Indian Cardamom Research Institute, Myladumparai.
		760-irrigated	Adaptable for Vandenumedu and Anamalai hills. Dry recovery: 22.5%, Essential oil: 9%.
Mudigree-1	Malabar	250-300	Released by RRS, Mudigree, suitable for Malned region of Karnataka State. Dry recovery: 20%, Essential oil: 8%.
PV-1	Malabar	500	Released by Cardamom Research Station, Pampadumpara. Dry recovery: 20%, Essential oil: 6.8%.
SKP-14	Malabar	430 rainfed; 590-irrigated	Released by ICRI, Regional Station, Saklespur.

2.4. Soil and climate

- Thick shady areas with loamy soil are ideal
- Grown at an elevation from 600 to 1500 m.
    - Malabar type elevation: 600-900 m
    - Mysore type elevation: 900-1200 m
    - Vazhukka type: 900-1200 m
- Heavy winds are unsuitable.
- Grown in forest loamy soils which are acidic with a pH range of 4.2-8.
- Rainfall is between 1500 and 4000 mm
- Temperature is 10-35°C

2.5. Season

- June-December.

2.6. Propagation

- Seedlings/suckers can be used for propagation.

2.6.1. Seed propagation

- Seed rate - 600g/ha (Fresh seeds).
- Treat with commercial grade nitric acid 25% for 10 minutes.
- Prepare the beds with equal quantity of well rotten cattle manure, wood ash and jungle soil.
- Sow the seeds in beds and cover with a thin layer of fine sand during August - October.
- Mulching and shading may be provided to seed beds.
- One year old seedlings with 3- 4 leaves are transplanted to secondary nursery.
- **Secondary nursery:** Prepare the beds and provide shade by erecting overhead pandal.
- 18-22 months old seedlings with 4-6 leaves are planted at a distance of 20 x 20 cm during March-May.
- Polybags of 20 x 20 cm size can also be used

2.6.2. Propagation from suckers

- Suckers from high yielding plants are planted in clonal gardens
- Suckers are planted adopting a spacing 1.8 x 0.6 m (6800 plants/ha of clonal nursery)
- Shade and irrigation is provided.
- 32-42 suckers are obtained from each planting unit in 12 months

2.7. Field preparation

- Dig pits of 60 x 60 x 60 cm size and fill with compost and top soil.
- Contour planting may be done in slopy areas.

2.8. Spacing

- Larger types: 2.5 x 2.0 m.
- Smaller types: 2.0 x 1.5 m.

2.9. Irrigation

- Grown as a rainfed crop, but provide sprinkler irrigation during summer for increased yields.

2.10. Manures and fertilizers

- Apply compost 25 t/ha; 75 kg N; 75 kg P_2O_5 and 150 kg K_2O/ha in two split doses during June-July and October-November.

2.11. After cultivation

- Shade trees like *Artocarpus fraxinifolius, Erythrina lithosperma, Diospyros ebenum* and *Lagerstroemia lanceolata* are grown.
- **Shade regulation**: Moderate shade: 50-60%; rainfed: 40-50%; irrigated: 55-60%
- Weed the field as and when necessary.
- At the end of monsoon rains, a light raking or digging and mulching is given around the plant to a radius of about 75 cm to conserve moisture during the dry period.
- Pollination can be enhanced by keeping bee hives at 10-15 nos/ha (*Apis cerana indica*)

2.12. Harvest

- Harvesting commences from the third year onwards.
- Harvesting is generally done once in a month.
- Pick only those fruits which are just ripe, but not fully ripe.

2.13. Yield

- Yield ranges from 200-250 kg/ha.

2.14. Curing of cardamom

- Capsules are picked when they are juicy.
- In earlier times, sulphuring is done to remove the moisture and bleaches them.
- Cardamoms are spread in bamboo trays one over other and sulphur is burnt underneath in special chambers.
- **Sun drying**
  - It takes 3-4 days for capsules to get dried and bleached.
  - Due to lack of regulation of heat, sun drying lacks green colouration of pods; they fetch low prices in the market.
- **Artificial curing**
  - It is a local device of heating on mud platform through wood fire.
  - In this device, there is hollow platform about 60 cm wide and 180-240 cm long running along one side of a room and hollow space underneath running from one side of the wall to the other opening at both the outside ends.

    - On one side, wood fire is lighted and the hot flames and smoke pass through the hollow platform.
    - Cardamom is placed on the smooth plastered platform and dried.
    - This is an unscientific method of curing.
- **Flue curing**
    - Required temperature is obtained by burning wood fuel in a fire chamber.
    - Hot air is passed through pipes or coils.
    - Smoke is allowed to escape through chimneys.
    - Room is provided with rafts and wire mesh.
    - Green cardamoms are spread at an interval of 30 cm.
    - Before spreading on the rafts or racks which arranged in tiers, the capsules are treated with 2% washing soda solution for 10 minutes.
    - To preserve the green colour, the capsules are treated with 2% sodium carbonate solution for 10 minutes.
    - After draining the solution, they are spread for drying.
    - Temperature is kept at 55°C for three hours.
    - At the end of curing, pods taken out of the chamber, cleaned and natural colour is retained.
    - White or bleached cardamom (Loss of green colour).
    - Bleached cardamom is sold as 'whole' in bottles in USA.

3

Clove (*Syzygium aromaticum* (L.) Merr. & L.M.Perry), Myrtaceae, 2n = 22

3.1. Origin and distribution

- Native to the Moluccas Islands of Indonesia.
- Although Indonesia is the largest producer of cloves, Zanzibar and Madagascar are the major exporters,
- In India, grown in the hilly tracts of Tamil Nadu, Kerala and Karnataka.

3.2. Economic importance

- Clove taste is imparted by eugenol and quantity of pice required is small.
- Pairs well with cinnamon, allspice, vanilla, red wine, basil, onion, citrus peel, star anise, and pepper corns.
- Oils, dried flower buds, leaves, and stems are used to make medicine.
- Applied directly to gums for toothache, pain control during dental work, and other dental related issues.
- Used in flavoring meats, curries and marinades, as well as fruits such as apples, pears, and rhubarb.
- Used as an ant and mosquito repellent.

3.3. Varieties

- Local cultivars are used.
- Penang, Amboyna and Zanzibar are some important locally identified selections.
- In addition to these selections, Burliar-1 and Odetham from Courtallum are grown in Fruit station at Kallar, Tamil Nadu.
- Some other famous varieties include Madagascar, Posi Posi, White, Afo, Kotok, and Siputi cloves.

3.3.1. Varieties popular in India

Penang: Cloves are plumpy and large in size and dark red-brown colour. They are commercially exported all over the east, but the demand is very high all across the globe. This spicy clove has uses in pomanders and culinary.

Zanzibar: Clove is smaller and leaner in shape as compared to the Penang cloves and has a black-brown color. This high quality clove works as a flavorful spice in savory and sauteed dishes. It is popular among the growers because of its high productivity and better adaptability.

Amboyan: Produces large-sized and plumpy cloves like the Penang. These pungent and aromatic cloves are popular in Indian savory recipes, tooth paste, cosmetics, and pomanders. But, this clove variety has slightly low productivity as compared to Zanzibar and they are also not very adaptable.

3.4. Soil and climate

- Humid tropical climate with an annual rainfall of 150-250 cm, temperature of 20 to 30°C and elevation up to 1000 m.
- Deep rich loams with high humus content and laterite soils.

3.5. Season

- June-December.

3.6. Planting

- Seeds and the sepals are reduced to triangular projections and this is known as mother clove which is used as a seed material.
- In addition to seeds, approach grafting is also used as propagation method.
- North and North Eastern slope is preferred.
- Two year old seedlings are planted in pits of 30 x 30 x 30 cm size filled with soil and FYM @ 10 kg/pit at a spacing of 6 m either way.
- Apply 50 g *Azospirillum* /pit before planting.

3.7. After cultivation

- Grow banana or erect pandal to provide shade during the initial establishment.
- Weed the basins and mulch with dried leaves.

3.8. Irrigation

- Frequent watering is essential in the initial stages.
- Irrigation should be given during summer months.

3.9. Manures and fertilizers

- **One year old plants**: FYM 15 kg, N 20 g, P_2O_5 20 g, K_2O 60 g/plant during June-July and September-October.

- **7 years old plants**: FYM 50 kg, N 300 g, P_2O_5 300 g and K_2O 960 g/plant.
- Apply 50 g each of *Azospirillum* and Phosphobacterium one month after manuring.

3.10. Harvest

- Bearing starts from sixth year onwards.
- Flower buds should be harvested when they are fully mature, but before opening.
- Buds are harvested as clusters, separated and dried in the sun for 5-7 days.

3.11. Yield

- 2-3 kg dried buds/tree.

4

Cinnamon (*Cinnamomum zeylanicum* J.Presl), Lauraceae, 2n= 24

4.1. Origin and distribution

- Originated in the central hills of Sri Lanka and Malabar Coast.
- Distributed in tropical and subtropical regions of North America, Central America, South America, Asia, Oceania
- Cultivated in Sri Lanka, Malagasy Republic and Seychelles.
- In India, it is grown in one or two locations in Kerala

4.2. Economic importance

- **Commercial products are**
  - **Quills**: Scraped peel of inner bark of mature cinnamon shoots, joined together with overlapping tubes, hollow of which has been filled with smaller pieces of cinnamon peels.
  - **Quillings**: Broken pieces and splits of all grades of cinnamon quills.
  - **Featherings**: Feather like pieces of inner bark consisting of shavings and small pieces of bark left over.
  - **Chips:** Rough unpeelable barks scraped off from the thicker stems.
  - **Leaf and bark oil:** Obtained by distilling the leaf and bark separately; used in perfumery and cosmetics
- Bark is a popular spice or powder with a delicate fragrance and a warm agreeable taste.
- Used in flavouring confectionary, liquors, pharmaceuticals and cosmetics.
- Helps in diabetics and digestion of sugar.
- Has astringent, stimulant and carminative properties and can check nausea and vomiting.

4.3. Varieties

- YCD 1, PPI-1, Nithyasree, Navasree, Konkan Tej, Sugandhini.
- In Sri Lanka, there are two types - sweet cinnamon and honey cinnamon.

- **Nithyasree** - Seedling selection from Indian collection. Duration: 3 years for first harvest. Yield: 200 kg dry quills/ha. Potential yield: 250 kg/ha.
- **Navasree** - Seedling selection from Indian collection. Duration: 3 years for first harvest. Yield: 200 kg dry quills/ha. Potential yield: 250 kg/ha.

4.4. Soil and climate

- Sandy loam or lateritic soils with high humus.
- Average temperature of 27°C.
- Grown up to an altitude of 800-1000 m asl and an annual rainfall of 150 to 250 cm.
- Hot and moist climate is suited; prolonged dry weather.

4.5. Season

- June-December

4.6. Propagation

- Seeds/Semi hardwood cuttings

4.7. Nursery

- Sown in beds in rows of 12 cm apart.
- July-August is the best season for sowing.
- From beds, 15 cm long seedlings are transplanted to polythene bags.

4.8. Planting

- Fill the pits of 60 cm^3 with top soil and FYM 10 kg.
- One year old seedlings or rooted cuttings are transplanted in pits at 2 x 2 m spacing under partial shade.

4.9. Irrigation

- Protective watering during summer.

4.10. Manures and fertilizers (per plant)

Manures and fertilizers	First year	Annual increase	Tenth year onwards
FYM	-	2 kg	20 kg
N	20 g	20 g	200 g
P	18 g	18 g	180 g
K	25 g	25 g	250 g

4.11. After cultivation

- Provide with temporary shade by erecting a small pandal.
- **Coppicing**: Young trees are cut close to the ground to produce more side shoots.

4.12.Harvest

- Harvesting starts from 4th or 5th year after planting.
- Shoots are cut for the extraction of bark once in May and again in November.
- Bark is peeled from the selected shoots of 18 to 24 months old, which are usually one metre long and 1 to 2 cm thick.
- Shoots ready for peeling are removed from the stumps and terminal ends of shoots are also removed.
- Peeling is done by knives after scraping off the outer bark.
- Harvested produce is called as quills.
- From leaves, bark and root, oil can be extracted by steam distillation.

4.13. Yield

- Dried bark: 100 g/bush; Leaf oil: 35 kg/ha/year
- Quills: 220-300 kg/ha for 10 or 11 years old; Quillings: 75 kg/ha.
- Leaves: 2.5 t/ha

5

Nutmeg (*Myristica fragrans* Hoult.) Myristicaceae, 2n = 42, 44

5.1. Origin and distribution

- Native to the Moluccas of Spice Islands of Indonesia.
- In India, cultivated in Thrissur, Ernakulam and Kottayam districts of Kerala, and parts of Kanyakumari and Thenkasi, Nilgiris (Kallar, Buraliar) districts in Tamil Nadu.

5.2. Economic importance

- Fleshy arils surrounding the nutmeg seed are the source of the spice mace.
- Used to flavour many kinds of baked goods, confectionaries, puddings, potatoes, meats, sausages, sauces, vegetables, and beverages such as eggnog.
- Fresh husks are used for making jelly.
- Nutmegs act as stimulant and carminative.

5.3. Varieties

- Vishwashree, Konkan Sugandha and Konkan Swad.
- IISR accessions - A 9-20, 22, 25, 69, 150, A 4-12, 22, 52, A11-23, 70.
- **Viswashree** - Clonal selection from elite trees. Duration: 6 years for first harvest. Yield: 480 kg mace/ha at 8th year; 3122 kg nuts/ha. Potential yield: 4800 kg mace/ha at 25th year; 31220 kg nuts/ha.
- **Keralashree** - Seedling selection from the seedlings raised from an elite mother tree from Burliar, Nilgiris of Tamil Nadu. Duration: 6 years for first harvest. Yield: 480 kg mace/ha at 8th year; 3122 kg nuts/ha. Potential yield: 4800 kg mace/ha at 25th year; 31220 kg nuts/ha

5.4. Soil and climate

- Friable, well drained clay, sandy loam, laterite to red soils.
- Grown up to an elevation of 1000 m with 150-250 cm of rainfall, humid tropical climate.
- River banks and sheltered valleys are also suited to this spice crop.

5.5. Propagation

- Seeds/grafts/budded plants.

5.5.1. Seed propagation

- Seeds are sown in beds at a spacing of 30 cm and 2.5-5.0 cm deep.
- After 90 DAS, transplant to poly bags (35 x 15 cm).
- Then, one year old seedlings are transplanted to bigger poly bags (35 x 20 cm).

5.5.2. Vegetative propagation

- Grafting (Approach graft, wedge graft, epicotyl grafting) or budding (Patch method) is recommended
- Use only orthotropic shoots as scion materials.
- Best season is October-January.

5.6. Season and planting

- Plant 18-24 months old seedlings during June-December in pits of 60 x 60 x 60 cm size filled with soil and FYM in equal parts.
- Spacing: 8 x 8m.

5.7. Manures and fertilizers

- **First year**: Apply FYM 15 kg, N 20 g, P_2O_5 20 g, K_2O 60 g/tree
- **Adult trees**: FYM 100 kg, N 400 g, P_2O_5 300 g, K_2O 1200 g/tree in June - July and September - October.
- Apply 50 g each of *Azospirillum* and Phosphobacterium one month after manuring.

5.8. Irrigation

- Irrigate once in 5-7 days during summer months.

5.9. After cultivation

- Plant banana or provide pandal for medium shade during the initial stages of growth.

Top working

- An important problem in nutmeg cultivation is the segregation of seedlings into male and female plants resulting in about 50% unproductive male trees.
- The only alternative is to adopt vegetative propagation either by top-working male plants or using budded or grafted plants.

5.10. Mixed crop

- Grown as mixed crop with arecanut and coconut.
- Nutmeg can be planted after every third row of arecanut.

5.11. Harvest

- Bearing starts from 6 to 7 years after planting.
- Mature fruits when they start splitting are harvested.
- Aril is known as mace and seeds are separated and dried.

5.12. Yield

- Fruits: 1000-2000/tree
- Dried nuts: 5-7 kg/tree
- Dried mace: 0.5-0.7 kg/tree

6

Turmeric (*Curcuma longa L.*) Zingiberaceae, 2n = 63

6.1. Origin and distribution

- Native of South and South East Asia, from there it is spread to Indonesia, China, Sri Lanka, Pakistan and Pacific Islands

6.2. Economic importance

- Used as condiments.
- Rhizomes are boiled in water, dried, ground into a deep orange yellow powder and used as a colouring and flavoring agent in many Asian cuisines, especially for curries, as well as for dyeing.
- Has a warm, bitter, black pepper like flavor and earthy, mustard like aroma.
- Yellow color is due to curcumin (3-9%)
- Anti-inflammatory and antioxidant properties
- Protect against heart disease
- May prevent cancer
- Ease symptoms of osteoarthritis
- Treat or prevent diabetes
- Treating depression
- Treating rheumatoid arthritis
- Improve skin health
- Protects body from free radicals
- Anti-aging supplement
- Prevent eye degeneration

6.3. Varieties

- **Local cultivar**: Salem and Erode turmeric (Tamil Nadu)
- Alleppey Finger (Kerala)
- Rajapore and Sangli turmeric (Maharashtra)

- Nizamabad Bulb (Telangana)
- In TamilNadu, the important varieties cultivated are Erode local, BSR 1, PTS-10, Roma, Suguna, Sudarsana and Salem local.
- Among these varieties, 70-75% is occupied by the local varieties.
- CO 1, CO 2, BSR 1, Swarna, Sugandham, BSR 2, Ranga, Rashmi, Rajendra Sonia, Krishna, Suroma, Allepy Finger Turmeric (AFT), IISR Prabha, IISR Prathiba, IISR Alleppey Supreme, IISR Kedaram, Sona, Varuna, BSR 3.
- Allepey Finger, Rajapore and Erode local are some of important exported varieties.
- Turmeric exported in the processed form is dry turmeric, fresh turmeric, turmeric powder and oleoresin.
- Alleppey Finger is known for its high content of curcumin - a yellow colouring substance. Its bright yellow colour has been preferred by spices importers in Europe and other continents.
- In Middle East, the UK, USA and Japan, some of the well accepted varieties are Alleppey Finger and Erode turmeric, Rajapore and Sangli turmeric and Nizamabad Bulb
- **IISR-Pragati** - Germplasm selection. Cultivated in Kerala, Tamil Nadu, Andhra Pradesh, Telangana, Karnataka and Chhattisgarh states. Crop duration: 180 days. Yield: 38.0 t/ha. Potential yield: 52 t/ha. Curcumin: 5.02%.
- **IISR-Kedaram** - AFT selection. Grown in Kerala, Karnataka, Maharashtra and North Bengal. Crop duration: 210 days. Yield: 5.28 t/ha dry turmeric. Potential yield: 29 t/ha fresh turmeric. Curcumin: 5.70%.
- **Prathibha** - Open pollinated progency selection. Adopted to all over India. Crop duration: 225 days. Yield: 39.12 tons of fresh rhizomes/ha. Curcumin: 6.52%.
- **Prabha** - Open pollinated progency selection. Adopted to all over India. Crop duration: 205 days. Yield: 37 tonnes of fresh rhizomes/ha. Curcumin: 6.52%.
- **Sudarshana** - Selection from the germplasm (PCT-14) collected from Singhat, Manipur. Cultivated in Kerala and Andhra Pradesh. Crop duration: 190 days. Yield: 28.8 tons of fresh rhizomes/ha. Potential yield: 54.9 tons of fresh rhizomes/ha. Curcumin: 7.9%.
- **Suvarna** - A selection from the germplasm (PCT-8) collected from Assam. Grown in Kerala, Karnataka and Andhra Pradesh. Crop duration: 200 days. Yield: 17.4 tons of fresh rhizomes/ha. Potential yield: 43.5 tons of fresh rhizomes/ha. Curcumin: 4.0%.

- **IISR-Alleppey Supreme** - AFT selection. Adopted to Kerala (Rainfed), Maharashtra, Karnataka and North Bengal (Irrigated). Crop duration: 210 days. Yield: 5.28 dry rhizome t/ha. Potential yield: 19.94 t/ha fresh. Curcumin: 5.5%.

6.4. Soil and climate

- A friable well drained light black, ashy loam or sandy, red loamy soil in wet or garden lands under tropical conditions is ideal.
- Annual rainfall of 1500 mm is ideal.
- Raised from sea level to 1500 m above mean sea level.

6.5. Season

- May-June

6.6. Propagation and planting

- Mother rhizome and finger rhizomes of 25-30 g weight at 2000 kg/ha.
- Dipped in phosalone 35 EC 2 ml/lit. or 0.3% copper oxychloride for 30 min or *Bacillus subtilis* at 10 g/kg and *T. asperellum* at 4g/kg.

6.7. Spacing

- 45 x 30 cm
- Rhizomes are to be dibbled at a depth of 4 cm.

6.8. Manures and fertilizers

- **Basal:** FYM at 25 t/ha, neem or groundnut cake at 200 kg/ha, 60 kg each of NPK/ha
- 30 kg of $FeSO_4$ and 15 kg of $ZnSO_4$, 10 kg each of *Azospirillum* and Phosphobacteria/ha
- **Top dressing:** 25 and 108 kg of N and K/ha respectively applied on 30, 60, 90, 120 and 150 days after planting.
- **Micronutrient:** Spray 375 g each of boron, iron and zinc during rhizome development stage + 375 g of urea in 250 lit of water/ha twice at 25 days interval.

6.9. Fertigation

- Fertigation at 150:60:108 kg of NPK/ha is applied throughout the cropping period once in three days.75% of the recommended dose of phosphorous is applied as basal dose.

- Water-soluble fertilizers like 19:19:19, mono ammonium phosphate (12:61:0), multi K (13:0:45) and urea are used.

6.10. Intercropping

- Onion, coriander and fenugreek are planted on sides of ridges 10 cm apart (250 kg seed onion/ha).
- Red gram and castor can also be planted at wider spacing. French bean, green gram, black gram also grown as intercrop.
- Turmeric can also be grown as an intercrop under coconut gardens.

6.11. After cultivation

- Earthing up at 2nd and 4th top dressings.
- Turmeric is a shade loving crop. It requires a minimum of 50% shade, which gave higher yield.
- Crops used for shading the turmeric are *Sesbania aculeata, S. grandiflora, Ricinus communis* and chillies.

6.12. Harvest

- Plants start lodging, yellowing and drying on crop maturity.
- Rhizomes are dug with spade or digging forks.

6.13. Yield

- Fresh rhizomes: 25-30 t/ha
- Cured rhizomes: 5-6 t/ha

6.14. Turmeric curing

- Rhizomes harvested, finger and mother rhizome are separated, cleaned and the green rhizomes are boiled in water.
- Rhizomes are boiled in water with galvanized iron or earthen vessels
- Stop boiling when the froths starting out and furnaces appear, giving out a typical turmeric odour.
- Boiled rhizomes are taken out for drying.
- They are spread in the sun for 10-15 days till they become hard and brittle and produce a metallic sound on breaking.
- Then, they cleaned and polished mechanically in drum rotated by hand or hexagonal drum mounted on a central axis and rotated by power.

- Curing and polished qualities vary with variety.
- Normally, the curing per cent varies from 16-20 per cent depending on the variety.
- Mother rhizomes will give higher curing per cent.

6.15. Turmeric products

1. **Whole turmeric**

- Cured and dried turmeric is packed in double burlap new gunny bags, which fumigated prior to packing.

2. **Turmeric powder**

- The ground turmeric powder will have a moisture level of above 12% on dry weight basis with free flow characteristics of the produce.
- Powder is packed in aluminium foil laminated pouches, which offers maximum protection against loss of volatile oil and increase of moisture.
- In bulk, turmeric powder is packed in fibre board drums, multiwall bags and tin containers.

3. **Volatile oil**

- It is derived by steam distillation of crushed turmeric rhizomes and is orange yellow liquid with an odour of turmeric.
- About 58% of oil is turmerones (3-4% curcuminoids) and 9% tertiary alcohol.

4. **Oleoresin**

- Turmeric oleoresin is extracted from ground turmeric by solvent extraction followed by vacuum concentration.
- The semi liquid viscous stuff contains both volatile aromatic principle and non-volatile acrid fraction
- Oleoresin contains 30-35% curcumin, 15-20% volatile oil.

5. **Curcumin**

- It is estimated by ASTA with acetone followed by measurement of absorption at 415 to 425 nm
- Curcumin is a crystallized dye consists of three pigments
  - Curcumin
  - Demethoxy curcumin
  - Bis-demethoxy curcumin
- The three can be separated by chromatography, but crystallized together as a mixture.
- Curcumin is an antioxidiant and therefore, anticancer neutraceutical.

6.16. Postharvest handling

- Curing involves boiling of rhizomes in fresh water and drying it in the sun.
- Cleaned rhizomes are boiled in copper or galvanized iron or earthen vessels, with water just enough to soak them till the fingers/mother rhizomes become soft.
- Cooked turmeric is taken out of the pan by lifting the troughs and draining the water into the pan itself.
- Same hot water in the pan can be used for boiling the next lot of raw turmeric which is already filled in the troughs.
- Alternatively, rhizomes may also be cooked using baskets with perforated bottom and sides.
- Cooking of turmeric is to be done within 2-3 days after harvest.
- Cooked fingers/mother rhizomes are spread separately in 5-7 cm thick layers on bamboo mats or cement floor under the sun for drying for 10-15 days.
- Dried material should be heaped or covered during night time.
- Artificial drying using cross-flow hot air at a maximum temperature of 60°C is also found good.
- Recovery of dry product varies from 20-25% depending upon the variety and the location where the crop is grown.
- Smoothening and polishing the outer surface by manual rubbing on a hard surface or mechanical rubbing improves the appearance.
- Improved method is by using hand-operated barrel or drum mounted on a central axis, the sides of which are made of expanded metal mesh.
- When the drum filled with turmeric is rotated, polishing is effected by abrasion of the surface against the mesh as well as by mutual rubbing against each other as they roll inside the drum.
- Turmeric is also polished in power-operated drums.
- In order to impart attractive yellow colour, turmeric suspension in water is added to the polishing drum at the last 10 minutes.
- When the rhizomes are uniformly coated with suspension they may be dried in the sun.

7

Ginger (*Zingiber officinale*) Zingiberaceae, 2n = 22

7.1. Origin and distribution

- Originated from Tropical South East Asia.
- Cultivated in India, China, Jamaica, Australia, Mexico, North and West Africa and Taiwan, Fiji, Mauritius, Brazil, Indonesia, Japan, Malaysia, Philippines, Thailand and Bangladesh.
- In India, cultivated in Gujarat, Kerala, Arunachal Pradesh, Odisha, Meghalaya, West Bengal, Tamil Nadu, Andhra Pradesh and Karnataka.
- Karnataka, Odisha, Assam, Meghalaya, Arunachal Pradesh and Gujarat together contribute 65 per cent to the country's total production.

7.2. Economic importance

- India, China and the Japanese people use ginger as food
- Used as drug in cold, cough, vomiting, dizziness, hypertension, impaired vision problems, digestion, fever and stomach diseases.
- Dry ginger and pepper along long pepper tea helps to remove mucus in the body for a long time and helps to protect health.
- Aromatic (0.25-3%) and volatile oil (1-2%) helps in preparation of ginger beer, ginger wine.
- Used for making biscuits, cakes, soup and pickle.
- Oleoresin content: 4-10%.

7.3. Varieties

- Rio de Janeiro, Assam, China, Nadia, Maran, Nadan, Suruchi, Suravi, Suprabha, Himagiri, IISR Varada, IISR Mahima and IISR Rejatha, Athira and Karthika.
- Kuruppam Padi, Thodupuzha, Calicut.

Variety	Fresh mean yield (t/ha)	Maturity (days)	Dry recovery (%)	Crude fibre (%)	Oleoresin (%)	Essential oil (%)
ICAR Indian Institute of Spices Research, Kozhikode - Kerala						
IISR Varada	22.6	200	20.7	4.5	6.7	1.8
IISR Mahima	23.2	200	23	3.3	4.5	1.7
IISR Rejatha	22.4	200	19	4	6.3	2.4
High Altitude Research Station, Odisha University of Agriculture and Technology, Pottangi-Odisha						
Suprabha	16.6	229	20.5	4.4	8.9	1.9
Suruchi	11.6	218	23.5	3.8	10	2
Suravi	17.5	235	23.5	4	10.2	2.1
Subhada	18	210	22.4	3.4	10.4	2
Y.S. Parmar University of Horticulture and Forestry, Nauni, Solan, Himachal Pradesh						
Himagiri	13.5	230	20.6	6.4	4.3	1.6
Kerala Agricultural University, Thrissur - Kerala						
Athira	21	220 – 240	22.6	3.4	6.8	3.1
Karthika	19	220 – 240	21.6	3.7	7.2	3.2
Aswathy	23	220 – 240	19.7	3.5	7.5	3.3

7.4. Season

- May-June
- In west coast, planted in May, December-January as irrigated crop.

7.5. Soil and climate

- A friable well drained sandy, lateritic loamy soil rich in humus.
- A day temperature of 25-35°C is ideal.
- Warm and humid conditions with 250 cm of annual rainfall.
- Grown as an irrigated crop in humid zones from sea level to an altitude of 1300-1600 m above msl.

7.6. Seed rate and treatment

- 1000-1200 kg bit rhizome/ha.
- Hilly areas: 2000 to 2500 kg/ha.
- Dip in 0.3% mancozeb and 0.1% malathion or 200 ppm streptocycline.
- Soaking in 5g *Bacillus subtilis* 10 days before sowing and dried in shade.

7.7. Spacing

- Irrigated crop:30 x 20 cm in ridges and furrows.
- Rainfed crop: Raised beds of 20 x 20 cm or 25 x 25 cm with a depth of 4-5 cm.

7.8. Manures and fertilizers

- Basal: FYM at 25-30 t/ha
- 75:50:50 kg of N, P and K/ha.
- Top dressing: 37.5: 12.5 kg of N and K/ha applied on 45 and 90 days after planting

7.9. After cultivation

- After each top dressing, earthing up is done.

7.10. Mulching

- 30 tons green leaves as mulch in three splits; 15 tons immediately after planting, 7.5 tons at 60 and 120 days after planting

7.11. Irrigation

- Watering at intervals of 15 days.
- Critical stages: Seed tuber germination and tuber production.

7.12. Weed management

- Weeding at 45-60 days interval,

7.13. Intercrop

- Cultivated as a single crop or intercrop with coffee, orange, banana, coconut plantations etc.

7.14. Harvest

- Harvest after 8-9 months when leaves start yellowing, wither and drying.

7.15. Curing ginger

- Green ginger is washed in water twice or thrice and is soaked in water for a day to prepare dry ginger.
- Next day, the skin of the rhizome is peeled off against coir or small cot or with sharp bamboo splinter.
- Peeling can be done safely by keeping the rhizomes inside rotating wire-mesh drum.
- Afterwards, the ginger is washed and dried in the sun for 10-12 days.
- Bleached ginger is obtained by soaking in 3% lime water for a few minutes and then, fumigated with sulphur and dried.
- Recovery of dry ginger is 18-20%.
- Moisture level is about 10 per cent.

7.16. Products from ginger

- **Ginger powder:** The dry ginger is pulverized to a mesh of 50-60. The powder is also used in distilleries for preparing ginger beer, brandy, wine etc.
- **Ginger oil:** This is obtained by steam distillation of dry ginger powder. Indian ginger contains 1.5-2.5% volatile oil. The oil is light yellow in colour.
- **Salted ginger:** For this purpose ginger rhizomes are harvested 4-5 months after sowing. The rhizomes are soaked in brine and citric acid for 14 days. Then, rhizomes are washed and preserved.

7.17. Yield

- About 12-15 t/ha.

8

Garlic (*Allium sativum*), Alliaceae, 2n=16

8.1. Origin and distribution

- Native land is Central Asia, Middle Asia and Northeastern Iran.
- China produces 80% of the world's supply
- India, Korea, Spain, Egypt, USA, Thailand and other countries are growing garlic.
- In India, cultivated in Gujarat, Haryana, Madhya Pradesh, Odisha, Maharashtra, Uttar Pradesh and Rajasthan
- India exports fresh and chilled garlic, dried garlic, dehydrated garlic flakes, dehydrated garlic powder, and garlic oil to Bahrain, Bangladesh, Germany, Japan.

8.2. Economic importance

- It contains amino acid allicin.
- Used both as a food flavoring and as a traditional medicine
- Allicin, antibacterial substance, the principle of amoebic dysentery.
- The garlic odour is due to diallyl disulfide.
- Consuming garlic on a daily basis (In food or raw) helps to lower cholesterol levels because of the anti-oxidant properties of allicin.
- Beneficial to regulate blood pressure and blood sugar levels.
- It is used as a condiment.

8.3. Varieties

- Ooty 1, Ooty 2, Singapore Red, Rajali, Tabiti, Creole, Madrasii, Italian, Godavari, Sweta, HG 1, HG 6, Pusaset 10, Agri Found White (G 41), Tamuna Safed (G1), Yamuna Safed (G 50), G 282, Agri Found Parvati.

8.3.1. Varieties developed by ICAR-DOGR

- **Bhima Omkar** - Recommended for cultivation in the states of Delhi, Gujarat, Haryana and Rajasthan. It matures in 120-135 days and average yield is 8-14 t/ha. It produces medium size compact white bulbs.

- **Bhima Purple** - This variety with attractive purple skinned bulbs has been recommended for cultivation in the states of Andhra Pradesh, Bihar, Delhi, Haryana, Karnataka, Maharashtra, Punjab and Uttar Pradesh. It matures in 120-135 days and the average yield is 6-7 t/ha.

8.3.2. Varieties from other organizations

- Garlic varieties developed by National Horticultural Research Development Foundation (NHRDF), Nashik; Mahatma Phule Krishi Vidyapeeth (MPKV), Rahuri; Chaudhary Charan Singh Agricultural University (CCSHAU), Hisar; Tamil Nadu Agricultural University (TNAU), Coimbatore; ICAR-Indian Agricultural Research Institute (IARI), New Delhi; ICAR-Vivekananda Parvatiya Krishi Anusandhan Sansthan (VPKAS), Almora and Junagadh Agricultural University (JAU), Junagadh are listed in Table 1.

Table 1. Garlic varieties for different regions of the country

Variety	Developed by	Colour	Region	Days to maturity	Yield potential (t/ha)
Short day type					
Agri Found White (G-41)	NHRDF	White	Madhya Pradesh and Maharashtra	130-135	12-14
Yamuna Safed (G-1)	NHRDF	White	All India	120-130	15-17
Yamuna Safed-2 (G-50)	NHRDF	White	Northern India	120-130	15-20
Yamuna Safed-3 (G-282)	NHRDF	Creamy white	Chhattisgarh, Gujarat, Haryana, Madhya Pradesh, Maharashtra, Punjab, Rajasthan and Uttar Pradesh	120-130	17-20
Yamuna Safed-4 (G-323)	NHRDF	White	North and Central India	165-175	16-17
Yamuna Safed-5 (G-189)	NHRDF	White	Andaman and Nicobar Islands, Arunachal Pradesh, Bihar, Delhi, Gujarat, Haryana, Jharkhand, Manipur, Meghalaya, Mizoram, Nagaland, Punjab, Rajasthan, Sikkim, Tarai region of Uttar Pradesh, Tripura and Uttarakhand	150-160	17-18
Godavari	MPKV	Purple	Maharashtra	140-145	10-11
Shweta	MPKV	White	Maharashtra	130-135	10-11
Phule Baswant	MPKV	White	Madhya Pradesh and Maharashtra	135-140	10-11
GG 4	JAU	White	Gujarat and Maharashtra	130-140	8-10
Intermediate type					
Ooty 1	TNAU	Dull white	Tamil Nadu	120-130	15-17
Long day type					
VL Garlic 1	ICAR-VPKAS	White	Bihar Himachal Pradesh, Jammu & Kashmir, Punjab, Uttarakhand and Uttar Pradesh	180-190	Hills:14-15 Plains: 9-10
VL Lahsun 2	ICAR-VPKAS	White	Himachal Pradesh, Jammu & Kashmir and Uttarakhand	190-200	Mid hills:14-16 Above mid hills: 24-26
Agri Found Parvati	NHRDF	Purple	Hills of Himachal Pradesh, Uttarakhand, Jammu & Kashmir and high altitudes of North eastern states like Sikkim *etc.*	165-175	17-18
Agri Found Parvati-2 (G-408)	NHRDF	White	Hills of Himachal Pradesh, Uttarakhand, Jammu & Kashmir and high altitudes of North eastern states like Sikkim *etc.*	165-175	17-22

8.4. Climate

- Grown at elevation of 1200-2000 masl.
- Requires short days, cool (12-18°C) moist period during vegetative growth.
- Short days are favourable for formation of bulbs.

8.5. Soil

- A well-drained friable (loamy) soil rich in organic matter with a pH range of 6-7 is suitable.
- Extreme acidic and heavy soils are not suited.

8.6. Season

- **Hilly areas**: 1st season during March-April; 2nd season during September-November.

8.7. Seed rate

- 500-600 kg cloves /ha having 8-10 mm dia meter.

8.8. Spacing

- Form ridges and furrows at 30 cm spacing or beds of convenient sizes.
- Cloves are planted at 15 x 10 cm spacing.

8.9. Irrigation

- Irrigate before and after planting
- Life irrigation is given on third day of planting and subsequently, at weekly intervals.

8.10. Manures and fertilizers

- Basal: Incorporate 50 t/ha of FYM.
- Apply *Azospirillum* at 2 kg and Phosphobacteria at 2 kg/ha.
- 40:75:75 kg/ha NPK, 50 $MgSO_4$ and 1 ton neem cake
- Top dressing: N at 35 kg/ha at 45 DAP.

8.11. Controlling rubberisation

- Avoid excessive application of N.
- Use ammonium sulphate or urea.
- Spray 1500 ppm of CCC on 30 DAP.
- Pendimethalin at 3-3.5 lit/ha after planting.
- Apply 0.2% boron + 0.1% sodium molybdate on 30, 60 and 90 DAP.

8.12. Weed management

- Weeding should be done on 30 and 60 DAP.

8.13. Harvest

- Harvest when turning of tops to yellowish or brownish colour and starts drying.

Splitting

- Delayed harvesting or orrigation after long spell of drought is the reason for splitting.

Sprouting of bulbs

- Excessive moisture or winter rains and supply of nitrogen.

8.14. Yield

- 8-12 t/ha

8.15. Curing

- Bulbs are cured for 7-8 days in shade either with tops or after cutting tops leaving 2.5 cm of the stalk.

8.16. Grading

- Big (30 mm diameter)
- Medium (25 mm)
- Small (12 mm)

9

Paprika *(Capsicum annuum var. longum L.)*, Solanaceae, 2n=24

9.1. Origin and distribution

- Originate from North America, in particular Central Mexico.

9.2. Economic importance

- Paprika contains capsanthin and capsorubin.
- Used to season and color rice, stews, and soups, such as goulash, and in the preparation of sausages, mixed with meats and other spices.
- There are six paprika American, Bulgarian, Hungarian, Moroccan, Spanish and Yugoslavian.
- In US, sprinkled raw on foods as a garnish.

9.3. Types and varieties

- **Types of bell pepper**
    - **Green pepper**: Can be a little bitter, but it is very useful, such a fruit reduces the level of cholesterol in the blood. And it has the least calories.
    - **Red pepper**: Is the sweetest and contains the most vitamin C.
    - **Orange pepper**: It is also very tasty, but it contains slightly less vitamin C than red.
    - **Yellow pepper**: It has the highest content of potassium.
    - **Dark purple** and almost black are also very helpful.
- **Varieties**
    - KTPL-19, Byadagi Kaddi, Byadagi Dabba, Paprika King.

9.4. Soil

- Well drained sandy loam or clay saline free soil with pH of 6.5-7.0.

9.5. Season

- June-July.

9.6. Seed rate

- 500 g/ha.

9.7. Nursery

- Prepare 10-12 beds of 7 m long, 1.2 m wide and 15 cm height
- Apply 15-20 kg well decomposed compost and 500 g of 15:15:15 NPK to each bed 15 to 20 days before sowing.
- Sow seeds in rows at 10 cm apart on 0.5 cm deep.

9.8. Spacing

- 60 x 45 cm

9.9. Manures and fertilizers

- **Basal**: Apply FYM 20-25 t/ha and 60, 100 and 60 kg NPK/ha.
- **Top dressing**: 20 kg N and 20 kg K/ha at 3 weeks after transplanting.
- 40 kg N and 40 kg K/ha at 6 weeks after transplanting.

9.10. Yield

- 25-35 t/ha.

10

Tamarind (*Tamarindus indica*), Caesalpiniaceae, 2n=24

10.1. Origin and distribution

- Native of Tropical Africa and cultivated throughout South East Asia, Australia, America and India.
- Cultivated in Tamil Nadu, Maharashtra, Karnataka, Andhra Pradesh, Madhya Pradesh and Odisha.

10.2. Economic importance

- The fruits are used as condiment or as acidulant.
- Fruits that contain a sweet, tangy pulp, which is used in cuisines around the world.
- Used as an acidic/flavouring agent in the Indian cookery.
- High in tartaric acid, sugar, B vitamins, and calcium.
- Used in traditional medicine and as a metal polish.
- Used for woodworking and oil can be extracted from the seeds.
- Seeds are astringent, aphrodisiac useful in giddiness, vertigo.

10.3. Varieties

- PKM 1, Urigam, Hasanur, Tumkur, Prathisthan, DTS 1, Yogeshwari.
- **Important varieties in India**
  - **PKM-1**: A clonal selection from the gene bank and it is early variety yielding 263 kg pods/tree with a pulp content of 39%. It can give 26 tonnes of pods/ha, if transplanted at a spacing of 10 m x 10 m.
  - **Urigam**: This is another local type providing very long pods, having sweet pulp.
  - **Prathisthan**: Variety released from Marathwada Agricultural University. Pulp has acid sweet taste, 60% pulp. 8% tartaric acid in pulp.
  - **Yogeshwari**: Variety released from Marathwada Agricultural University. High yielding type with red pulp, regular bearer with 11.00% acidity.

10.4. Soil and climate

- Grown on variety of soils ranging from poor degraded, eroded, gravelly, saline and alkaline soils.
- Productivity is higher in red loamy, deep well drained soils.
- Maximum temperature: 36-47.5°C; minimum temperature: 0-17.5°C.
- Rainfall: 750-1900 mm.
- Altitude: up to 100 m above msl.

10.5. Season

- June-December.

10.6. Propagation

- Seeds/grafts

10.6.1. Nursery

- Sown in nursery beds in March-April.
- Soaking of seeds in 10 % cow urine or in cow dung solution (500 g in 10 litre of water) for 24 hrs.
- Two year old seedlings are transplanted to the main field.

10.6.2. Vegetative propagation

- Softwood grafting: March-April
- Air layering: Shoots treated with IBA @ 4000 ppm.

10.7. Spacing

- 8-10 x 8-10 m is adopted.

10.8. Planting

- Planted in the pits of 1 x 1 x 1 m filled with FYM and top soil.
- Add 50 g of methyl parathion 1.3% dust in the pit.
- Immediately after planting, support the graft with stakes.

10.9. Irrigation

- Once in seven days.

10.10. Manures and fertilizers

- Apply 200:150:250 g of NPK/tree/year along with 25 kg of FYM and 2 kg of neem cake.

10.11. Intercropping

- Grow leguminous crops, short duration vegetables, annual drumstick, sesamum and sorghum up to four years.

10.12. Training and pruning

- Early training is necessary to form a high head and uniform scaffold branches in all directions.
- Removal of rootstock sprouts, dried, diseased and criss-cross branches.

10.13. Harvest

- Bearing starts from 4th year onwards and economic yield from 9th year onwards.
- Harvest in March-April every year.

10.14. Yield

- 150-200 kg of fruits/tree/year.

11

Vanilla (*Vanilla planifolia* Andrews) Orchidaceae, 2n=28

11.1. Origin and distribution

- Native of Atlantic Coast from Mexico to Brazil.
- Vanilla comes from orchids of the genus vanilla.
- Originally came from Mesoamerica, including parts of modern day Mexico and Guatemala.
- Major vanilla producing countries are Java, Mauritius Tahiti, Seychelles, Madagascar, Indonesia, Mexico, Cameroon and Reunion.
- Madagascar and Indonesia produce two thirds of the world's supply of vanilla.
- In India, vanilla cultivation is gaining importance in Kerala, Karnataka and Tamil Nadu since early 1990s.

11.2. Economic importance

- Second most expensive spice after saffron
- Used in both commercial and domestic baking, perfume manufacture and aromatherapy.
- To treat intestinal gas and fever and to increase sexual desire (As an aphrodisiac).
- As a flavouring agent, used in the preparation of ice creams, milk, beverages, candies, confectioneries and various bakery items.

11.3. Varieties

- **Madagascar**: The Madagascar Bourbon vanilla bean is what most of us associate with the typical vanilla extract. It has a sweet, classic flavor that is great for baking or using in your favorite comfort foods.
- **Mexican**: Mexican vanilla beans are an interesting variety. They have the smooth, classic flavor of vanilla but with an added kick of spice. It is an excellent choice for using in chocolate dishes, cinnamon-flavored dishes, or barbecue sauces.

- **Indonesian**: This variety of vanilla has a milder, earthy flavor with an added touch of smoke. It is a good option for baking or for pairing with chocolate.
- **Tahitian**: Tahitian variety is fun. It has a fruity flavor with a hint of cherry undertones. It is an excellent option for use in ice cream, paired with fruit, use din puddings, or in beverages.
- **Indian**: Indian variety is known for having a bolder flavor. Because of the boldness, it is a great option to be paired with chocolate.
- **Tonga**: Tonga vanilla has less spice and more of a woody flavor to it. It is great when paired with dishes that highlight raisins and figs.

11.4. Climate

- Humid tropical climate with an annual rainfall of 150-300 cm
- Elevation: 700-1500 m above msl
- Temperature: 21-32°C.

11.5. Soil

- Lateritic loamy soil rich in humus having good drainage.

11.6. Season

- 6 months after planting standards i.e., August-November.

11.7. Standards

- *Glyricidia maculata, Erythrina indica, Plumeria alba* and *Casuarina equisetifolia.*
- Planting during onset of rain i.e., after summer during May and June.
- Pit size 30 cm^3 for standards

11.8. Propagation

- Stem cuttings of 60-120 cm.
- 1m cutting gave the best performance.

11.9. Spacing

- Plains: 2.0 to 3.0 m x 2 to 1.5 m
- Hills: 1.5 x 1.5 m

11.10. Planting

- It requires partial shade of about 50% sunlight.
- Place 2 nodes of unrooted cuttings of 60-120 cm below the soil surface.

11.11. Training

- Train the vines to a height of 1.2 to 1.5 m.
- Then, trained horizontally or allowed to grow downward towards the ground.
- Horizontally trained vines are coiled round the pole connecting the two supporting trees.
- Vines trained to grow downward is allowed to touch the soil and allowed to root and again brought back upward on the same supporting tree and the same procedure is repeated.
- Mulch the vine with pruned vegetation 2-3 times in a year.

11.12. Manures and fertilizers

- FYM 112g/vine/year in two doses during May-June.
- Apply 40 to 60: 20 to 30: 60 to 100 g of NPK/vine/year in 2 to 3 splits.
- Spray 1% solution of 17: 17: 17 NPK mixture once in a month for boosting growth and flower.
- Apply *Azospirillum* and phosphobacteria at 50 g/vine.

11.13.Flowering and pollination

- Flowering starts in December-January from third year onwards.
- Pinching of top 7.5-10 cm of vine at 6-8 months before flowering seasons for flower bud initiation.
- Pruning of older fruiting branches also encourages flower production.
- Flowers start opening from 4 am – 1 pm.
- Hand pollination must be done on the same day.
- A skilled worker can pollinate 1000-1500 flowers in a day.
- In Mexico, the insect Mellipona and humming birds are used for pollination.

11.14. Harvesting

- Pods are harvested in 6 to 9 months after flowering when distal end of pod turns yellow.
- Matured beans change colour from green to pale yellow.
- Harvested by cutting with a knife.

11.15. Yield

- Average cured bean yield is 300 to 600 kg/ha/year.
- 6 kg of green pods produces 1 kg of cured beans.
- Economic life of vine is 12-14 years.

11.16. Vanillism

- Labourers working in plantation of vanilla develop severe headache or gastric trouble.
- This is owing to the presence of calcium oxalate in vines.

11.17. Curing and drying

- The well matured and ripened pod is harvested.
- There are different methods of processing.
    1. Peruvian curing is done by hot water. The pods are dipped in boiling water and the pods are hung in the open and allowed drying for 20 days.
    2. Guiana process: Pods are dried in sun.
    3. Mexican process: Harvested pods kept under shade till shrivel.
    4. The best quality pods are 17-25 cm long, vanillin 1.48-2.90%.

12

All Spice (*Pimenta dioica* (L) Merr), Myrtaceae, 2n=22

12.1. Origin and distribution

- Indigenous to West Indies (Jamaica), Central and South America. Jamaica is the main producer of allspice.
- In India, there are few trees in Maharashtra, Tamil Nadu, Karnataka and Kerala.
- Available in the Mahendragiri hills of Kanniyakumari, and Kallar and Burliar of the Nilgiris of Tamil Nadu; Ambalavayal in Kerala.

12.2. Economic importance

- It combines the odour and flavour of nutmeg, clove, cinnamon and pepper and hence, called all spice.
- Berry, berry oil, oleoresin, leaf oil are the products of economic use.
- Used in food industry (65 to 70%), domestic use (5 to 10%), production of berry oil (20 to 25%), extraction of oleoresin (1 to 2%), and pharmaceutical and perfume industry.
- Has, anti-microbial, insecticidal, nematicidal, anti-oxidant and deodorant properties.
- Dry unripe fruit is the spice of commerce.

12.3. Soil and climate

- Deep rich loams with high humus content and lateritic soils
- Altitude: 1000 m above msl
- Annual rainfall: 100-200 cm
- Ideal temperature: 27°C

12.4. Propagation

12.4.1. Seeds

- Soak seeds in water overnight, rubbing in a sieve, washing with clean water and drying in shade.

- Sow the seeds in nursery beds of 1.2 m width incorporated with organic matter or a mixture of sand and coir dust or coir dust alone and mulched.

12.4.2. Vegetative propagation

- Air layering of soft wood and semi hard wood shoots in January.

12.5. Spacing and planting

- 6 x 6 m.
- Plant 9-10 months old seedlings of 25-30 cm height.
- Since dioecious, keep a male: female ratio of 1:10 in plantation.

12.6. Manures and fertilizers

- First year:Apply FYM at 10 kg + 20:180:50 g of NPK/tree in shallow trenches about 1-1.5 m away from the tree.
- 15 years old trees and more: FYM at 50 kg + 300:250:750 g of NPK/tree is applied as two split doses.

12.7. Growth regulator

- Soil application of paclobutrazol (1.25g/tree) induces flowering, increased number of panicles, flowers and setting of berries.

12.8. Harvesting

- Clonally propagated plants: Flowering start in three years
- Seed propagated plants: 6 years.
- Fully developed green berries are harvested.

12.9. Yield

- 50-60 kg dry berries/year/tree.

Section II
Production Technology of Condiments

1

Coriander (*Coriandrum sativum* L.) Apiaceae, 2n=22

1.1. Origin and distribution

- Native to Southern Europe and Northern Africa to Southwestern Asia.
- Rajasthan, Andhra Pradesh, Tamil Nadu, Odisha, Karnataka and Madhya Pradesh are major producers of coriander in India.

1.2. Economic importance

- All parts of plant are edible, but fresh leaves and dried seeds are used in cooking.
- May help to lower blood sugar.
- Rich in immune boosting antioxidants.
- May benefit heart health.
- May protect brain health.
- May promote digestion and gut health.
- May fight infections.
- May protect your skin.
- Easy to add to your diet.
- Used for digestion problems including upset stomach, loss of appetite, hernia, nausea, diarrhea, bowel spasms, and intestinal gas.
- Also used to treat measles, hemorrhoids, toothaches, worms, and joint pain

1.3. Varieties

- CO 1, CO 2, CO 3, CO (Cr) 4, Karan, CIMPOS-33, CS 2

Variety	Characteristics
CO 1	It was released by TNAU, Coimbatore. Plant is taller in height with many umbels per plant. It is suitable for green and grains. Duration is 110 days, while yield is 500 kg per ha.
CO 2	Released by TNAU, Coimbatore. High yield, dual purpose variety, tolerant to drought, oil 0.3%. Duration: 90-110 days. Yield: 600-700 kg per ha.

CO 3	Released by TNAU, Coimbatore. High yield, dual purpose, grain is medium sized; seed oil content is 0.38-0.41%. Duration is 103 days and yield is 400 kg per ha.
Gujarat Corinader-1	Released by GAU, Jagudan. High yield, more number of branches, seeds are bolder with greenish color. Duration: 112 days. Yield: 1100 kg per ha.
Gujarat Coriander-2	Released by GAU, Jagudan. High yield, more branches, dense foliage, umbels large size, grain purpose variety, bold seeds, no lodging. Duration is 110-115 days. Yield is 1500 kg per ha.
Rajendra Swati	Released by RAU, Dholi. Yield potential is high, suitable for intercropping, fine seeded, rich in essential oil and resistant to stem gall disease. Duration is 110 days. Yield is 1200-1400 kg per ha.
Rcr-41	Released by RAU, Jobner. It is tall, erect, suitable for irrigated areas, and resistant to stem gall. Duration is 130-140 days. Yield is 1200 kg per ha and it is a high yielding variety.
Swati	Released by APAU, Guntur. Semi erect and suitable for delayed sowing. Duration is 80-90 days and yield is 885 kg per ha. Also, high yielding.
Sadhana	Released by APAU, Guntur. High yielding and suitable for rainfed areas. It is semi erect, resistant to aphid and mites. Duration is 95-105 days. Yield is 1000 kg/ha.
Sudha (LCC-128)	Medium duration variety with 80-98 days duration. Grain is oblong shaped, medium sized. Yield potential is 750-1200 kg/ha. Essential oil content is between 0.36% and 0.40%.
Sindhu	Medium duration variety with 95-100 days duration. Grain is medium in size with an yield of 1050 kg/ha.
APHU Dhania-1 (LCC-170)	Medium duration variety with 85-90 days duration. Grain is bold, oblong shaped. Yield potential is of 850-1200 kg/ha. Essential oil content is 0.40 %.
Suguna (LCC-236)	Medium duration variety with 90-95 days duration. Grain is slender, oval shaped. Yield potential is 750-1350 kg/ha. This variety is having high volatile oil content (0.52 %).
Suruchi (LCC-234)	It is a high yielding leaf variety suitable for off-season production, comes to harvesting in 35 to 55 days after sowing. This variety gives an average yield of 15-18 t/ha of grain in rabi season and 2.5-4.5 t/ha greens in off-season under 50% shade net. Herb volatile oil content is 0.15%.
Susthira (LCC-219)	It is a high yielding medium duration variety (85-90 days), gives stable yield under rainfed areas. Yield potential is 1.20 to 1.43 t/ha under rainfed conditions and 1.20 to 1.75 t/ha under irrigated conditions. This is variety having high volatile oil content (0.59%).

1.4. Soil and climate

- Well drained silt or loamy soils.
- Rainfed: Clay soil with pH of 6-8.
- Temperature range of 20-25°C.
- Cool and comparatively dry, frost free climate.

1.5. Season

- June-July and October-November

1.6. Seed

- Irrigated crop: 10-12 kg/ha
- Rainfed crop: 20-25 kg/ha
- Split open seeds into halves before sowing for more germination.
- Technically, the full seed is called schizocarp.
- Mericarps are used as seeds.
- Seeds are grouped into two *viz.* leafy type and grain types.
- Seed types classified into small seeded and big seeded types.
- The small seeded types are thin pericarp skinned fruits.
- This is the high oil yielding type.
- The Bulgarian and Russian types are high oil yielding types.
- The major volatile oil is d-linalool or coriandrol.

1.6.1. Seed treatment

- Soak the seeds in water for 12 hrs.
- Treat with *Azospirillum* at 1.5 kg /ha + *Trichoderma asperellum* at 50 kg/ha.
- Presowing hardening with potassium dihydrogen phosphate at 10 g/litre of water for 16 hrs.

1.7. Field preparation and sowing

- Add FYM 10 t/ha before last ploughing.
- Form beds and channels
- Spacing: 20 x 15 cm.

1.8. Manures and fertilizers

- Basal: FYM 10 t/ha; 10 kg N, 40 kg P and 20 kg K for rainfed and irrigated crops.
- Top dressing: Top dressing with 10 kg N/ha on 30 DAS for irrigated crop.

1.9. Irrigation

- Irrigation at 7-10 days interval.

1.10. After cultivation

- Spray pre-emergence herbicide fluchloralin at 700 ml in 500 litres of water/ha.

- Also pendimethalin at 7 kg/ha.
- Weeding is done 30 days after sowing.
- Spray CCC at 250 ppm one month after sowing for inducing drought tolerance in rainfed crops.

1.11. Harvest

- Plants are pulled just when the fruits are fully ripe, but green and start drying.
- Dried and threshed with sticks, winnowed and cleaned.
- For leaf, pull out the plants when they are 30-40 days old.

1.12. Yield

- Rainfed grain yield: 300-400 kg/ha
- Irrigated: 500-600 kg/ha
- Leaf yield: 6-7t/ha

2

Fenugreek (*Trigonella foenum graecum L.*), (*T. corniculata*), Fabaceae, 2n=16

2.1. Origin and distribution

- Native to South Eastern Europe, West Asia and Mediterranean region.
- Cultivated in Central and Southeastern Europe, Western Asia, India, and Northern Africa.
- In India, cultivated in Rajasthan, Madhya Pradesh, Gujarat, Uttar Pradesh, Maharashtra and Punjab.

2.2. Economic importance

- Culinary use in smaller quantities.
- Used as an herb (Dried or fresh leaves), spice (Seeds) and vegetable (Fresh leaves, sprouts, and micro greens).
- Seeds are used both whole and powdered in the preparation of pickles, vegetable dishes, dal and spice mixes such as panch phoron and sambar powder.
- Balance cholesterol.
- Soothe upset stomach and digestive problems.
- Reduce menstrual cramps.
- Seeds are aphrodisiac effect.
- Reduce appetite.
- Reduce fat mass.
- Maintain liver and kidney health.
- Soothe muscle pain.
- Promote lactation of women.
- Reduce fever.
- Diosgenin present in the seed used in sex hormone synthesis.
- Algerian and Morocco types contain high level of diosgenin.

2.3. Varieties

- CO 1, CO 2, Pusa Early Bunching, Lam Selection 1, Rajendra Kranti, Hissar Sonali, RMT 1, RMT 143, RF 101, RF 125, RF 35.

Variety	Pedigree / Parentage	Average yield (kg/ha)	Duration (days)	Salient features
CO 1	Reselection from TG-2356 introduced from North India	680	80-85	A quick growing, dual purpose, early maturing variety tolerant to root rot disease. Seeds contain 21.7% protein.
CO 2	Selection from CF 390	480	85-90	Short duration dual purpose variety, field tolerant to *Rhizoctonia* root rot disease, suitable for both kharif and rabi season. Early maturity, short duration.
Rajendra Kanti	Pure line selection from Reghunathpur collection	1300		Medium sized bushy plant; early maturity, suitable for intercropping in both kharif and rabi season, field tolerant to *Cercospora* leaf spot, powdery mildew and aphids.
RMT. 1	Pure line selection from Nagpur local	1400		Vigorous semi erect medium sized, moderately branched growth habit, medium sized, bold and attractive typically yellow coloured grains, moderately resistant to root knot nematode and powdery mildew and aphids
Lam Sel.1	Selection from germplasm collection of Uttar Pradesh	740		Dual purpose variety early maturing, bushy type and medium height 94 cm, more number of branches and green matter. When cultivated for green leaf purpose, it gives an average green yield of 12 t/ha. Field tolerant to major pests and diseases.
Hisar Sonali	Pure line selection from germpalsm	1700		Tall and bushy vigorous growing variety, dual purpose variety, late maturity (140-145 days), suitable for cultivation under irrigated condition. Moderately resistant to root rot and aphids.
Hisar Suvarna	Pureline selection from local germplasm	1600		A quick growing, erect and tall, dual purpose, medium maturity (130-140days), moderately resistant to Cercospora and powdery mildew. Wider adaptability, suitable for cultivation throughout the country.

Hisar Madhavi	Pureline selection from local germplasm of UP	1900		A quick growing, erect and tall, dual purpose, medium maturity (130-140 days) moderately resistant to powdery mildew and to downy mildew. Suitable for both irrigated and rainfed condition.
Hisar Muktha	Pureline selection, natural green seed coated mutant line from UP	2000		A quick growing seed type variety, medium maturity (135-140 days), moderately resistant to powdery mildew and to downy mildew. Erect and tall plants. Wide adaptability. Suitable for both irrigated and rainfed condition
RMT 303	Mutation breeding from variety RMT 1	1900		Medium maturity variety (145- 150 days), seeds bold with typical yellow colour, less susceptible to powdery mildew
RMT 305	Mutation breeding from variety RMT 1	1300		First determinate type, early maturing, wider adaptability, resistant to powdery mildew and root knot nematodes. Seeds bold, attractive and yellow, duration 120-125 days.
Guj Methi 1	Recurrent selection based on pure line selection from J. Fenu 102	1864		The first variety from Gujarat released for the state. Plants dwarf.
RMT 143		1600	140-150	Moderately resistant to powdery mildew, seeds bold yellow colour, suitable for heavier soils.
Pant Ragini	Selection from local germplasm	1200	170-175	A dual purpose tall bushy type, resistant to downy mildew and root rot, medium maturity. Seed contain 2-2.5% essential oil
AM-01-35	Selection from local germplasm	1720		Dual purpose, tolerant to powdery mildew

2.4. Soil

- A rich well drained loamy soil.
- pH 6.0-7.0
- Tolerant to salinity.

2.5. Climate

- Cool and comparatively dry, frost free climate.
- Tropical and temperate spice.
- Elevation up to 2000 m

2.6. Season

- June-July and October-December

2.7. Field preparation and sowing

- Seed rate: 20 kg/ha
- Form beds and channels of 3.5 x 1.5 m.
- Spacing: 20 x 15 cm.
- Cycocel: 50-100 ppm improves germination.
- Ascorbic acid at 250-450 ppm improves growth, flowering and seed number.

2.8. Manures and fertilizers

- Basal: FYM 20-25 t/ha and N, P, K at 30:25:40 kg/ha.
- Top dressing: 20 kg of N at 30 DAS.

2.9. Irrigation

- First irrigation is given immediately after sowing, second on tthird day and subsequently at 7-10 days interval.

2.10. After cultivation

- Spray pre-emergence herbicide fluchloralin at 700 ml in 500 litres of water/ha.
- Plants are thinned at 20-25 DAS.
- One pinching at a height of about 4" will encourage branching.

2.11. Crop duration

- Greens: 20-25 days
- Grains: 90-100 days

2.12. Yield

- Greens: 4-5 t/ha
- Grains: 500-700 kg/ha

3

Fennel (*Foeniculum vulgare Mill.*) Apiaceae, 2n=22

3.1. Origin and distribution

- Originated in the Southern Europe, Mediterranean region and grows throughout Northern, Eastern, and Western hemispheres specifically in Asia, Rumania, North America, Europe, India, Russia, Germany and France.
- In India, cultivated in Rajasthan, Andhra Pradesh, Punjab, Madhya Pradesh, Uttar Pradesh, Maharashtra, Gujarat, Karnataka, and Haryana.

3.2. Economic importance

- The leaf stalks are used both as vegetable and salad.
- A common practice in most Indian households is to have few fennel seeds at the end of every meal to freshen the mouth.
- Used for various digestive problems including heartburn, intestinal gas, bloating, loss of appetite and colic in infants.
- Source of copper, potassium, calcium, zinc, manganese, vitamin C, iron, selenium and magnesium.
- The seeds are used as condiment
- Seeds yield 4-5% volatile with anethole as major constituent.
- *F. vulgare* sub *capillaccum* has two distinct varieties
  - i) *F. vulgare* sub *capillaccum* var *vulgare*-Cultivated or wild yielding bitter fennel oil
  - ii) *F. vulgare* sub *capillaccum* var *dulce* - Cultivated yielding sweet or roman or Florence fennel oil.

3.3. Varieties

- CO 1, UF 32, PF 35 and Gujarat Fennel 1.
- **Ajmer Fennel 1**: Tolerant to *Ramularia* and *Alternaria* blight. The seeds bold, fragrant, volatile oil 1.6%, 2-2.2 t/ha. Suitable for all over the country in irrigated condition.

- **Ajmer Fennel 2 (AF-2)**: Developed at NRCSS, Ajmer. It is suitable for both early sowing and as rabi crop. Plants are erect and tall, bearing large size umbels. Average yield is 1.79 t/ha during rabi crop.
- **Gujarat Fennel 12:** Synchronized maturity
- **Gujarat Fennel 2**: High yielding variety developed at Spice Research Centre, Gujarat Agricultural University through pedigree selection of local genotype. This variety is suitable for cultivation during rabi season under irrigated condition. It produces yield of 1.94 t/ha.
- **Hisar Swarup**: Developed at Haryana Agricultural University, Hisar. This is a mass selection from indigenous germplasm of Haryana and has a yield potential of 1.6 t/ha.
- **HF 143**: Developed at HAU Hisar, Haryana. Average yield is 1.78 t/ha.
- **CO 1:** Developed by Tamil NaduAgricultural University, Coimbatore through reselection from PF-35. Suitable for intercropping and border cropping with chilli and turmeric. It matures in 210-220 days and gives an average yield of 570 kg/ha.
- **Rajendra Saurbha**: Developed by SKN College of Agriculture (RAU), Jobner. It gives 1.8-1.9 t/ha seed yield.
- **RF 101:** It bears large umbels with long bold grains. The average yield is 1.55 t/ha
- **RF 125:** Bears compact umbels and long bold grains. Average yield is 1.55 t/ha
- **RF 143:** Developed by RAU, Jobner through recurrent selection from individual plant progeny. It gives 1.2 t seed yield per hectare.
- **RF 205:** Developed by RAU, Jobner. It gives maximum 2.94 t/ha seed yield.
- **RF 178:** Developed by RAU, Jobner. It gives 1.8-1.9 t/ha seed yield.
- **RF 281:** Developed by RAU, Jobner. It gives 1.8-1.9 t/ha seed yield.

3.5 Soil and climate

- Well drained loamy soil or black or sandy soil rich in organic matter.
- Comes up well in fairly mild with frost free climate, cold weather crop.
- Elevation up to 1825 m.

3.6. Season

- Hills: March-April
- Plains: October-November.

3.7. Field preparation and sowing

- Seed rate: For direct sowing - 9-12 kg/ha.
- For transplanting: 3-4 kg/ha
- Seeds are soaked in water for 5 days for quick germination.
- Nursery area: 100 sq.m.
- Form beds and channels.

3.8. Spacing

- 4-5 weeks old seedlings are transplanted at 60 x 30 cm spacing.
- 2 seedlings/hill.

3.9. Irrigation

- First irrigation should be given immediately after sowing, second on the third day and subsequent irrigations at 7-10 days intervals.

3.10. Manures and fertilizers

- Basal dressing: 25:10:0 kg/ha NPK
- 12-15 tons FYM as basal.
- Top dressing: 25kg N/ha after sowing.

3.11. After cultivation

- In direct sown crop, thin 4-5 weeks old seedlings to 25-30 cm spacing.
- Pre-emergence application of pendimethalin at 1 kg a.i./ha.

3.12. Harvest

- Matures in 7-8 months.
- Harvest the umbels 4-5 times at 10-15 days interval, dry in sun for 4-5 days and seeds are threshed.

3.13 .Yield

- 500-1200 kg/ha/year.

4

Cumin (*Cuminum cyminum* L.) Apiaceae, 2n=14

4. 1. Origin and distribution

- Native to Egypt, Syria, Middle East and stretching east to India.
- In India, cumin is mainly cultivated in Western Indian states like Rajasthan and Gujarat.

4.2. Economic importance

- Important spice used in Indian kitchens for flavoring various food preparations.
- The flavor is due to chief constituent cuminol (Cuminaldehyde)
- Volatile oil is up to 2.5-3.5 %.
- Used in various ayurvedic medicines especially for the conditions like obesity, stomach pain and dyspesia.
- Many Latino, Vietnamese, United States and Thai foods recipes have cumin and many chilli recipe includes cumin.
- In India, cumin is a traditional ingredient not only curry, but also kormas, masalas, soups and other recipes.
- Used in some cheeses like Leyden cheese as well as some French breads.
- Curry powder is not the only blend in which cumin is found; achiote, chilli powder, adobos, sofrito, garam masala and bahaarat all owe their distinct ethnic flavors partially to cumin.
- Used whole or ground in some pastries and pickles.
- A mix of cumin, garlic, salt, and chilli powder on grilled corn is delicious.

4.3. Varieties

- **RZ 19**: Tall variety of cumin with erect stem, pink flowers and bold pubescent grains; matures in 120-140 days with an average yield of 560 kg/ha.

- **RZ 209**: Erect with pink flowers and bold, grey, pubescent grains; matures in 140-150 days with an average yield of 650 kg/ha.
- **GC 1**: Erect with pink flowers and bold, linear, oblong, ash brown colour grains; matures in 105-110 days with an average yield of 700 kg/ha.

4.4. Climate

- Moderate sub-tropical mild climate is ideal
- Moderately cool and dry climate is best.
- Does not stand high humidity and heavy rainfall.

4.5. Season

- November-December.

4.6. Soil

- Well drained, loamy soils rich in organic matter.

4.7. Propagation

- Seed propagation
- Seed rate of 12-15 kg/ha
- Sown by broadcasting or line sowing at 10cm depth with 20-30 cm spacing in rows, 0-15 cm plant to plant.

4.8. Manures and fertilizers

- FYM: 10-15 tons/ha
- 20kg P_2O_5/ha as basal application; 40-50 kg N/ha on 30 DAS.

4.9. Weed management

- Weeding at 30 and 60 DAS.
- Preemergence herbicide application of oxadiazon at 0.5-1.0 kg/ha and fluchloralin or pendimethalin at 1.0 kg/ha.

4.10. Irrigation

- Light irrigation is done soon after sowing and thereafter, second irrigation at 8-10 days after first irrigation.
- Subsequent irrigation at 15-25 interval.
- Consumptive use of water is 227 mm.

4.11. Harvesting

- Done by cutting the plants with sickle and stacked on clean threshing floor for sundrying.
- After drying, seeds are separated by light beating with sticks and winnowing.

4.12. Yield

- Average yield of 500 kg/ha

5

Indian Mustard (*Brassica juncea* (L.) Czern), Brassicaceae (Cruciferae), 2n=36

5.1. Origin and distribution

- True or black mustard: *Brassica nigra* Koch-Self sterile
- White mustard (Yellow mustard) : *Brassica compestris,* Syn: *Sinapsis alba*-Self sterile
- Indian or brown mustard: *Brassica juncea* (L.) Czern
- These three species are commonly used as condiment.
- White mustard is indigenous to Southern Europe, whereas brown mustard is from China introduced to Northern India.
- Black mustard is endemic in Southern Mediterranean region.
- White mustard is widely cultivated in Australia, China, Chili, Denmark, Italy, Japan, UK, Netherlands, North Africa, Canada and USA.
- The Indian mustard is fertile.
- Third most important edible oilseed crops of the world after soybean and oil palm (37 to 49 %).

5.2. Economic importance

- Seed and oil are used as condiment in the preparation of pickles, curries, vegetables, hair oils, medicines and manufacture of greases.
- Oil cake is used as cattle feed and manure.
- In the tanning industry, mustard oil is used for softening leather.
- Used in the manufacture of mayonnaise, dried or dehydrated mustard leaves, whole mustard seeds.
- Whole mustard is used as a flavouring agent in Indian cooking, whereas ground mustard provides flavour and consistency in Bengali fish curries.
- Flour has preservative and antioxidant properties in addition to providing flavour and colour.
- The mustard flour of commerce is a mixture of brown or black and white mustard.

- Volatile oil of brown mustard is allyl isothiocyanate.
- Volatile oil of white/yellow mustard is acrinyl isothiocyanate.
- Volatile oil of black mustard allyl isothiocyanate.
- The volatile oil of mustard on hydrolysis is the glucosinolates (Sinigrin) which breaks to enzyme myrosin.

5.3. Varieties

5.3.1. Toria varieties

- PBT 37, TL 15, TL 17, Raya, RLM 619, PBR 91, PBR 97, PBR 210

5.3.2. Gobhi sarson (Canola type)

- Oil is good for health of human.
- GSL 1, PGSH 51, Hyola PAC 401, GSC 6

5.3.3. Indian mustard

- RH 0749, RH 0406, T 59 (Varuna), Pusa Agrani, Pusa Mustard 21, Pusa Mustard 24, NPJ 112, Pusa Mustard 26, Pusa Mustard 28.

NRCDR 2

- Plant height: 165-212 cm
- Average seed yield: 1.95-2.63 t/ha
- Oil content: 36.5-42.5%
- Days to maturity: 131-156 days

NRCHB 506 hybrid

- Recommended for Rajasthan and Uttar Pradesh
- Plant height: 180-205 cm
- Average seed yield: 1.55-2.54 t/ha
- Oil content: 38.6-42.5%
- Days to maturity: 127-148 days

NRCDR 601

- Recommended for Delhi, Haryana, J&K, Punjab and Rajasthan
- Plant height: 161-210 cm
- Average seed yield: 1.94-2.62 t/ha
- Oil content: 38.7-41.6%
- Days to maturity: 137-151 days

NRCHB 101

- Recommended for M.P., U.P., Uttarakhand, Rajasthan, Bihar, J&K, West Bengal, Odisha, Chhattisgarh, Manipur.

- Plant height: 170-200 cm
- Average seed yield: 1.38-1.50 t/ha
- Oil content: 34.6-42.1%
- Days to maturity: 105-135 days

DRMRIJ-31 (Giriraj)

- Recommended for Delhi, Haryana, Jammu & Kashmir, Punjab and parts of Rajasthan
- Plant height: 180-210 cm
- Average seed yield: 2.23-2.75 t/ha
- Oil content: 39-42.6%
- Days to maturity: 137-153 days

DRMR 150-35

- Recommended for Bihar, Jharkhand, West Bengal, Odisha, Chhattisgarh and Manipur
- Early sown rainfed condition
- Average seed yield: 1.83 t/ha
- Oil content: 39.8%
- Days to maturity: 114 days (86-140 days)

DRMR 1165-40

- Recommended for Rajasthan, Delhi, Haryana, Punjab and Jammu & Kashmir
- Plant height: 177-196 cm
- Average seed yield: 2.2-2.6 t/ha
- Oil content: 40-42.5%
- Days to maturity: 133-151 days

NRCYS-05-02 (Yellow Sarson)

- Plant height: 110-120 cm
- Average seed yield: 1.24-1.71 t/ha
- Oil content: 38.2-46.5%
- Days to maturity: 94-181 days

5.4. Climate

- Requires cool and dry weather and a fair supply of soil moisture during the growing period and a dry clear weather at the time of maturity.

5.5. Soil

- Prefers sandy loamy or clayey loam soil with neutral pH of 6.5-7.5.
- Does not tolerate water logging conditions or heavy soils.

5.6. Season

- Rabi season - September to October to February to March.
- Rainy season - July to November in South India.
- Toria: Mid to the last week of September.
- Sarson and rai: First fortnight of October.

5.7. Seed rate and spacing

- Seed rate: 6.5-7.5 kg/ha.
- Toria: In rows of 30 cm apart; Sarson and rai: 45 cm apart.
- Thinning is done three weeks after sowing to maintain a plant to plant distance of 10 to 15 cm.

5.8. Manures and fertilizers

- Apply 15-20 tons of FYM or compost at the time of field preparation.
- 60-90 kg N, 60 kg P_2O_5 and 40 kg K_2O/ha.

5.9. Irrigation

- Two irrigations at pre bloom and pod filling stages.

5.10. Weed management

- Application of nitrofen at 1-1.5 kg a.i./ha or isoproturon 1 kg a.i./ha as pre-emergence spray.

5.11. Rotation and mixed cropping

- Rotation with maize, cotton, pearl millet, pulses.

5.12. Harvesting and threshing

- Harvest when pods turn yellowish brown; just before pods open to avoid shattering losses
- Sarson is less liable to shattering as compared to toria and mustard.
- Stacked in threshing floor for 5 days before threshing.
- Cleaned seed must be dried in the sun for 4 to 5 days to reduce moisture to 8%.

5.13. Yield

- Rapeseed: 1.4-2 t/ha
- Mustard: 2.0-2.5 t/ha.

6

Curry Leaf (Murraya koenigii Linn. Sprengal), Rutaceae, 2n=18

6.1. Origin and distribution

- The plant originated in the Tarai region of Uttar Pradesh, India, and at present it is cultivated in Burma, Ceylon, China, Australia and the Pacific Islands.

6.2. Economic importance

- Curry leaves are a rich source of vitamin A, vitamin B, vitamin C, vitamin B2, calcium, and iron, apart from a heavy distinctive odor and pungent taste.
- It helps in the treatment of dysentery, diarrhea, diabetes, morning sickness, and nausea by adding curry leaves to your meals.

6.3. Varieties

- Sen Kaampa, Dharwad-1 and Dharwad-2

6.4. Soil and climate

- Red sandy loam soils with good drainage are ideal for better leaf yield.
- Optimum temperature is 26° to 37°C.

6.5. Season and planting

- Immediately after collection of fruits, the seeds should be pulped and sown in nursery beds or poly bags.
- One year old seedlings are suitable for planting.
- One seedling is planted at the centre of the pit.

6.6. Spacing

- Pit size of 30 x 30 x 30 cm is dug one to two months before planting at a spacing of 1.2 to 1.5 m.

6.7. Irrigation

- Immediately after planting, the pits are irrigated.
- On the third day, the second irrigation is given and the subsequent irrigation is given once in a week.

6.8. Manures and fertilizers

- FYM: 20 t/ha
- After each harvest, 20 kg of FYM/plant is applied and mixed with soil.

6.9. Inter cultivation

- Periodical hoeing and in the first year, intercrop like pulses can be grown.
- After attaining 1 m height, the terminal bud is cut off to encourage basal branching.
- In total, 5-6 branches are maintained per bush.

6.10. Plant protection

6.10.1. Pests

Citrus butterfly

- Hand picking and destruction of the larvae and spray malathion at 1 ml/lit.

Psyllid bug and scale

- Spraying of dimethoate at 1 ml/lit.

6.10.2. Diseases

Leaf spot

- Spraying of carbendazim at 1 g/lit of water.
- Spraying sulphur compounds should be avoided.

6.11. Harvest and yield

- At the end of first year, 250-400 kg of leaves/ha can be harvested.
- II year: Once in 4 months, every time 1800 kg/ha which accounted for 5400 kg/ha/year
- III year: 5400 kg/ha
- IV year: 2500 kg/ha once in 3 months which accounted for 10,000 kg/ha/year
- V year onwards: 5000 kg/ha once in 3 months which accounted for 20,000 kg/ha/year

7

Pest of Spices and Condiments

7.1. Pepper

7.1.1. Pollu beetle: *Longitarsus nigripennis*

Symptoms

- Grub feeds on growing tip, tender stem, leaf petiole, spikes and berry.
- Tunneling of tender stem causes drying up of shoot.
- Spikes tunneled by grubs turn black and cause dropping of spikes due to rotting.
- Grubs bore into the berries and feed on entire internal contents.
- Exit holes could be seen on dry berries and turn dark - called 'Pollu' berries.

Management

- Regulation of shade in the plantation.
- Spraying quinalphos (0.05%) during June/July and September/October or neem gold (0.6%) during August and September.

7.1.2. Top shoot borer: *Laspeyresia hemidoxa*

Symptoms

- Larvae bore into tender terminal shoots and feed on internal tissues resulting in blackening and decaying of affected shoots.

Management

- Spray quinalphos (0.05%) on tender terminal shoots at monthly intervals (During July-October).

7.1.3. Marginal gall thrips: *Liothrips karnyi*

Symptoms

- Downward and inward curling of leaves resulted in the formation of marginal leaf galls.
- Later, infested leaves become crinkled and malformed.
- In severe cases, whole plant becomes stunted, affecting adversely the formation of spikes.

Management

- Spray dimethoate (0.05%) during emergence of new flushes in young vines and cuttings in nursery.

7.1.4. Pepper mussel scale: *Lepidosaphes piperis*

Symptoms

- Sucking leaf tip, leaf and stem of suckers.
- Infected parts become yellowish and later dry up.

Management

- Spray dimethoate 0.1% followed by one more spray on 21st day of first spray.

7.1.5. White tailed mealy bug: *Ferrisia virgata*

Symptoms

- Mealy bug found in cluster on the terminal shoots, leaves, berries and suck sap.
- Yellowing, withering and drying of plants or shedding of berries.

Management

- Spray dimethoate 2 ml/lit.

7.2. Cardamom

7.2.1. Thrips: *Sciothrips cardamom*

Symptoms

- Panicles become stunted, shedding of flowers and immature capsules.
- Formation of corky encrustations on pods resulting in malformed and shriveled condition.
- Such pods lack their fine aroma and the seeds are also poorly developed.

Management

- Spray quinalphos 0.025% or phosalone 0.07% during March, April, May, August and September.

7.2.2. Shoot, panicle and capsule borer: *Dichocrocis (Conogethes) punctiferalis*

Symptoms

- Early stage larva bores unopened leaf buds and feeds on leaf tissue.
- Also bore panicles leading to drying up
- Feed on young seeds inside rendering the capsules empty.

- Late stage larvae bore pseudostem and feed central core of stem
- Drying of terminal leaf and produce 'dead heart' symptom.

Management

- Remove tillers showing 'dead heart' symptoms.
- Spraying fenthion 0.075%

7.2.3. Beetle borer: *Onthophagus* spp.

Symptoms

- Larva makes neat, circular tiny borc hole and at right angles to pericarp mostly on ventral side.
- Matured appearance of young immature capsules, decay and drop off.
- Adults bore flowers and feed on tissues.

Management

- Judicious shade regulation.
- Spraying of quinalphos at 2ml/lit.

7.2.4. Cardamom aphid: *Pentalonia nigronervosa f. caladii*

Symptoms

- Nymphs and adults suck plant sap.
- Colonies of aphids are seen under conccalcd conditions inside leaf sheaths of the older pseudostems.
- Vector of mosaic or katte virus of cardamom.

Management

- Removal of colocasia and other hosts in the vicinity.
- Remove partly dried and decayed pseudostems which harbour aphids
- Spraying with dimethoate at 2 ml/lit at 2 weeks interval in November and April.

7.2.5. Hairy caterpillar: *Eupterote cardamomi, E. canarica, E. fabia*

Symptoms

- Caterpillars are gregarious in habit and congregate on trunks of shade trees during day time and drop down on cardamom plants during night time.
- Defoliating whole cardamom plants during October-December.

Management

- Caterpillars that congregate at base of shade trees can be collected and killed.

- Installing light traps.
- Spray monocrotophos or quinalphos at 3 ml/lit.

7.3. Turmeric and ginger

7.3.1. Shoot borer: *Conogethes punctiferalis*

Symptoms

- Yellowing and drying of leaves of infested pseudostems.
- Presence of a bore-hole on pseudostem through which frass is exuded and withered.

Management

- Spraying malathion 0.1% at 30 days interval during July to October.

7.3.2. Rhizome scale: *Aspidiella hartii*

Symptoms

- Plants look devitalized, pale and withered before drying completely.
- At harvest, minute yellowish crawlers can be seen moving in large numbers.

Management

- Treat seed material with quinalphos 0.075% (20-30 minutes) before storage and sowing.
- Apply well rotten sheep manure or poultry manure at 10 t/ha in two splits followed by drenching dimethoate 30 EC at 2 ml or phosalone 35 EC at 2 ml /lit of water.

7.3.3. Leaf roller: *Udaspes folus*

Symptoms

- Leaves become folded or rolled longitudinally; complete defoliation.

Management

- Spray carbaryl (0.1%) or dimethoate (0.05%).

7.3.4. Thrips: *Panchaetothrips indicus*

Symptoms

- Leaves become rolled up, turn pale and gradually dry up

Management

- Spray quinalphos 0.025% or phosalone 0.07%.

8

Diseases of Spices and Condiments

8.1. Pepper

8.1.1. Phytophthora foot rot / quick wilt: *Phytophthora capsici*

Symptoms

- **Die back**
- Discoloration occurs and rotting progress upwards and downwards resulting in die-back.
- Lateral branches of affected vines break off at nodes and fall off.
- **Foot rot or collar rot**
- Stem near ground level get infected and rotting and death of vine within 2-3 weeks.
- Affected portion emits bad odour.
- **Root rot**
- Infection starts at main root or at feeder root.
- Leaves become yellow and defoliate.

Management

- Soil drenching with 1% bordeaux mixture after removal of affected plant.
- Spraying with 1% bordeaux mixture or copper oxychloride 0.25%
- Soil application of neem cake and *Trichoderma asperellum* or *Bacillus subtilis*

8.1.2. Slow decline/slow wilt: *Fusarium, Rhizoctonia* and *Diplodia* along with nematode infection

Symptoms

- Yellowing of lower leaves which progress upwards.
- Affected leaves become flaccid and fall off.
- Tip burn of leaves and die-back of twigs appear.
- Affected vines gradually die.

Management

- Soil drenching with copper oxy chloride in May-June and September-October.

8.1.3. Anthracnose/Pollu disease: *Gloeosporium gloeosporioides*

Symptoms

- Circular or irregular grey spots appear in leaves.
- Concentric rings of acervuli appear on upper surface of the leaves.
- Berry becomes brown in colour and extends downwards from upper portion, and become chaffy (Pollu means hallow thing)

Management

  - Three rounds of spray with bordeaux mixture 1%

8.2. Cardamom

8.2.1. Damping off/rhizome rot/clamp rot: *Pythium vexans, Fusarium oxysporum, Rhizoctonia solani, Phytophthora* spp.

Symptoms

- Infected leaves become pale, yellow and ultimately young leaves die.
- Older leaves die prematurely and new shoots are weak.
- Rhizomes rot at the base of the stem.

Management

- Drenching nursery with copper oxy chloride 0.25% or bordeaux mixture 0.5%, before 15 days of sowing.
- Soil drenching with B.M 1% or PCNB 1.0%

8.2.2. Azhukal disease/capsule rot/fruit rot: *Phytophthora parasitica var. nicotianae/ Phytophthora palmivora*

Symptoms

- Large circular, irregular, water soaked spots with black colour appear on leaves.
- Grey patches of irregular spots with brown margin at base of leaf sheath.
- Basal portion rots and the pseudostem break away at the collar region.
- Rotting of rhizomes.
- Small light brown lesions appear in green tender fruits which fall in 3-6 days.

Management

- Three sprays with bordeaux mixture 1% in May, June, July
- Soil drench with bordeaux mixture 1% or copper oxy chloride 0.25%

8.2.3. Chenthal disease/leaf blight: *Colletotrichum gloeosporioides*

Symptoms

- Elongated, water soaked lesions on leaf upper surface.
- Spots become brown to dark with pale yellow hole.
- Leaves wither and pseudostem wilts.
- Flowers fail to develop.
- Inflorescence dry up starting from tip downwards.
- Affected garden shows burnt appearance.

Management

- Three sprays with carbendazim 0.1% or mancozeb 0.2% or copper oxy chloride 0.25% at 30 days interval.

8.3. Clove

8.3.1. Seedling wilt: *Cylindrocladium* sp., *Fusarium* sp., *Colletotrichum* sp. *Rhizoctonia* sp. and *Trichoderma* sp.

Symptoms

- Seedling wilt in nurseries and causes 5 to 40% death of seedlings.
- Leaves of affected seedlings loose natural lustre, tend to droop and ultimately die.
- Root system and collar region of seedling decay.

Management

- Drenched with copper oxy chloride 0.25%.

8.3.2. Leaf rot: *Cylindrocladium quinquiseptatum*

Symptoms

- Dark spots at leaf margin and spreads sometimes with no definite pattern.
- Rotting may be in whole leaf or at tip resulting in defoliation.

Management

- Sprayed with carbendazim 0.1%.

8.4. Nutmeg

8.4.1. Leaf spot and shot hole: *Colletotrichum gloeosporioides*

Symptoms

- Sunken spots surrounded by a yellow halo.
- Central portion of necrotic region drops off resulting in shot hole symptoms.
- Dieback of mature branches.

Management

- Spraying 1% bordeaux mixture two or three times during rainy season.

8.4.2. Fruit rot: *Colletotrichum gloeosporioides, Botryodiplodia theobromae*

Symptoms

- Water soaked lesions on fruits, tissues become discoloured and disintegrated.
- Premature splitting of the pericarp and rotting of mace and seed.

Management

- Spraying 1% bordeaux mixture

8.5. Turmeric

8.5.1. Leaf spot: *Colletotrichum capsici*

Symptoms

- Elliptic or oblong spots on leaves with yellow halo.
- Centre of spots are greyish white or black dots.
- Finally, leaves dry up and give a scorched appearance.

Management

- Spray mancozeb 0.25% or copper oxy chloride 0.25% at 15 days interval

8.5.2. Leaf blotch: *Taphrina maculans*

Symptoms

- Spots of 1-2 mm diameter appear on both sides of leaf.
- Leaf shows a reddish brown appearance.
- Spots coalesce to form irregular bigger patches.

Management

- Spray with mancozeb 0.25% or copper oxy chloride 0.25% at 15 days interval.

8.5.3. Rhizome rot: *Pythium aphanidermatum*

Symptoms

- Basal portion of shoots appear watery and soft.

- Reduced root system.
- Leaves exhibit gradual drying along margin.
- Infected rhizomes soft, rotted with different shades of brown

Management

- Rhizome dip in copper oxy chloride or zineb (0.3%) for 30 minutes before planting
- Soil drenching with copper oxy chloride (0.25%) in and around affected plants
- Soil application of *Bacillus subtilis* talc formulation (2.5 kg/ha)

8.6. Ginger

8.6.1. Soft rot or rhizome rot: *Pythium aphanidermatum/ P. vexans/ P. myriotylum*

Symptoms

- Infection starts at collar region of pseudo stem and progresses upwards as well as downwards.
- Becomes water soaked and rotting spreads to the rhizome resulting in soft rot.
- Yellowing of leaves spreads from lower region upwards, followed by drooping, withering and drying of pseudostem.

Management

- Treatment of rhizomes with mancozeb 0.3% for 30 minutes before storage and once again before planting.
- Removal of affected clumps and drenching surrounding beds with mancozeb 0.3%.

8.6.2. Bacterial wilt: *Ralstonia solanacearum*

Symptoms

- Water soaked spots appear at collar region of pseudo stem and progresses upwards and downwards.
- Mild drooping and curling of leaf margins of lower leaves which spreads upwards
- Yellowing starts from the lowermost leaves and gradually progresses to the upper leaves.
- Plants exhibit severe yellowing and wilting symptoms.
- Affected pseudo stem and rhizome when pressed extrudes milky ooze from vascular strands.

Management

- Treat rhizomes with streptocycline 200 ppm for 30 minutes and shade dried before planting.
- Beds should be drenched with bordeaux mixture 1% or copper oxy chloride 0.2%.

8.6.3. Leaf spot: *Phyllosticta zingiberi*

Symptoms

- Water soaked spot and later turns as a white spot surrounded by dark brown margins and yellow halo.
- Lesions enlarge and adjacent lesions coalesce to form necrotic areas.

Management

- Spraying copper oxy chloride 0.25% or mancozeb 0.2%.

8.7. Vanilla

8.7.1. Stem rot: *Fusarium* sp.

Symptoms

- Water soaked lesions giving stem a brown coloured appearance resulting in rotting of tissues, leaves turning yellowish and drying off.

Management

- Spray 0.2% carbendazim, repeat spray after one month.

8.7.2. Root rot: *Fusarium* sp., *Sclerotium* sp.

Symptoms

- Browning of roots leading to rotting and decay.
- Yellowing of leaves, stem and leaves becoming flaccid, shrivelled and giving a drooping appearance.

Management

- Soil drench and spray with 0.2% copper oxy chloride at 2 or 3 litres per plant.

8.7.3. Bean rot: *Phytophthora* sp.

Symptoms

- Rotting and shedding of beans

Management

- Spray 0.1% carbendazim and 1% bordeaux mixture alternatively at weekly intervals

8.7.5. Shoot tip rot: *Fusarium, Sclerotium*

Symptoms

- Decaying of shoot tip and leaves at the tip

Management

- Spray 0.1% carbendazim at 15 days interval

8.7.6. Vanilla mosaic: Vanilla mosaic virus

Symptoms

- Mosaic pattern or mottling on foliage.
- Stunting sterility and leaf distortion

Management

- Uproot and destroy affected plant

8.8. Coriander

8.8.1. Wilt disease: *Fusarium oxysporum* f.sp. *corianderii*

Symptoms

- Lower leaves turn yellow and later dry
- Affected plants show wilt symptom
- Vascular discolouration of roots of affected plants.

Management

- Seed treatment with carbendazim at 2g/kg of seeds or *Bacillus subtilis* at 10g/kg of seeds
- Soil drenching with carbendazim 0.1% or copper oxy chloride 0.25%

8.8.2. Powdery mildew: *Erysiphe polygoni*

Symptoms

- Small, white spots appear on leaf surface and later, cover the entire surface
- Powdery growth can be seen on petiole and umbels
- Under severe infection, plants present an ashy white appearance

Management

- Spray with wettable sulphur 0.2%.

8.8.3. Stem gall: *Protomyces macrosporus*

Symptoms

- Galls appear on the leaves and stems.
- Shape of coriander seeds changes.

Management

- Treating seeds with 2 g bavistin/ kg of seeds.

8.9. Fenugreek

8.9.1. Powdery mildew: *Erysiphe polygoni* and *Leveillula taurica*

Symptoms

- Powdery white growth on both surfaces of leaves.

Management

- Dusting sulphur at 25 kg/ha or spraying wettable sulphur 0.25% at 20 days interval.

Section III

Facts and Figures of Medicinal and Aromatic Crops

1

Tea (*Camellia sinensis, Thea sinensis*) Camelliaceae (Theaceae), 2n=30

1.1. Origin and distribution

- The oldest known beverage, tea is native of China (Between China and Burma) in South East Asia.
- Known to the Chinese as early as 2737BC, but it was a popular drink in England in 1664 AD.
- Original species, which produce tea, were *C.assamica* (Assam jats), *C.sinensis* (China jats) and their natural hybrid, *C.assamica* subspecies *lasiocalyx* (Indo China or cambod type).
- **Chinese (Small leaf) tea - *C. sinensis* var. *sinensis***: Originated in southern China possibly with hybridization of unknown wild tea relatives
- **Chinese Western Yunnan Assam (Large leaf) tea**
- **Indian Assam (Large leaf) tea- *C. sinensis* var. *assamica***: Originated from *Camellia taliensis* in the area where southwestern China, Indo-Burma, and Tibet meet
- **Chinese Southern Yunnan Assam (Large leaf) tea**: Hybridized with the closely related species *Camellia taliensis*
- **Difference between Assam and China types**

Assam	China
It is a small tree 1.5-1.8 m height	It is a shrub 90-150 m height
Few robust branches	Branches abundant and whippy shoots
Large, glossy leaves 8-20	Small leathery erect leaves
Light to medium green	Dark green colour
High yield and medium quality	Low yield but good quality
Susceptible to drought and frost	Hardy and resistant
Sparse flowering	Profuse flowering

- Commercial cultivation of tea started in India from British era when Robert Bruce in 1823 discovered tea plants growing wild in Upper Brahmaputra valley.

- In May 1838, first Indian tea from Assam was sent to England for public sale.
- India is world's second largest tea producer after China.
- India is the largest producer and consumer of black tea in the world.
- Grown in Assam, West Bengal, Tamil Nadu, and Kerala which accounts for about 95% of total tea production.

1.2. Economic importance

- Rich source of flavonoids, bioactive compounds that can lessen oxidative stress, relieve inflammation, and provide other health benefits
- Green teas contain the polyphenol, catechin.
- Contains antioxidants.
- Has less caffeine than coffee.
- May reduce your risk of heart attack and stroke.
- May help with weight loss.
- Help to protect your bones.
- May keep your smile bright.
- Boost immune system, battle cancer
- Soothe digestive system, calorie free
- Lower blood pressure, protect teeth

1.3. Varieties

- Pandian, Sundaram, Golconda, Jayaram, Evergreen, Athrey, Brookeland, BSS 1, BSS 2, BSS 3, BSS 4, BSS 5, Biclonal seed stocks

Special characters	Clone	Originators
Wind Tolerance	UPASI-2, UPASI-10	UPASI-TRF India
Drought resistance	UPASI-9	UPASI-TRF India
Frost resistance	B-26	HPKV-TES India
Smallest leaf	CH-1	IHBT India
Biggest Leaf	Betjan	BETJAN T.E, India
Blister blight tolerance	TRI-2043,	TRI, Sri Lanka
High pubescence	TRI-2043	TRI, Sri Lanka
High anthocyanin pigmentation	TRI-2025	TRI, Sri Lanka
High tolerance to pH	TN-14-3	TRF, Kenya
Poor fermenter	12/2	TRF, Kenya
Mite tolerance	7/9	TRF, Kenya
Scale insect tolerance	TN-14-3	TRF, Kenya
High polyphenol content (53.7%)	Luxi white tea	TRI, China

High amino acid content (6.5%)	Anji white tea	TRI, China
Low caffeine content (0.14%)	Guangdong tea	TRI, China
High caffeine content (6.96%)	Wild tea at Yunnan	TRI, China
Water logging tolerance	TV-9	TES, India

UPASI - TRF: United Planters Association of Southern India - Tea Research Foundation, Valparai, Tamil Nadu.
TRI - Tea Research Institute, Sri lanka
TRI - Tea Research Institute, China
TRF - Tea Research Foundation, Kenya

- **Improved clones**

Clone	Important features
UPASI 1 (Ever green)	Hardy, quality-above average
UPASI 2 (Jayaram)	Hardy, quality-above average, tolerant to drought and wind
UPASI 3 (Sundaram)	Natural triploid quality clones and very high yielding
UPASI 6 (Brooklands)	Suited to mid and higher elevations
UPASI 8 (Golconda)	Suited to all elevations, high yielding
UPASI 9 (Arthrey)	Fairly tolerant to drought and withstand slightly high pH, high yielding
UPASI 10 (Pandian)	Hardy, quality-above average, tolerant to drought and wind
UPASI 14 (Singara)	Suited to higher elevations, high yield
UPASI 15 (Spring field)	Flushes throughout the year
UPASI 17 (Swarna)	Flourishing well at mid and high elevations
UPASI 24	Hardy
UPASI 25	High yielding
UPASI 16	High yielding
UPASI 27	Drought tolerant
UPASI 28 (UPASI 10 x TRI2025)	Biclonal, good strength and high quality

1.4. Soil and climate

- Well drained soil with rich organic matter and pH of 4.5 to 6.5.
- Grows at 650-2000 m above msl
- Optimum temperature: 20-27°C.

1.5. Nursery

- Well drained and deep loam soil with pH of 4.5 to 4.8.
- Form beds of one metre width and 8 cm height and of a convenient length.
- Drench with 2% solution of aluminium sulphate applied at 10 litres/2.5 sq.m of area.

- One more layer of soil of 8 cm height is spread and again drenched with equal quantity of water twice.
- **Preparation of sleeves**
  - Polythene sleeves of 150 or 200 gauge and 10 cm width and 30-45 cm length may be used.
  - Drainage holes may be provided at the bottom.
  - Lower 3/4 of the sleeves should be filled with 1:3 sand and soil mixture and top 1/4 with 1:1 sand and soil mixture and staked in rows.
  - Overhead shade is provided.
- **Selection of mother bush and its treatment**
  - Healthy and vigorously growing high yielding bushes should be selected.
  - Apply 40 g of young tea mixture + 60:90 NK mixture up to 5 years.
  - Following mixture has to apply before taking the cuttings.
  - 0.5% $AlSO_4$ + 1% $MgSO_4$ (Before 3 weeks); 2% $ZnSO_4$ (Before 2 weeks); 1% urea (Before 1 week)
- **Preparation of cuttings**
  - In April-May and August-September, semi hard-wood cuttings are prepared with one leaf and an internode with a slanting cut at the bottom.
- **Planting of cuttings**
  - Sleeves are watered thoroughly and holes are made in the soil.
  - Cuttings are inserted in the hole and the soil around is pressed firmly to avoid airspace followed by watering.
  - Small polythene tents may be provided which maintain high humidity and regulate the temperature inside.
  - Cuttings may take 10-12 weeks for rooting.
  - After 90 days i.e. when all the cuttings have rooted, polythene tent may be removed gradually over a period of 10-15 days.
- **Manuring of nursery**
  - Apply 30 g of nursery mixture consisting of ammonium phosphate (20:20), potassium sulphate or MOP, magnesium sulphate, zinc sulphate at 35: 15 or 12: 15:3 parts by weight (Total 80 parts by weight) dissolved in 10 litres of water for 4 sq.m area at fortnightly intervals.
- **Hardening of the cuttings**
  - 4-6 months old cuttings should be removed from shade gradually in stages over a period of 4-6 weeks.

1.6. Methods of planting

- **Single hedge system:** Spacing adopted is 1.20 x 0.75 m accommodating 10,800 plants/ha.
- **Double hedge system:** Spacing adopted is 1.35 x 0.75 x 0.75 m accommodating 13,200 plants/ha.

1.7. Season and planting

- May-June or September-October
- Sleeves should be opened lengthwise without injuring roots and planted in pit and soil is gently pressed.

1.8. Irrigation

- Sub-soil irrigation may be given for young tea seedlings during summer months.

1.9. Manures and fertilizers

- At 2 months after planting, apply P at 80-100 kg/ha as rock phosphate once in a year at 15-25 cm depth up to the first pruning and thereafter once in two years.
- N: K ratio of 2:3 should be adapted for the first 3 years and a ratio of 1:1 thereafter.

Year of application	Total weight kg/ha/annum		No. of applications	Qty/plant (g)	
	N	K		Ammonium sulphate	Urea
I year	180	270	5	13	27
II year	240	360	6	23	15
III year	300	450	6	29	18
IV year onwards	300	300	6	33	19

1.10. After cultivation

- Spraying glyphosate 1.75 litres + kaoline 2 litres + 2 kg of wetting agent in 450 litres water followed by gramoxone 500 ml in 200 litres of water to control weeds.

1.11. Training young tea

- **Centering:** To induce more laterals, centering should be done 3-5 months after planting.
- Main leader stem should be cut, leaving 8-10 matured leaves.

- **Tipping**: Done at a height of 35 cm from the ground level; second tipping at 60 cm.

1.12. Pruning

- Done in April-May or August-September to maintain convenient height and remove dead and diseased branches.
- Area to be pruned every year = $\frac{\text{Total extent of the garden}}{\text{Pruning cycle}}$

  Pruning interval – (Elevation in feet / 1000) + 1
- **Types of pruning**
- **Rejuvenation pruning**: Whole bush should be cut near the ground level less than 30 cm.
- **Hard pruning:** Done at 30 to 45 cm (12" to 18") for proper spread of bushes.
- **Medium pruning:** To stimulate new wood and to maintain the foliage at lower levels less than 60 cm.
- **Light pruning:** Raising height by an inch or less to manageable heights for plucking (Less than 65 cm).
- **Skiffing**: Removal of only top 5-8 cm new growth to obtain a uniform level (More than 65 cm).
- **Fringe pruning**: All the leaves below the pruning level including straggling branches and periperal branches.
- **Bagjan pruning**: To maintain the bush at constant height for number of years and this causes ends of all branches to get thickened and knotted. Shoots yield poor quality tea.

1.13. Shade regulation

- Pollarding of shade trees should be done prior to heavy rains at a height of 8-10 m from the ground
- **Annual lopping:** Cutting the erect type branches on the laterals in shade trees before monsoon season.

1.14. Crop duration and harvest

- Plucking commences when the tea bush is 3 years old.
- Plucking of extreme tip of the growing branch consists of an unopened bud together with two leaves is popularly known as "Two leaves and a bud", while fine plucking is anything less than this.
- In South India, plucking continues throughout the year at weekly intervals during March - May and at intervals of 10-14 days during the other months.

- **Rush period:** Harvesting is done at 7 to 10 days interval.
- **Lean period:** Harvesting is done at 10-15 days interval.

1.15. Yield

- Yield of green leaves is 10 t/ha.

1.16. Manufacture of tea

- This processes involves
  - Withering
  - Rolling
  - Fermentation
  - **Withering**
    - Leaves undergo dehydration by spreading in tiers of gunny piece.
    - Temperature of the air is adjusted by fans and the leaves attain a flaccid condition.
    - Withered leaves will reach a silk hand kerchief to touch.
  - **Rolling**
    - It is carried on by machines and the cells in the leaves broken and the sap containing enzymes for fermentation.
    - Leaves attain a twist characterises of the final product.
    - During the breaking of the leaf, the polyphenol contact with the enzyme polyphenol oxidase with presence of oxygen and changed to the aflavins and the arubigins teas are rolled 2-3 times for 30-40 minutes.
  - **Fermentation**
    - It starts from rolling to entry of drier, and oxygen and temperature are major factors.
    - Fermentation time is two hours; high quality and colour can be obtained.
    - Low temperature develops poor liquor and higher temperature for dull liquor
  - **Drying**
    - To arrest fermentation, remove the moisture in the leaf to preserve the liquor quality, brightness.
    - Drying is carried at one or two times in a special oven called drier at 36-155°C for 30-40 minutes.
    - Final drying is done at 22-24°C before packing.

- **There are two methods of processing**
  - o CTC (Cutting, tearing and curing)
  - o Orthodox manufacture.

1.17. Grades of tree

- After drying, the teas are sorted into different grades. The dried broken leaves sorted into different grades.
  - o Orange Pekoe (OP)
  - o Broken Orange pekoe (BOP)
  - o Orange Fannings (OF)
  - o Broken Pekoe (BP)
  - o Pekoe (P)
  - o Broken Pekoe Souchong (BPS)
  - o Pekoe Fannings and
  - o Dust.
- Three orange grades are finer portions.
- Two fannings are grey small pieces.
- Broken pekoe, pekoe, broken pekoe souchong are coarser grades.
- Duster is the smallest grade.

1.18. Blending

- Before tea reaches to consumers, it is blended.
- **Green tea**: The leaves are heated after harvest to inactivate the oxidising enzymes.
- **Oolong tea**: Leaf is slightly withered and light fermentation is allowed. Some oxidation is permitted to inactivate the enzymes.
- **Brick tea**: Teas are made into solid cakes or bricks.
- Coarse leaves are heaped on a cement floor to generate heat and a black fungus grows for imparting a peculiar taste.
- In marketing, teas are judged by the quality and appearance of leaves.
- Judgment is known as tea tasting and this is done by special type of men as tea tasters.
- Assam type or collar type not to exceed 30 cm and china type 225 cm.

2

Coffee (*Coffea canephora/Coffea arabica)*, Rubiaceae

2.1. Chromosome number

- Basic chromosome number for the genus Coffea is n=11
- *Coffea canephora*-diploid (2n=2x=22)
- *Coffea arabica* (2n=4x=44)

2.2. Origin and distribution

- Original domesticated coffee plant is said to have been from Harar and native population is derived from Ethiopia (Abyssinia Plateau) with distinct nearby populations in Sudan and Kenya.
- From where, it was taken to Arabia in 11th century.
- From Arabia, its seeds were brought to India by Baba Budan in the 17th Century and were raised in the Baba Budan Hills of Karnataka.
- British planters established large coffee estates near Chikmagalur (Karnataka) in 1826, in Manantody (Wynad) and Shevoroys in 1830 and Nilgiris in 1839.
- Almost the entire production is shared by Karnataka, Kerala and Tamil Nadu, smaller production in Andhra Pradesh, north eastern states and Madhya Pradesh.
- Karnataka is the largest producer accounting for 70% of total coffee production and 60% of the area
- Kozhikode, Waynad, Malappuram, Kollam, Kannur and Palakkad are the chief producing districts in Kerala.
- Tamil Nadu is the third largest producer where India's 6 % coffee is produced.
- About half of Tamil Nadu's coffee is produced in Nilgiri district.
- Other districts are Theni, Tirunelveli, Salem and Coimbatore.

2.3. Types of coffee

- ***Coffee arabica*** and ***Coffee robusta*** are the two main varieties accounting for 49% and 51% area, respectively.
- Indian coffee has created a niche in the international market
- Indian Robusta earning high premium value as highly preferred for good blending quality.

Arabica	**Robusta**
Mild coffee	More strength
More aromatic	Used in making various blends
Higher market value	High premium
Grown in higher altitudes with cool and equable temperature, ranging between 15 and 25°C	Grown in hot and humid climate with temperature ranging from 20 to 30°C
Arabica requires more care and nurture and more suitable for large holdings	Suitable irrespective of size of the farm
Susceptible to pests and diseases such as white stem borer, leaf rust	Less susceptible
Requires more shade	Requires moderate shade
Harvest between November to January	Harvest between December to February.

2.4. Economic importance

- Indian coffee is the most extraordinary of beverages, offering intriguing subtlety and stimulating intensity.
- Typically mild and not too acidic, these coffees possess an exotic fullbodied taste and a fine aroma.
- Indian coffee has a unique historic aroma and flavour.
- Caffeine, most widely consumed psychoactive substance, is the best known ingredient of coffee.
- Boosts physical performance.
- May help to lose in weight.
- Helps to burn fat.
- Helps to focus and stay alert.
- Lowers risk of death.
- Reduces risk of cancers.
- Reduces risk of stroke.
- Reduces risk of Parkinson's disease.
- May lower risk of Type II diabetes.
- Protects the brain.
- Brightens the mood, helps to fight depression and lowers risk of suicide.

2.5. Varieties

- **Arabica varieties:** Sln 795, Sln 7, Sln 9, Sln 10, Cauvery and its selections and HRC (Hawaian Red Cuturra), Chandragiri and San Roman
- **Robusta varieties:** Sln 274, Sln 270, Sln 3

Varieties	Parentage	Special characteristics
Selection 1 (S 288)	Tetraploid hybrid from S 26 (Natural cross - *C. liberica* x *C. arabica*)	Resistance to leaf rust race 1 and 2, high yielding. Wider adaptability
Selection 3 (S 795)	Cross between S 288 x Kent	Resistance to leaf rust race 1 and 2,700-1200 kg/ha. Bold fruits 75% A grade
Selection 4	Cioccie, Agaro and S 12 Kaffa	Leaf rust resistant
Selection 5	Devamachy x S 881	Small, oblong fruits, 900-1100 kg/ha
Selection 6	S 274 (Robusta) x Kent	High A grade beans, 900-1000 kg/ha
Selection 7	San Ramon - a short internode arabica	Dwarf in stature, segregates to tall by 30%
Selection 8	Pure line selection of Hibrido de Timor	High resistance to leaf rust, drooping branches
Selection 9	Sln 8 x Tafarikela	Drought hardy, suitable to different coffee zones
Selection 10	Double cross hybrid, Catura x (S 795 x HDT)	Resistant to leaf rust
Selection 11	Progency of *C. liberica* x *C. eugenoides*	Field resistance to rust and drought hardiness
Cauvery	Catimor x Hibrido de Timor	Plants are dwarf, suitable for high density planting. Yield - 3000 kg/ha. More A grade beans with superior cup quality.
Selection 12	Cross Caturra x HDT, followed by selfing	Precocious, suitable for close planting, resistant to leaf rust

2.6. Soil and climate

- Deep, friable, open textured rich in plant nutrients with plenty of humus and pH of 4.5 to 6.5

Varieties	Elevation (m)	Rainfall (mm)	Distribution
Robusta	500-1000	1000-2000	Blossom shower: February-March
Arabica	800-1650	1600-2500	Blossom shower: March-April

- Backing shower during April-May is required for both the varieties

2.7. Season

- June-December

2.8. Seeds and sowing

- Propagated by seeds

2.9. Nursery

- Sound fruits are depulped, sieved and mixed with sieved wood ash and dried in shade.
- Graded to remove all cut, triangular and elephant beans
- Select light loamy soil of good drainage with high organic matter.
- Form raised beds of 15 cm height, 1m width and at convenient length.
- Incorporate 30-40 kg of well rotten compost, 2 kg of finely sieved agricultural lime and 400 g of rock phosphate to a bed of 1 x 6 m size.
- **Sowing**: Treat with *Azospirillum* and Phosphobacterium
- Sown in February-March in the bed 1.5-2.5 cm apart with the flat side down wards in regular rows.
- Cover with a thin layer of fine soil and a layer of paddy straw.
- Water the beds daily and protect from direct sunlight by an overhead pandal.
- Seeds germinate about 28-35 days. The germination is epigeal and the testa stage is called coffees.
- After 45 days, they are transplanted to a secondary nursery beds for raising bag nursery.
- **Bag nursery:** Polythene bags with holes and filled with a mixture of jungle soil, FYM and sand in 6:2:1.
- Seedlings are planted in polythene bags.
- An area of 12 x 8 m can accommodate 5000 seedlings.

2.10. Preparation of field

- Selective felling may be done while retaining a number of desirable shade trees.
- Terracing should be done in deep sloppy areas.
- After the summer showers, pits of 45 x 45 x 45 cm are dug.
- Pits are left open for weathering and filled with top soil along with 500 g of rock phosphate and heaped.
- Planting is done along the contour in sloppy areas.

2.11. Spacing

- Arabica coffee: 1.5 to 2.0 m either way.
- Dwarf varieties: Sanraman - 1 x 1 m.
- Robusta coffee: 3 x 3 m.

2.12. Planting shade trees

- *Grevillea robusta, Albizia lebbeck, Leucaena glauca, Erythrina lithosperma, Glyricidia maculata* are planted during June when south west monsoon rain commences.
- Two metre long stakes are planted for every two plants of coffee.
- During summer, stem of young *Erythrina lithosperma* are painted with diluted lime or wrapped in agave leaves or polythene sheets in order to prevent them from sun scorch.
- Regulate shade by cutting criss-cross branches during monsoon season.

2.13. Irrigation

- Sprinkler irrigation during March - April increases blossoming and results in higher yields.

2.14. Fertilizer application

Species	Pre - blossom March N:P:K (kg/ha)	Post - blossom May N:P:K (kg/ha)	Mid - monsoon August N:P:K (kg/ha)	Post - monsoon October N:P:K (kg/ha)	Total (kg/ha)
Arabica					
1st year after planting	15:10:15	15:10:15	-	15:10:15	45:30:45
2nd and 3rd year	20:10:20	20:10:20	-	20:15:20	60:45:60
4th year	30:20:30	20:20:20	-	30:20:30	80:60:80
5 years and above for less than one ton/ha crop	40:30:40	40:30:40	-	40:30:40	140:90:120
For one ton/ha and above	40:30:40	40:30:40	40:30:40	40:30:40	160:120:160
Robusta					
For less than one ton/ha crop	40:30:40	-	-	40:30:40	80:60:80
For 1 ton/ha and above	40:30:40	40:30:40	-	40:30:40	120:90:120

2.15. After cultivation

- Weeding and mulching should be done as and when necessary.
- Digging is done to a depth of 30 cm towards the end of monsoon (October-November).
- **Cover digging**: Weeds and vegetative debris are completely buried in soil and stumps are removed.
- In sloppy areas, dig trenches on the contour 45 cm wide and 30 cm deep of any convenient length.
- Prune water shoots and disease affected shoots.

2.16. Training and pruning

- Arabica is grown in single stem.
- When the stem matures, it is topped at a height of one metre.
- Timely topping helps to develop frame work on the bush which will carry equal distributed crown flush.
- Centering is started in early period of the coffee and that is needed in 2 and 3 years to removal of the vegetative growth with half foot circle or cuts to the first node of each primary from the main stem from crown downwards.
- Unproductive wood between all primaries and secondaries are removed.

2.16.1. Rejuvenation pruning

- It includes parrot pole pruning done once in 4 or 5 years; to bring back old coffee to shape
- This pruning is the removal of dead, exhausted, dried and worn out branches.
- This is carried out after harvest.

2.16.2. Flowering

- Under natural condition, there exists a dormancy of flower bud opening.
- Flower buds in coffee will not open unless the receipt of rain.
- Only after receipt of the rain, all the flower buds in the plantation will open simultaneously.
- This type of flowering is called gregarious flowering.

2.17. Harvest

- Starts in November and harvesting extends up to February.
- Coffee fruits should be harvested as and when they become ripe.
- When gently squeezing the fruits, beans inside come out easily.
- Dried separately as cherry.
- **Fly picking:** Small scale picking of ripe berries during October to February
- **Main picking:** Well-formed and ripened and bulk of the berries are harvested during December
- **Stripping:** Picking of all the berries left irrespective of ripening
- **Gleanings:** Collection of fruits that have been dropped during harvesting.

2.18. Processing of coffee

- Coffee is processed in two ways
  - Wet processing by which plantation or parchment is prepared.

    - Dry method by which cherry coffee is prepared.
- Market is preferable for parchment coffee.
- Cherry coffee is with mucilage and fruit in dry form.
- Preparation of parchment and cherry picking the fruits in ideal stage are important part of processing.

2.19. Preparation of parchment coffee

- Well ripened berries are chosen for preparing parchment coffee.
- The steps involved in parchment coffee are
    - Pulping
    - Fermentation
    - Washing and trampling
    - Drying
- **Pulping**
    - Fruits are pulped on the same day to avoid fermentation
    - Pulping is the removal of mucilaginous flesh and the peel and freezing the berries inside by pulpers.
    - Pulping requires pulping equipment and adequate water is fed to the pulper through siphon and to ensure uniform feeding to separate lights and floats from fruits.
    - Uniform feeding ensures removal of the peel and prevents ants, choking of pulps etc.
    - The pulped parchment should be sieved to eliminate unpulped and fruit peel
    - The lights may be separated from well filled mature heavy beans at every stage.
    - At pre-fermentation stage, the coffee is passed in a channel with 4% slope.
    - The channel is cemented and smoothened to avoid residue sticks to rough surface.
    - The peel separated by pulping may be led away from vats into collection pits to eliminate microbial decomposition of the peel.
- **Fermentation**
    - It may be natural and by this method, mucilage breaks down in arabica and requires 24-36 hours.
    - Over fermentation lead to foxy beans and under-fermention, the sticky mucilage left on the parchment.

- The leads to mistiness in the produce.
- **Enzymatic:** Pectinolytic enzyme is used to hasten the fermentation.
- The dose of requirement is 350-750g/ton for 12 hours at 20°C.
- Treatment with alkali: Removal of mucilage by treatmenting with sodium hydroxide 1 kg/10 lit water takes one hour for arabica and 1½ h for robusta.
- Parchment is no longer slimy and makes rattling noise, clear water is let in and the parchment is washed clean with three or four changes of water.

- **Washing and trampling**
  - After loosening the mucilaginous coat, they are trampled under feet and washed in water repeatedly until the berries are free and clean. Washing produces spots on the dry parchment and it is called foxiness in coffee.
- **Drying**
  - Surface drying in trays with wire mesh bottom.
  - The parchment is spread in trays to thickness of 4 to 7 cm.
  - At the right stage of dryness, the parchment becomes crambly.
  - The cleaned and dried beans are known as parchment or parchment coffee.
  - Seven days of drying is the correct stage.

2.20. Grading

- Beans are graded through number of sieves of dimensions
- 'AA' grade refers to monsoon arabica, sieve with 7.2 mm round hole.
- For robusta, 7.0 mm round holes.
- Cherry coffee: Fruits are dried in the sun. This is called cherry or native coffee.
- **Defects of bean**
- Elephant bean
  - In normal bean, two seeds will be produced.
  - However in some berries, one develops at the expense of the other that will result in one is bigger and another smaller.
  - The bigger been is called elephant bean.

- Pea berry
  - In some beans, due to defective pollination or fertilization, only one seed develops with normal shape. Such beans are called pea berry.
- Triage
  - Normal, defective and broken bits of coffee is known as triage.
- Blacks
  - Three fourth of bean is black in colour.

2.21. Yield

- 750-1000 kg dry parchment/ha

3

Coconut (*Cocos nucifera L.*) Arecaceae, 2n=32

3.1. Origin and distribution

- Origin is in South Asia – Malaysia, Pacific.
- In the Pacific, cultivated on islands in Southeast Asia - Philippines, Malaysia, Indonesia.
- In the Indian Ocean, centre of cultivation is southern periphery of India including Sri Lanka, Maldives and Laccadives.
- In India, coconut palms are grown on the entire coastal belt of Kerala, Karnataka and Tamil Nadu followed by Goa, Maharashtra, Andhra Pradesh, and Odisha.
- Kerala, Tamil Nadu, Karnataka and Andhra Pradesh are the major coconut producing states accounting for more than 90% of area and production

3.2. Economic importance

- Supplies food, fuel, drink, oil and shelter.
- Wet kernel, desiccated copra tender, coconut water, cream flour, oil, cake, toddy, husk, fibre, shell, coir, wood and leaves are used for one purpose or other.
- Hence, it is Kalpa Vriksha (Tree of Heaven)

3.3. Varieties

- There are only two distinct varieties of coconut, the tall and the dwarf.
- Tall cultivars that are extensively grown are the West Coast Tall and East Coast Tall.
- Dwarf variety is shorter in stature and its life span is short as compared to the tall.

3.3.1. Tall varieties characteristics

- Long lived palm, living generally to an age of about 80 to 90 years

- Palms thrive well under different soil conditions varying from littoral sands to red loams and laterites.
- Palms grow well up to an altitude of 1000m above the sea level.
- It is fairly resistant to diseases and pests.
- Tree attains a height of about 15m to 18m or more
- It begins to bear in about 8 to 10 years after planting
- Nut is medium to big in size varying in shape from spheroid to linear-oblong and with colors varying from green, yellow and orange to shades of brown.
- About 6,000 nuts yield a ton of copra.
- **Suitable varieties for Tamil Nadu**: West Coast Tall, Chandrakalpa or Lakshadweep Ordinary (LCT), VPM-3 (Andaman Ordinary), East Coast Tall, Aliyar Nagar 1, Kera Chandra (Philippines Ordinary)
- **Suitable varieties for Kerala**: West Coast Tall, Chandrakalpa or Lakshadweep Ordinary (LCT), Philippines Ordinary (Kerachandra), VPM-3 (Andaman Ordinary), Kera Sagara (Seychelles)
- **Suitable varieties for Karnataka**: West Coast Tall, Tiptur Tall (TPT), Chandrakalpa or Lakshadweep Ordinary (LCT), VPM-3 (Andaman Ordinary), Kera Chandra (Philippines Ordinary)

3.3.2. Dwarf varieties characteristics

- Dwarf coconut is small in stature (5-7m) and commences bearing earlier than the tall variety
- Dwarf coconut palms flower as early as the third year after planting and come to regular bearing in the ninth year.
- Average life span is 40-50 years.
- Dwarf or short variety are producing green, orange and yellow nuts.
- Susceptible to drought.
- Nuts are small in size and ovoid or round in shape
- Nut weighs about 3 oz (85 gm) with 65 per cent oil content.
- **Suitable varieties for Tamil Nadu**: Chowghat Orange Dwarf (COD), Chowghat Green Dwarf (CGD)
- **Suitable varieties for Kerala**: Chowghat Orange Dwarf (COD), Chowghat Green Dwarf (CGD)
- **Suitable varieties for Karnataka**: Chowghat Orange Dwarf (COD)

3.3.3. Hybrid varieties characteristics

- Hybrids are the intervarietal crosses of two morphological forms of coconut.

- They show earliness in flowering and give increased yield, higher quantity and better quality of copra and oil when compared to the parents.
- When the tall is used as female, they are called TxD hybrid, while the reciprocal is known as DxT hybrid.
- Besides, intervarietal hybrids like Tall x Tall and Dwarf x Dwarf are also produced.
- Hybrids perform well under good management conditions including nutrient management and irrigation.
- **Suitable hybrids for Tamil Nadu:** Kerasankara (WCT x COD), Chandrasankara (COD x WCT), Kerasoubhagya (WCT x SSAT), VHC 1 (ECT x MGD), VHC 2 (ECT x MYD), VHC 3 (ECT x MOD)
- **Suitable hybrids for Kerala**: Kerasankara (WCT x COD), Chandrasankara (COD x WCT), Chandralaksha (LCT x COD), Keraganga (WCT x GBGD), Lakshaganga (LCT x GBGD), Anandaganga (ADOT x GBGD), Kerasree (WCT x MYD), Kerasoubhagya (WCT x SSAT)
- **Suitable varieties for Karnataka**: Kerasankara (WCT x COD), Chandrasankara (COD x WCT), Chandralaksha (LCT x COD), Kerasoubhagya (WCT x SSAT)

List of tall varieties	List of dwarf varieties (Tender coconut)	List of hybrid varieties
West Coast Tall	Chowghat Orange Dwarf (COD)	Kerasankara (WCT x COD)
East Coast Tall	Chowghat Green Dwarf (CGD)	Chandrasankara (COD x WCT)
Chandrakalpa or Lakshadweep Ordinary (LCT)	Chowghat Yellow Dwarf (CYD)	Chandralaksha (LCT x COD)
Phillippines (Ordinary Kerachandra)	Gangabandom	Keraganga (WCT x GBGD)
VPM - 3 (Andaman Ordinary)	Malaysian Dwarf Yellow	Lakshaganga (LCT x GBGD)
Aliyar Nagar 1 (ALR 1)	Strait Settlement Dwarf Green	Anandaganga (ADOT x GBGD)
Aliyar Nagar 2 (ALR 2)	Kalpa Raksha	Kerasree (WCT x MYD)
Tiptur Tall		Kerasoubhagya (WCT x SSAT)
Kera Sagara (Seychelles)		VHC 1 (ECT x MGD)
Benavali Green Round (Pratap)		VHC 2 (ECT x MYD)
Phillipines Tall (Chandrathara)		VHC 3 (ECT x MOD)
Assam Tall (Kamaroopa)		Gadavani Ganda (East Coast Tall x Gangabandom)
Kalpadhenu		Kalpa Samrudhi
Kalpa Pratiba		Kalpha Shankara (Chowghat Green Dwarf x West Coast Tall)
Kalpa Mitra		
Kerakeralam		

3.4. Soil and climate

- Light sandy to heavy soils with a pH of 5.2 to 8.0.
- Proper drainage and good water holding capacity.
- Avoid rock or any hard substratum within 2 m of surface.
- Altitude: 600 to 900 m
- Rainfall: 200 cm/year, 180-300 cm is ideal.
- Mean annual temperature of 27°C is the best

3.5. Mother palm and seed nut selection

- Seed gardens should have palms with a high proportion of heavy bearers.
- Free from the pest and disease incidence.
- Avoid trees growing closer to households, cattle shed and compost pits.

3.5.1. Features of superior mother palm

- **Regular bearer**: Should produce on an average one leaf and an inflorescence in its axil every month.
- Straight stout trunk with even growth and closely spaced leaf scars.
- Spherical or semi spherical crown.
- More collar girth at the fifth months from the time of germination in the nursery are indicator for early flowering and high yield.
- High rate of leaf (More than 30 fully opened leaves) and spathe production (12 inflorescences)
- Short and stout petiole and wide leaf base firmly attached to the stem,
- Short and stout inflorescence stalk with bunches resting on leaf petioles of lower whorl and more number of female flowers (25 or more)
- Age of the palm from 25 to 40 years.
- Even trees with 15 years age can be selected, if it is high yielding and has stabilized yield (Eg. Chowghat dwarf).
- Choose high yielding mother palms giving not less than 100 nuts/palm/annum under irrigated condition (70-80 nuts/annum under rainfed conditions) with copra output of 20 kg/palm.
- Husked nuts should weigh not less than 650-700g.
- Mean copra content of 150 g/nut or more
- Select high yielding and root wilt disease free West Coast Tall (WCT), Chowghat Green Dwarf (CGD) and Chowghat Orange Dwarf (COD) palms.

- **Avoid palms which have**
  - Long, thin and pendulous inflorescence stalks.
  - Narrow crown with upward branching are poor yielder.
  - Long, narrow, small sized or barren nuts
  - Alternate bearing tendency
  - Shedding of immature nuts in large numbers

3.5.2. Harvesting and storage of seed nut

- Medium sized, round or oblong mature nuts, showing a resonant and ringing sound, are harvested when at least one nut in the oldest bunch starts becoming dry.
- Harvested during February - August in Tamil Nadu and December to May in Kerala
- In tall varieties, seed nut matures at 11-12 months
- In dwarf varieties, nuts will mature in 10-11 months
- Thin husked nuts with an average of 2.9 cm thickness and below, sprout earlier and produce more number of leaves.
- Using a rope, harvested bunches are lowered to ground to avoid injury
- Seed nuts of tall varieties and hybrids are air cured for one month, followed by sand curing for two months.
- For dwarf varieties, air curing should be less than one month, followed by sand curing for two months.
- Tall variety nuts are stored up to two months after harvest and dwarfs are sown within 15 days.
- For storing, arrange the seed nuts with the stalk end up over a 8 cm layer of sand in a shed and cover with sand to prevent drying of nut water.
- Up to five layers of nuts can be arranged one over the other.

3.6. Nursery management

- Select well drained plot with coarse texture soil near water source for irrigation.
- Should be a open space with artificial shade or in adult coconut garden.
- Plant seed nuts in a long and narrow bed at 30 x 30 cm spacing either horizontally or vertically in deep trenches with 20-25 cm depth.
- Five rows of nuts may be planted in each bed accommodating 50 nuts per row.
- Irrigate nursery beds once in three days.

- Seed nuts start germination 6 - 8 weeks after planting and continue up to six months.
- Grow sunhemp 2 times and harvest at flowering (Used for green manuring for adult coconut), followed by one hand weeding at 6th month was found effective in weed management.
- Provide shade to the nursery by raising *Sesbania* or *Leucaena* on the sides of beds.
- Remove nuts which do not germinate 5 months after sowing.

3.6.1. Selection and harvesting of seedlings

- Select seedlings, having good girth of 10 cm at collar and early splitting of leaflets (Minimum of 6 leaves), at 9 to 12 MAS with short and broad leaf stalks.
- Eliminate the seedlings which are deformed or having stunted growth.
- Remove the seedlings from the nursery by lifting with spade.
- Do not pull out the seedlings by pulling leaves or stem.

3.7. Planting seasons

- June-July, December-January

3.8. Spacing

3.8.1. Tamil Nadu

Particulars	Spacing
Tall varieties	7.5 x 7.5 m (25' x 25')
Dwarf varieties	6.5 x 6.5 m (20' x 20')
Hybrids	8.5 x 8.5 m (26' x 26')

- For planting in field border as a single row, adopt 20' spacing between plants.

3.8.2. Kerala

Planting system	Spacing	Plants/ha
Triangular	7.6 m	198
Square	7.6 to 9 m (7.6 x 7.6m, 8 x 8m, 9 x 9 m)	170-120
Single hedge	5 m in the rows and 9 m between the rows or 6.5m in rows and 9 m between rows	220
Double hedge	5 x 5 m in rows and 9 m between pairs of rows or 6.5 x 6.5 in rows and 9 m between pairs of rows	280

3.8.3. Karnataka

Varieties	Spacing	Plants/ha
Tall varieties (Southern region and hilly region) Tall varieties (Coastal region)	9 x 9 m, 8.2 x 8.2 m	125-149
Dwarf and hybrids (Southern dry region and hilly region) Dwarf and hybrids (Coastal region)	7.5 x 7.5 m, 7 x 7 m	178-204

3.9. Planting

3.9.1. Tamil Nadu

- Dig pit of 3' x 3' x 3' and fill to a height of 2' (60 cm) with FYM, red earth and sand in equal proportions.
- At centre of pit, plant seedling after removing all roots.
- Press the soil well around the seedling and provide shade using coconut or palmyrah leaves.
- Remove soil covering the collar region.
- As the seedlings grow and form stem, fill up the pits gradually by cutting the sides.

3.9.2. Kerala

Soil types	Pit size
Loamy soil	1 x 1 x 1 m
Laterite soil	1.2 x 1.2 x 1.2 m
Sandy soil	0.75 x 0.75 x 0.75 m

- Pits may be filled up with topsoil and powdered cow dung/compost to a height 60 cm
- Burial of two layers of husks in floor of pits will be useful for moisture conservation.
- Sprinkle carbaryl 10% dust on the husk to prevent colonization by termites.

3.9.3. Karnataka

- Pits of 1 m cube (3.3 ft) with proper spacing
- Fill the pits with green leaves, compost/FYM and top soil.
- Plant seedlings in centre of the pit and press the soil firmly around the plant and provide support.
- Deep placing of nut up to 45 cm is preferred.

3.10. Water management

- From 5th year onwards, adopt the following irrigation schedule for drip and basin irrigation.
- Form a circle of 200 cm radius in root zone around base of coconut tree up to a depth of 100 cm.
- Provide a set of four drippers set equidistant in a circle 100 cm away from base of tree and discharging water at 30 l/h for 2.5 h with a irrigation frequency of 8 days.

3.10.1. Western region of Tamil Nadu

Months	Normal condition (for best yield)	Moderate water scarcity condition	Severe water scarcity condition
A. Drip irrigation			
February to May	65 lit/day	45 lit/day	22 lit/day
January, August and September	55 lit/day	35 lit/day	18 lit/day
June and July, October to December	45 lit/day	30 lit/day	15 lit/day
B. Basin irrigation			
February to May	410 lit/6 days*		
January, August and September	410 lit/7 days*		
June and July, October to December	410 lit/9 days*		

3.10.2. Eastern region of Tamil Nadu

Months	Normal condition (for best yield)	Moderate water scarcity condition	Severe water scarcity condition
A. Drip irrigation			
March-September	80 lit/day	55 lit/day	27 lit/day
October-February	50 lit/day	35 lit/day	18 lit/day
B. Basin irrigation			
March-September	410 lit/5 days*		
October-February	410 lit/8 days*		

* Quantity of water to be applied in the basin.

Add 30-40% of the above quantity of water (135-165 litres/palm) to meet the conveyance loss.

3.11. Drought management and soil moisture conservation

3.11.1. Mulching with coconut husks/ leaves/ coir pith

- Apply coconut husks with convex surface facing upwards (100 nos.) or dried coconut leaves (15 nos) or coir pith up to a height of 10 cm in the basin of 1.8 m radius around the palms.

3.11.2. Burial of coconut husk or coir pith

- Bury 100 husks with concave surface facing upwards or 25 kg of coir pith/ palm in circular trenches which are 30 cm width and 60 cm depth at 1.5 m radius
- Also buried in trenches (45 cm deep and 150 cm width) at 3 m distance from palm in between two rows of coconut.

3.12. Manures and fertilizers

Age (Years)	FYM (kg/tree)	Urea (kg/tree)	Super Phosphate (kg/tree)	Muriate of Potash (kg/tree)
1	10	0.308 (140 g N)	0.500 (80 g P_2O_5)	0.480 (300 g K_2O)
2	20	0.616 (280 g N)	1.000 (160 g P_2O_5)	0.960 (600 g K_2O)
3	30	0.924 (420 g N)	1.500 (240 g P_2O_5)	1.440 (900 g K_2O)
4	40	1.23 (560 g N)	2.000 (320 g P_2O_5)	1.920 (1200 g K_2O)
5th year onwards	50	1.23 (560 g N)	2.000 (320 g P_2O_5)	1.920 (1200 g K_2O)

- Apply in circular basins at 1.8 m distance from base of palm, incorporate and irrigate.
- Apply in two split doses, in June-July and in December - January.

3.13. Weed management

Broad leaved weeds	Grasses and sedges
Pre emergence spraying of atrazine at 1.0 kg a.i./ ha	Post emergence spraying glyphosate at 10 ml and 20 g ammonium sulphate/litre of water

3.14. Inter cropping

3.14.1. Tamil Nadu

Below 7 years of age (up to 5 years)	7-20 years of age	Above 20 years of age (Sunlight transmission of above 50%)
Groundnut, sesame, sunflower, tapioca, turmeric and banana can be grown Avoid crops like paddy and sugarcane	Green manure crops and fodder crops (Napier grass and Guinea grass)	*Annuals:* Groundnut, bhendi, turmeric, tapioca, sweet potato, sirukizhangu, elephant foot yam, ginger *Biennials:* Banana varieties like Poovan and Monthan Cocoa*, pepper* (Panniyur 1 or Panniyur 2 or Panniyur 5 or Karimunda), nutmeg* and vanilla*

*Suitable in Pollachi tract and Kanyakumari district.

3.14.2. Karnataka

- Grown during the pre-bearing or early stage of growth

- Dry land paddy, millets, groundnut, cowpea, chilli, wheat, potato, tuber crops and rhizomatous crops.
- In older plantations, elephant foot yam, cassava, colocasia, greater and lesser yams, sweet potato, ginger and turmeric are grown.
- Under rainfed conditions, French bean, fodder grass (Hybrid napier, guinea grass and *Stylosanthes gracilis*), sunflower, cowpea and finger millet are preferred.

3.14.3. Kerala

- ***Cereals:*** Rice, maize
- ***Legumes and pulses:*** Groundnut, horse gram, cowpea
- ***Tubers:*** Tapioca, sweet potato, yams, colocasia
- ***Spices and condiments:*** Ginger, turmeric, chilli, pepper, nutmeg, cinnamon, clove
- ***Fruit plants:*** Banana, pineapple, papaya (Banana variety Poovan is recommended in Kuttanad at 3 suckers/clump).
- ***Beverage crop:*** Cocoa
- ***Fodder grasses:*** Hybrid napier, guinea grass

3.15. Multi storied cropping

- In normal coconut grooves, 78% area is not effectively used.
- Hence, this space can be economically utilized by growing multitier cropping.
- Under irrigation, four crop architecture i.e., coconut, pepper, cocoa and pineapple is followed.
- **Top floor** - Coconut spaced at 7.5 to 8.0 m apart
- **Second floor** - Pepper vine trained up to 8.0m on the coconut palm trunk
- **First floor** - Cocoa branch spread confined to a height up to 3.5 m
- **Ground floor** - Pineapple

3.16. Mixed cropping

- Growing banana, papaya, pineapple, arecanut, mango, jack fruit etc., along with coconut is common in small farmers holdings.
- Rubber, clove, nutmeg and cinnamon are also planted in centre of four coconut palms spaced 7.5 m apart.
- 1000 banana/ha (125 palm/ha) is found to be the most important intercrop for coconut garden.

3.17. Multiple cropping systems

- Coconut + Banana + Sirukizhangu + Bhendi are suitable system for the eastern region.
- Banana, pepper, cocoa, nutmeg, vanilla are suitable in the western region.
- **Perennials:** Cocoa, nutmeg, pepper, clove, lemon grass and cinnamon.
- **Annuals: Kharif -** Rice, maize, groundnut, ginger, turmeric, chilli, yams, colocasia, red gram, vegetables, sweet potato, tapioca, banana, pineapple, papaya and fodder grass.
- **Rabi:** Sesame, horse gram, red gram, vegetables, cowpea, sweet potato and banana.
- **Summer:** Vegetables

3.18. Green manure and cover crops

3.18.1. Green manures

- *Crotalaria juncea* (sunnhemp), *Tephrosia purpurea* (kolinji), *Gliricidia maculata, Indigofera hirsuta, Pueraria phaseoloides.*

3.18.2. Cover crops

- *Calapagonium muconoides, Stylosanthes gracilis*

3.18.3. Shade cum green manure shrub

- *Tephrosia candida*
- Sow green manure and cover crop during April - May with the onset of pre monsoon rains and *in situ* incorporation during August-September.

3.19. Harvesting

- Twelve months old nuts are harvested at 30-45 days.
- For tender nut purposes, 7 to 8 months old nuts are harvested.
- Nuts which are 11 months old give fibre of good quality.
- On an average, eight harvests are taken.

3.20. Yield

Variety	Nut yield (No's / tree / year)	Earliness (Year)
Hybrid	100	3-5
Tall	60-80	6-7
Dwarf (Tender coconut)	70-90	4-5

3.21. Special problems in coconut

3.21.1. Rejuvenation of existing garden

1. **Thinning of thickly populated gardens**

- Cutting and removal of trees in thick plantation, yield could be increased.
- Maintain 175 palms/ha.

2. **Ensuring adequate manuring and irrigation**

- Adopt as per recommendation.

3.21.2. Pencil point disorder (Micronutrient deficiency)

- Apply 225 g each of borax, zinc sulphate, manganese sulphate, ferrous sulphate, copper sulphate and 10 g of ammonium molybdate dissolved in 10 litres of water and poured in the basin of 1.8 m radius.

3.21.3. Button shedding

- Provide regular irrigation, recommended nutrition to reduce the button shedding.

3.21.4. Barren nuts

- Apply extra 2 kg of K_2O with 200 g of borax/palm over and above the usual dosage of fertilizer.

3.21.5. TNAU MN mixture at 1 kg/tree/year

- Enriched FYM prepared at 1:10 ratio of MN mixture and FYM, mixed at friable moisture and incubated for one month in shade.

3.22. Food products

- **Coconut water:** It is a liquid endosperm or the tender coconut (Seven to eight months old) and used as a drink.
    - It has caloric value of 17.4/100 g of water.
    - The percentage of arginine, alanine and cystine in protein are higher.
- **Nut meat or kernel**: The kernel of seven to eight months old nut consumed along with tender coconut water.
    - At this stage, kernel is soft with high content of protein and sugar.
    - The ripe kernel is derived from 10-12 months old nut.
- **Coconut milk:** It is an emulsion of coconut oil in water with soluble component of the meats.
    - It is milk like liquid obtained from the grafting of the meat.

- **Coconut flour**: It is the partially soluble coconut flour and is used in bakery. It contains 7-8 per cent protein, 94% oil and 17 per cent crude fibre.
- **Edible copra:** Copra is available in two different forms of ball copra and cup copra
  - Ball copra is made from matured (>12 months) whole nuts and when it is split in to two halves, it is known as cup copra.
  - The Lakshadweep Micro is the suitable variety for ball copra.
  - Fresh meat is processed and converted into chips in Philippines.
- **Sweet sap**: A sugar containing juice or sap is obtained by tapping the opened spadix.
  - The word tapping includes extraction and manipulation for stimulating the sap-yielding palms to exude juice from a part (Spadix).
  - An average of 18 litres per month sweet sap (Pathaneer) is tapped from a single tree.
- **Coconut vinegar** of the sap is not added with calcium chloride; the sap becomes unpalatable and is allowed to ferment for 24 hours.
  - After the period, acetic fermentation sets in converting the alcohol into acetic acid.
- **Coconut oil:** The dried copra on crushing gives 60-70% oil and 33-40% cakes.
  - The oil is edible and used for soap making
  - The cake is utilized as a cattle feed.
- **Miscellaneous products**
- **Coconut leaves**: The plated leaves are used for cling, huts, fencing and shedding purposes.
- **Coir and fibre**
  - These are obtained from the husk.
  - The fibre after cleaning is dried in the sun and spun into yarn litter by hand or by machines.
  - The yarn is used in ropes, mats, mattings, nets, bags etc.

4

Arecanut (*Areca catechu*) Arecaceae, 2n=32

4.1. Origin and distribution

- Native to Indian ocean, Malaysia or Philippines
- Grows in tropical Pacific, Asia, and parts of east Africa.
- Areca is derived from a local name from Malabar Coast of India and catechu is from Malay name caccu.
- Karnataka (40%), Kerala (25%), Assam (20%), Tamil Nadu, Meghalaya and West Bengal are major arecanut growing states.
- India is the largest producer and consumer.

4.2. Economic importance

- Chewing nut, popularly known as betel nut or supari.
- In India, it is used in religious rites of birth, marriage, and nuptial; linked with religious practices.
- Offered to Gods in veneration in the form of *taamboola*, which consists of one arecanut placed over two betel leaves.

4.3. Varieties

- Mangala, Sumangala, Subamangala, Mohitnagar, Srimangala, Sreevarathan, Samruthi (Andaman), Hirehalli Dwarf, VTLAH 1, 2 and Thirthahalli Dwarf.

Varieties	Characteristics
South Kanara	Mostly grown in South Kanara district of Karnataka and Kassaragod district of Kerala. Large nuts and uniform bearing. Average yield is 1.5 chali/palm/year (7 kg ripe nuts).
Thirthahalli dwarf	Grown extensively in Maland area of Karnataka. It is preferred for tender nut processing and not as dry nut.
Sree Varjdhan or Rotha	Grows in coastal Maharashtra. Nuts are oval in shape and the yield is 1.5 kg chali (7 kg ripe nuts) per palm per year. Kernal colour when cut is marble white. Its endosperm is tastier than other varieties. It starts bearing after 6-7 years of planting.
Mettupalayam Local	Grown widely in Mettupalayam area of Tamil Nadu. Nut size is very small.

Kahikuchi	Grown in North East India.
Mohitnagar	Grown in the Northern part of West Bengal. Nut is very big and uniform and much similar to Kahikuchi. At Kidu farm in Karnataka, it yielded 3.7 kg chali (19.5 kg ripe nuts) per palm per year.
Mangala (VTL-3)	Selection from China and released for cultivation by Arecanut Research Station, Vittal. Semi-tall type and bearing starts after 3-5 years of planting. Early bearing, early stabilization, high fruit set and yield. Average yield is 2 kg chali (10 kg ripe nuts) per palm per year.
Sumangala (VTL-11)	Introduced from Indonesia. Average yield is 3.3 kg chali (17.5 kg ripe nuts) /palm /year. Tolerant to burrowing nematode, *Radopholus similes.*
Sree mangala (VTL-17)	Selection from Singapore. Average yield is 3.1 kg chali (15.6 kg raw nuts)/palm/year. Tolerates *Radopholus similes.*
Swarnamangala (VTL-12)	Trees are semi tall to tall, stem sturdy with shorter internodes. Bearing-by 4th year. Potential yield (kg chali/palm/year) - 6.28. Average yield (kg chali/palm/year) - 3.88. Recommended for rrigated areas of Karnataka and Kerala
Vittal Areca Hybrid-1 (VTLAH-1)	Hybrid between Hirehalli Dwarf x Sumangala. Average yield (kg chali/palm/year) - 2.54. Recommended for Coastal Karnataka and Kerala.
SAS-I	Tall tree. Round and medium nuts. Average yield is 4.60 kg chali/palm/year. Recommended for Uttara Kannada
Sreevardhana	Tall tree. Round and medium nuts. Average yield is 2.00 kg chali/palm/year. Recommended for Coastal areas

4. 4. Soil and climate

- Well drained soils of laterite, alluvial, red loam.
- Adequate protection from exposure to South-Western sun is essential to avoid sun scorch.
- Quick growing shade trees have to be planted on the southern and western sides well in advance of planting seedlings. The fruit trees like mango, jackfruit and coconuts may be grown. On the initial stage, banana may be grown before taking up arecanut planting.
- Sensitive to moisture deficit.
- Temperature to a maximum of 40°C, optimum 14-36°C.
- Altitude up to 1000 m above msl.
- Rainfall: 750-4500 mm.

4.5. Season

- June-December
- October

4.6. Nursery

- Seed nuts from outstanding mother palms are sown at 5-6 cm apart in sand beds under partial shade with their stalk end pointing upwards.

- Propagation only through seeds. There are four stages.
  - Selection of mother palm
  - Selection of seed nuts
  - Selection of techniques in germinating and raising seedlings
  - Selection of seedlings
- Age at bearing, regular bearing, higher number of leaves on the crown, shorter internodes and high fruit set.
- Position of bunch in the tree, position of nuts in the bunch weight of nuts with in a bunch, maturity of nuts and floating habit are some factors.
- Fully ripe nuts from middle portion of the middle bunch, the second and third bunch.
- Nuts float vertically with calyx and upwards in water are vigorous and high per cent germination.
- After two to three leaves, sprouts are transplanted to 30 x 10 cm polythene bag filled with forest soil and are allowed to grow for 12 to 18 months under partial shade.
- Also transplanted in secondary nursery beds with a spacing of 30 cm on either side.

4.6.1. Primary and secondary nurseries

- A seed nut takes 53-94 days for completion of germination.
- Nuts sown in nursery beds in sand/soil/burnt earth are vigorous.
- Such type of sowing in sand bed is primary nursery which takes six months and seedlings with two or three leaves are transplanted in secondary nursery in beds of convenient length with 35-45 cm spacing.
- Nursery may be provided with shade using coconut/arecanut leaves.

4.7. Planting

- Select dwarf and compact seedlings with more number of leaves.
- 12-18 month old seedlings with 5 or above leaves are planted in pits of 90 x 90 x 90 cm at a spacing of 2.75 m either way and covered with soil to the collar level and pressed around.
- Provide shade in southern and western side during summer months.
- Growing banana or other crops in advance may also provide shade.

4.8. Irrigation

- Irrigate weekly once during November-February, once in 4 days during March-May.
- For flood irrigation, 175 lit/ tree/ day.
- In drip irrigation, 16-20 lit/ tree/ day.

4.9. Manures and fertilizers

- **5 years and above**: 10-15 kg of FYM or green leaf; 100:40:140 g of NPK/ tree/year.
- **Less than five years old**: Half of the above dose is recommended.
- Manures are applied during January-February after North East monsoon in a basin of 0.50-1.00 m radius around tree to a depth of 15-20 cm.

4.10. Multiple cropping

- Growing of more crops of shorter duration in arecanut garden to utilize the land effectively is multiple cropping.
- **Inter cropping**
  - Growing two or more crops simultaneously of annuals or biennials in the inter spaces of arecanut.
  - Crops like paddy, sorghum, cowpea, vegetables, yams, pineapple, banana are grown.
  - Crops like yams, elephant foot yam, taro, tapioca, pineapple and banana are raised in pits or trenches.
  - Other crops like turmeric, ginger, arrow root, chillies are planted in raised beds.
- **Mixed cropping**
  - Growing perennial crops like cocoa, coffee, cardamom, black pepper, clove, cinnamon, nutmeg in the inter space of arecanut.

4.11. After cultivation

- Weeding is done twice or thrice a year by spade digging.
- Wherever the land is sloppy, terracing has to be done to prevent soil erosion.

4.12. Harvest

- Bearing starts after 5 years of planting.
- Harvest when three quarter nuts ripe.
- Number of harvests: 3 to 5/year.

4.13. Yield

- An average of about 1250 kg/ha can be obtained.

4.14. Chemical composition

- Major chemical constituents are polyphenols, fat, polysaccharides, fibre and protein.
- Alkaloid is a minor constituent and arecoline is major alkaloid and minor ones are arecaine, guvacoline and guvacine.

4.15. Processing

- **Fresh nuts**
  - In Kerala, the fresh fruits are stored by steeping in water, such water preserved nuts is known as neetadaka.
- **Dried ripe nuts**
  - Dried whole nut is known as chali or kottapak.
  - Ripe nuts are dried in the sun for 35-40 days.
- **Kalipak**
  - Nuts of 6-7 months maturity is soft and finger can be pressed into it.
  - Processing involves
    - Dehusking
    - Cutting the soft nuts into pieces
    - Boil nut pieces with water
    - Kali coating using the extract of tender nut boiled in water.
    - Drying.
- **Supari**
  - There are two types; one from chali and the other from kalipak
- **Uses of leaf sheath**
  - For making food plates

5

Rubber (*Hevea brasiliensis*) Euphorbiaceae, 2n=36

5.1. Origin and distribution

- Indigenous tree of tropical rainforest of Amazon basin of Brazil and introduced in Asia in 1876
- Malaysia, Indonesia, Thailand and Africa are the main rubber producing countries.

5.2. Economic importance

- Natural rubber is available from *Manihot glaziovii* (Cera rubber), *Ficus elastica* (India rubber), *Castiolla elastica* (Panama rubber), *Parthenium argenatum* (Guayul), *Taraxacum koksaghyz* and *Hevea brasiliensis* (Para rubber).
- *Hevea brasiliensis* is the most important commercial source of natural rubber.
- After proper chemical treatment, rubber wood provides enough strength and durability of any semi-hard wood available in India
- Used for the manufacture of useful articles like door and window components, furniture, wall paneling, interior decoration, tool handles etc.
- Consists of polymers of the organic compound isoprene, with minor impurities of other organic compounds, plus water.
- Rubber is harvested as latex
- Tapping: Collecting latex, a sticky, milky colloid, by making incisions in the bark
- Latex is refined into rubber.
- Rubber is useful because of its invaluable qualities such as elasticity, resistance to water and non-conduction of electricity.
- Useful substance for waterproofing, insulating and in making motor car and pneumatic tyres (Heating the rubber and mixing it with sulphur makes a stronger substance).

5.3. Varieties

- Tjir 1, PB 86, BD 5, BD 10, PR 17, GT 1, RRII 105, RRIM 600, PB 28/59, PB 217, PB 235, RRIM 703, RRII 5, PCK-1, 2 and PB 260
- **Clones recommended by Rubber Board of India**

Name	Parents	Characters
Primary clones		
TJIR-1	-	Indonesian clone, yield 930 kg/ha per year. Susceptible to Phytophthra, Oidium and pink disease.
G.T.1	-	Indonesian clone, yield 1360 kg/ha year.Tolerant to Phytophthora, pink disease and brown bast.
G.I.1	-	Malaysian clone, yield 1130 kg/ha year. Susceptible to brown bast, possess drought tolerance.
P.B. 86	-	Malayasian clone.Yield is 1130 kg per ha/year.
Secondary clones		
PRIM-600	TJIR-1 x P.B. 86	Developed by Rubber Research Institute of Malaya (RRIM). Yield is 1317 kg/ha/year. Susceptible to Phytophthora and pink disease.
RRIM-628	TJIR.1x RRIM.527	Yield is 1051 kg/ha/year. Susceptible to brown bast and poor yield during summer.
Tertiary clones		
RRIM-703	RRIM.600xRRIM.500	Yield is 1725 kg/ha/year. Susceptible to brown bast and wind damage.

- Other important clones recommended for South Indian conditions are PR.107, PB.5/51, RRIM-118, RRII-203 and RRII-208.

5.4. Soil and climate

- Deep and lateritic fertile, red lateritic or clay loam, fairly rich soil with acidic pH of 4.5 to 6.0 and highly deficient in available phosphorous.
- Tropical climate with annual rainfall of 2000-4500 mm.
- Elevation- 300-500 m above msl
- Temperature - 20 to 34°C with 80% RH.
- Avoid heavy winds.

5.5. Season

- June-July

5.6. Propagation

- Propagated by green budding, brown budding, crown budding and modified forkert method.

5.7. Nursery

- Bed size: 60-120 cm width and of convenient length.
- Spacing: Seedling stumps - 23 x 23 cm, 30 x 30 cm and 34 x 20 cm
- Budded stumps - 30 x 30 cm
- Stump budding - 60 x 60 cm
- Bud wood nursery - 60 x 90 cm or 60 x 120 cm.

5.8. Planting

- Pit size of 1 m^3 are dug and filled up with top soil and compost

Planting material	Spacing (m)	Population/ha
Budded plants		
Hilly areas	6.7 x 3.4	445
Plains	4.9 x 4.9	420
Seedlings		
Hilly areas	6.1 x 3.0	539
Plains	4.6 x 4.6	479

- Germinated seeds are sown *in situ* in the pits.
- Healthy ones are retained.

5.9. Manures and fertilizers

- **For immature rubber trees:** Apply 12 kg FYM and 125 g of rock phosphate/ pit before planting.
- Apply 10:10:4:1.5 NPK and Mg as per schedule given below:

Months after planting	Period of application	Quantity/plant	
		10:10:4 (g)	12:12:6 (kg)
3	September - October	225	190
9	April - May	450	380
15	September - October	450	380
21	April - May	450	480
27	September - October	550	480
33	April - May	550	380
39	September - October	450	380

- **5th year till the tree is ready for tapping**: Apply 400 kg of mixture/ ha in 2 doses, once in April-May and another in September-October.
- **Matured rubber trees under tapping:** Apply NPK 12:6:6 grade mixtures at 900 g/tree (400 kg/ha) every year in two split doses.
- Add 10 kg magnesium sulphate for every 100 kg of the above mixture.

5.10. After cultivation

- Growing *Pueraria phaseoloides, Calopagonium mucunoides, Centrosema pubescens* and *Desmodium ovalifolium* and *in situ* incorporation.

5.11. Weed management

- *Eupatorium, Lantana camara, Mimosa pudica* are some weeds.
- As a pre emergence, simazine at 5 kg/acre; 2, 4, D at 0.75-2 kg in 450 litres of water may be sprayed.

5.12. Tapping

- It is the periodical removing of thin slices of bark to extract rubber latex.
- Rubber trees attain tappable stage in about seven years
- Trees must attain a girth of 55 cm at a height of 50 cm from the ground.
- Budded tree reaches the girth of about 50 cm at a height of 125-150 cm from the bud union.
- Following is an arbitrary system of calculating the relative intensity of tapping.

S/2 d/2	Half spiral, alternate day for 6 months and rested for 3 months	100% intensity
S/2 d/26 m /9	Half spiral, alternate day for 6 months and rested for 3 months	67% intensity
S /2 d/3	Half spiral, third daily	67% intensity
S/2 d/31m/2	Half spiral, daily for one month and rested for next month	100% intensity
S /1 d/4	Full spiral, fourth daily	100% intensity
V /2 d/212 m/16	Half circumstances and cut alternate day for 12 months and rested for the next 4 months	75% intensity

- **Tyre rubber**
  - Tyre manufacturing industry consumes about 50% of natural rubber.
  - Cost of the rubber is reduced by blending lower grade scraps and latex rubber and
  - Viscosity of the rubber is kept low by addition of extender oil or plastics.

5.13. Ethrel treatment

- To increase latex yield of trees.
- Apply at 5% a.i. concentration with a brush below the tapping cut to a width of 5 cm after light scraping of the outer bark.

- First application may be done after a drought period during pre-monsoon showers.
- Subsequently in September and November.

5.14. Processing of rubber

- Product from the rubber tree is latex, a milky white dispersion of rubber in water which is harvested by the process of tapping.
- Latex flows out from the trees on tapping are channeled into containers such as coconut shell or polythene cups.
- Latex collected in the cups is transferred to clean buckets 2-3 hours after tapping
- Anti-coagulants like ammonia or formalin or sodium sulphite are added to prevent the coagulation of latex.
- It is diluted to consistency of 12-15% of rubber.
- Metrolac is a special hydrometer used for the measurement or percentage of rubber after dilution for precipitation of particles strained through 60 mesh screens for second time.
- Then, it is poured into coagulating tank or aluminium pans, and acetic acid and formic acid are used for coagulation.
- Slow coagulation produces a soft rubber which is easy to work on the rollers.

5.15. Marketable forms of natural rubber

- Important forms are
  - Ribbed smoked sheets (RSS)
  - Crebe rubber
  - Preserved latex and latex concentrates
  - Block rubber or crumb rubber.
- **Ribbed smoked sheets**: Latex coagulated into thin slabs of coagulum is sheeted through a set of smooth rollers followed by a grooved set and dried to get ribbed smoked sheet.
  - Based on drying method, sheet rubbers are grubbed into ribbed smoked sheets and air dried sheets.
  - About 67% of the rubber is produced in sheet form.
- **Crepe rubber**: when coagulated latex or any form of field coagulation is passed several times through minimum of 3 mills with heavy rolls
  - This lace like rubber is air dried and is called as crepe rubber.
  - It is classified into sole crepe, pale latex crepe (PLC), estate brown crepe (EBC), remilled crepe and smoked blanket crepe.

- **Dryness (Brown bast)**
  - Syndrome characterized by prolonged dripping of latex with the gradual decline in volume yield, pre coagulation of latex and partial or complete drying of tapping area (10-25%).
  - Reduce tapping intensity or give a tapping rest for 3 to 12 months.

5.16. Yield

- Higher yield after 14 years of planting.
- Seedling trees: Annual yield of rubber is 375 kg/ha
- Budded plants: 800-1000 kg/ha.

6

Cashewnut (*Anacardium occidentale* L.), Anacardiaceae, 2n=42

6.1. Origin and distribution

- Native of America and Brazil.
- Introduced in India during later half of the Sixteenth Century.
- In 2017, Vietnam, India, and Ivory Coast were the major producers.
- Cultivated in Andhra Pradesh, Goa, Karnataka, Kerala, Maharashtra, Odisha, Tamil Nadu and West Bengal; also in few pockets of Assam, Chhattisgarh, Gujarat, Meghalaya, Nagaland and Tripura.
- Cashew is called gold mine of waste lands.

6.2. Economic importance

- Commonly eaten as food; rich source of vitamin C and sugar.
- Shell oil can be used in lubricants, water proofing, paints and starting in World War II, arms production.
- Heat resistant paint is prepared from shell oil which is used in automobile industries. Wood varnish prepared from shell oil is substitute for French polish.
- Tannin extracted from cashew testa may be used in leather industry.
- Fruit is light reddish to yellow fruit and pulp can be processed into a sweet, astringent fruit drink or distilled into liquor.
- Used in medicine; stomach and intestinal (Gastro-intestinal) ailments.
- Applied directly to the skin as a skin stimulant and to seal (Cauterize) ulcers, warts, and corns.

6.3. Varieties

- Tamil Nadu: VRI-1,VRI-2,VRI-3,VRI-4,VRI (CW) H1
- Kerala:Akshaya (H-7-6), Amrutha (H-1597), Anagha (H-8-1), Anakkayam-1 (BLA-139-1), Dhana (H-1608), Dharasree (H-3-17), K-22-1, Kanaka (H-1598), Madakkathara-1 (BLA-39-4), Madakkathara-2 (NDR-2-1), Priyanka (H-1591), Sulabha (K-10-2)

- Andhra Pradesh: BPP-1, BPP-2, BPP-3, BPP-4, BPP-5, BPP-6, BPP-8 (H2/16)
- Karnataka: Chintamani-1, NRCC-1, NRCC-2, Ullal-1, Ullal-2, Ullal-3, Ullal-4, UN-50
- Maharashtra: Vengurla-1, Vengurla-2, Vengurla-3, Vengurla-4, Vengurla-5, Vengurla-6, Vengurla-7
- Goa: Goa-1
- Odisha: Bhubaneswar-1
- West Bengal: Jhargram-1

Andhra Pradesh

- BPP 1: Medium size with an average weight of 4-5g. Average yield is about 8-10 kg per tree with W 400 kernel grade.
- BPP 2: Average yield of this variety is 8-11 kg/tree. Nuts are comparatively smaller in size, with an average nut weight of 3-4g.
- BPP 3: Average yield is 9-11 kg/tree with 4-5g of nut weight.

Tamil Nadu

- VRI 1: Native to Cuddalore District of Tamil Nadu. Average yield of 6-8 kg/tree. Kernel grade is W 320 and has 4-5g of average nut weight.
- VRI 2: National variety was collected from the Changelpattu district of Tamil Nadu.Well adapted to a wide range of soils and yields 6-8 kg fruits/tree with 4-6g of nut size.

Odisha

- BH8 5: Early flowering variety produces nuts of 6-8 g weight. Having an average yield rate of 9-10 kg per tree and tree flowers between December to February.

West Bengal

- Jhargram 1: Nut weight of this variety is 3-5g, whereas the yield of this variety is 8-9 kg/tree.
- Jhargram 2: Mid-season flowering variety has a yield of 12-14 kg/tree. Average nut weight is 8-10g, with a kernel grade of W 180.

Maharashtra

- Vengurla 1: Has an average yield of 19-20 kg on every tree. This early flowering variety produces nuts with 6-8g weight.
- Vengurla 2: Has a nut weight of 4-5g with an average yield of 24-25 kg/tree. Known for its short fruiting and flowering phase, its kernel grade is W 320.

Goa

- Goa 2: With an average yield of 5-7 kg/tree. Average weight is 9-10 g. This variety is selected from the Ela village of Goa that has a W 210 kernel grade.
- Goa 1: Developed by the ICAR Research Complex of Goa. Average yield is 7-8 kg/tree. The tree belongs to the Balli village of Goa that produces nuts with 7-8g weight.

Kerala

- Sulabha: Produces bold nuts with a mean weight of 9.8 g with a kernel grade of W 210.
- BLA 39-4: Flowering is between November to January. Average yield is 13-14 kg/tree. Producing medium sized nuts with 6-8g weight and kernel grade of W 280.
- Dhana: Variety has a yield of 17-18 kg/tree and has 2-3 g of the kernel weight. This hybrid bears nut with a kernel grade of W 210.
- Akshaya: Variety has a yield of 9-11 kg/tree and bears nuts with a mean weight of 10-11g. This hybrid produces nuts with W 180 kernel grade.

Karnataka

- Ullal 2: This late flowering variety produces medium sized nuts with 5-6g weight. Average yield is 8-9 kg/tree with W 240 kernel grade.
- Ullal 1: Selected from the Taliparamba district of Kerala and has an average yield of 14-16 kg/tree. Nut weight is 6-8g with a W 210 kernel grade.
- H 130: This hybrid is early flowering with a long fruiting duration. It produces big kernels and bold sized nuts with 12-13g weight.
- Bhaskara: Released in 2006 in Karnataka, this is a mid-season flowering variety. Its nut and kernel weights are 7-8 g and 2-3 g, respectively.
- NRCC Selection 1: This late flowering variety has an average yield of 9-10 kg/tree. Medium sized nuts have a 7-8g weight with a W 210 kernel grade.

6.4. Soil and climate

- Red sandy laterite loam
- A temperature between 7°C and 49°C, the tree performs satisfactorily.
- A wind velocity of 30 km/h is harmful during blossoming.
- Rainfall of 50-250 cm per annum.

6.5. Season

- June-December

6.6. Propagation

- Soft wood grafting, air layering and epicotyl grafting

6.7. Preparation of field

- Pits of 45 x 45 x 45 cm size are dug and filled up with a mixture of soil + 10 kg FYM + 1 kg neem cake.

6.8. Spacing and planting

- About 200 plants/ha
- Adopt a spacing of 7 m either way
- **High density planting**: Spacing of 5 x 4 m accommodating 500 plants/ha

6.9. Manures and fertilizers (per tree)

Manures & fertilizers	I year old	II year old	III year old	IV year old	V year onwards
FYM (kg)	10	20	20	30	50
N (g)	70	140	210	280	500
P (g)	40	80	120	160	200
K (g)	60	120	180	240	300

- Applied 100:125:250 g NPK/tree in 2 equal split doses during June-July and October-November under east coast area.

6.10. Intercropping

- Groundnut or pulses or minor millets till the trees reach bearing age.
- Tapioca, chilli, ginger

6.11. Irrigation

- Grown as a rainfed crop.
- Irrigation once from flinching to fruit maturity stage.

6.12. Training and pruning

- Prune the interlocking branches during July-August to maintain the frame.
- Dried twigs and branches should be removed every year.

6.13. Rejuvenation of old cashew orchard

- **Top working:** Cut leaving a stump of 1-3 m height from ground level.

6.14. Harvest

- Starts yielding 3rd year onwards.

- Peak picking at March and May.
- Good nuts are grey green, smooth and well filled.
- After picking, nuts are separated from apple and dried for two to three days to bring down the moisture content to 10 to 12 %.
- Properly dried nuts are packed in alkathene bags. This will keep for 6 months.

6.15. Processing

- This consists of roasting, shelling, extracting the oil, peeling, grading and packing.

1. Roasting

- It adds flavor and taste to the kernels and makes shells brittle and extraction of kernels easier
- Under roasting or over roasting affects recovery of kernels.
- **Open pan system**
  - In shallow iron pan or earthen pots, one kg nuts are kept and heated over open fire.
  - During the roasting, shell liquid and smoke come off and causes irritation on the skin of the personnel engaged.
  - Roasted nuts are spread on the floor and covered with earth which would absorb oil adhering to the roasted nuts.
- **Sun drying**
  - Nuts are dried under sun for 2-3 days and they lose the moisture and become brittle ending for shelling.
  - This process is followed at Panruti areas of Tamil Nadu which is a major cashew growing location.
- **Improved methods of roasting cashewnuts**
  - Continuous roasting process
  - Principle involved is as in the open pan system.
  - In this process, single or double walled metallic drums are used.
  - In the case of single walled, there is no provision to recover shell oil, whereas in the double walled, the smoke produced during roasting escape *via* inter space between the two walls of the drum.
  - In the single walled drum, the gases produced are escaped through chimney provided at the lower end and there is provision for collecting liquid.

- **Oil bath process**
  - In this method, nuts are kept in the wire trays and passed through a bath of cashew shell maintained at temperature of 200°C to 202°C for a period of three minutes there by the shell oil is recovered from the shell.

2. **Shelling**

- After roasting, shelling is carried out by the human labour and cracked open with wooden mallet and care has to be taken that the inner kernel is intact.

3. **Peeling**

- Extracted kernels have thin outer skin which is brown and pink in colour.
- Removal of the skin is done by hand with help of safety pin or hand knife.
- Peeling made easier when the kernels given to a heat treatment for about 4 hours in drying room.
- After peeling, the kernels are spread on the cement floor to absorb moisture to less brittle.
- Afterwards, the kernels are kept in the ventilated room called sweating chamber for 2-3 hrs.

4. **Grading**

- It is done based on count number of kernels per kg.
- Kernels which have not split are separated as wholer.
- They are separated into six grades 210, 240, 280, 320, 400 and 450 whole nuts per kg.
- Broken and split kernels are classified as standard, scorched pieces, splits, small pieces and each grade is separately packed.

5. **Packing**

- Packing is done in tins.
- Each tin contains 10 kg of kernels.

6.16. Yield

- 3-4 kg/tree/year.
- 375 kg of nuts and 1250 kg of cashew apple/ha.

7

Cocoa (*Theobroma cacao*) Malvaceae, 2n=20

7.1. Origin and distribution

- Native to Tropical America, the Amazon Basin.
- Domesticated in Mexico and Central and South America.
- Cultivated in Karnataka, Kerala and Tamil Nadu mainly as intercrop and multi-storey crop with arecanut and coconut from 1970 onwards.

7.2. Economic importance

- Cocoa is a cauliflorus beverage crop.
- Chocolate, hot chocolate, cocoa powder and dry powder are made
- Cocoa butter (the fat) can be extracted.
- Used in Mesoamerican foods including tejate and Mexican drink that also includes maize.
- Contains antioxidants called flavonoids.
- Lowers blood pressure and promotes inflammation or blockage of blood vessels.

7.3. Varieties

- Three varietal types in cocoa namely Criollo, Forestero and Trinitario.
- Forestero types are known to perform well under Indian conditions.
- **Kerala Agricultural University, Thrissur**
- CCRP-1, CCRP-2, CCRP-3, CCRP-4, CCRP-5, CCRP-6 and CCRP-7 and 3 hybrids CCRP-8, CCRP-9, CCRP-10.
- **ICAR-Central Plantation Crops Research Institute, Vittal**
- Clone: VTLCC-1; hybrids: VTLCH 1, VTLCH 2, VTLCH 3 and VTLCH 4.

7.3.1. Varieties released by CPCRI

- **NC-45/53**
  - Selection from Nigerian clone
  - Early, heavy bearer, self and cross compatible
  - Pods (no/tree/year) - 75, pod weight (g) - 321, beans/ pod - 37, single dry bean weight (g) - 1.05
  - Average dry bean yield (kg/tree/year) - 1.33, potential yield (kg/tree/ year) - 2.5, yield (kg/ha) - 911
  - Recommended for arecanut and coconut gardens of Karnataka, Kerala, Tamil Nadu
- **II-67 x NC-42/94**
  - Hybrid between Malaysian and Nigerian clone
  - Early, heavy bearer, suitable for water limited conditions
  - Pods (no/tree/year) - 40, pod weight (g) - 429.5, beans/ pod - 43, single dry bean weight (g) - 1.01
  - Average dry bean yield (kg/tree/year)- 1.25, yield (kg/ ha) - 856
  - Recommended for rainfed and irrigated areca, coconut gardens of Karnataka, Kerala, Tamil Nadu, Andhra Pradesh, Maharashtra, Goa.

7.3.2. Commercial cocoa has two major varieties, Criollo and Forestero.

Character	Criollo	Forestero
Pod colour	Dark red	Yellow
Other pod characters	Rough surface, ridges prominent, pronounced point and thin walled	Smooth, inconspicuous ridges, thick walled, melon shaped with round end
Flavour and aroma	Blend flavour	Harsh flavour, bitter taste.
Adaptability in India	Poor adaptability and less yield potential and hence, discourage for commercial cultivation	Good adaptability and high yielding and hence, recommended for commercial cultivation.

7.3.3. Varieties released by KAU

CCRP-1	Local population (SPS)
CCRP-4	Local population (SPS)
CCRP-5	Selection from germplasm IV
CCRP-6	Selection from germplasm VI
CCRP-7	Selection from germplasm VI
CCRP-2	Poly clonal selection for high yield
CCRP-3	Poly clonal selection for high yield

CCRP-11	Poly clonal selection for bold bean size and high yield
CCRP-12	Poly clonal selection for bold bean size and high yield
CCRP-13	Poly clonal selection for bold bean size and high yield
CCRP-14	Poly clonal selection for bold bean size and high yield
CCRP-15	Poly clonal selection for bold bean size and high yield

7.4. Climate

- Cultivated at altitudes up to 300-1600 m above msl
- Annual rainfall of 1000 to 2500 mm, RH of 80% and 15-35°C
- Grown as intercrop in coconut and arecanut gardens.

7.5. Soil

- Alluvial, clay red laterite soils with pH of 4.5-8.0.

7.6. Propagation

- Propagated by seeds.
- **Potting media**: Red soil + sand + FYM at 2:1:1 + super phosphate (5 kg/ ton)
- **Vegetative propagation**

Vegetative propagation	Rootstock	Scions	Success (%)
Budding: Patch budding	10-12 month old seedlings	Bud patch of 2.5 cm length and 0.5 cm width from the bud wood	85
Grafting: Soft wood gratting	3-4 month old seedlings	Scion stick of 12-15 cm length with 2-3 buds	70

7.7. Planting

- Cocoa is a shade loving tropical rain forest crop.
- Hence, it requires in the young stage 50 per cent and the grown phase needs 70 per cent shade.
- If it is grown as a pure crop, necessitates primary shade crops like banana in the young stage and in the grown up stage trees like *Erythrina lithosperma* may be tried.
- However, under Indian conditions, cocoa is raised as inter crop in the plantation of coconut, arecanut and rubber.
- Pits of 50 x 50 x 50 cm are dug, after one month refilled with topsoil and 15-20 kg of FYM.
- Plant vigorous and healthy 4-6 month old seedling or grafts or buddings of 60 cm height in the centre of pit.

- Polythene strip tied over graft joint should be removed and the joint should be above the soil.
- Requires 50% shade and later, the shade requirement is about 40%.
- Grown as a mixed crop in spices and rubber and intercrop in coconut and arecanut gardens.

Main crop	Spacing (m)	Cocoa as inter crop
Coconut	7.5 x 7.5	One row of cocoa plants at 3 m interval at the centre of two coconut rows and one cocoa at the centre of two coconut plants along the coconut row; 500 plants/ha
Arecanut	2.7 x 2.7	One row of cocoa plants at 2.7 m interval at the center of two arecanut rows. Likewise, alternate gaps of arecanut rows should be filled in; 686 plants/ha.
Oil palm	4.5 x 4.5	Five cocoa plants between four oilpalm plants; 400 plants/ha.

7.8. Irrigation

- During summer, irrigation at weekly intervals.
- In mixed crop with arecanut, irrigated once in a week during November-December, once in 6 days during January-March and once in 4-5 days during April-May with 175 litres of water.

Age of plant	Water requirement (Litre/plant/day)
1st year	3-5
2nd year	10
3rd year and later	20-25

7.9. Manures and fertilizers

- In the initial stage, FYM at a dose 15-20 kg/plant may be applied.
- Apply in two equal splits during April-May and September- October.

Fertilizer (g/plant)	I year	II year	III year onwards
Urea	72	144	220
Rock phosphate	65	130	200
Muriate of potash	77	154	230

7.10. Fertigation

- Drip irrigation is done once in 2 days.
- 100:40:140 kg of N, P and K are applied through drip irrigation at weekly intervals.

Phases of crop	Months	Weeks	Number of weeks	*Water soluble fertilizers (NPK - %)
1st season flower initiation	January and February	1-8	8	12.5
Pod set	March	9-13	5	10.0
Pod development	April and May	14-22	9	12.5
Pod maturity, harvest and 2nd season flower initiation	June and July	23-30	8	20.0
Pod set	August	31-35	5	12.5
Pod development	September and October	36-43	8	15.0
Pod maturity and harvest	November and December	44-52	9	17.5
Total				100

***Water soluble fertilizers:** All 19 (19% N: 19 % P: 19 % K); Mono-ammonium phosphate (12:61:00 % NPK); Water soluble form of N and K fertilizers containing 13% N and 45% K (Multi K) and urea (46% N).

7.11. Training and pruning

- First pruning is done after the main harvest (April to July) and just before the rainy season.
- Second pruning during November-December.
- Young plants develop a jorquette at 1 m height; increasing light intensity decreases jorquette height.
- Strongest of re-growing chupon can be selected and all others removed.
- In due course, this chupon will produce a jorquette at a higher level.
- Vegetatively propagated plants form a jorquette at ground level.
- Fan branches should be limited to 3-4 to allow more light and decrease canopy humidity.
- Basal chupons should be removed at regular intervals and all lower branches that form or bend below the jorquette should be trimmed off.
- Further, all branches within 60 cm of the jorquette, all old and diseased branches growing into the centre of the tree canopy should be removed
- **Pruning of grafted plants**: Pruning is done after first year of planting in August-September to obtain a supporting framework of one or more upward growing main stems.
- Drooping or inward growing branches are to be removed.

- Secondary pruning is suggested to develop well shaped canopy and desired canopy should be maintained in umbrella shaped form with about 3.8 to 4.2 m spread and 2.7 m height.
- Apply Bordeaux paste immediately after the pruning.

7.12. Harvesting

- Pods take about 140-160 days to ripen.
- Pods turn yellow or orange.
- Economic yield starts from 5th year.
- Yield: 1-2 kg of pods/tree/year; 50-70 pods/tree/year.
- Two main crops in a year (September-January and June-August).

7.13. Processing

- Pods are heaped together and kept under shade for 5-6 days to ensure uniform ripening to reduce acidity and helps in development of a better flavor suited for chocolate making.
- Pods are broken, husks removed and the white yellowish seed masses are heaped together for fermentation in box or basket for 5 to 7 days.
- On an average, 10-12 pods give 1 kg wet beans and 3 kg of wet beans (From 30-36 pods) give 1 kg of fermented and dried beans.
- **Sweating box method of fermentation**: Boxes of 60 x 60 x 45 cm made of wood with reapers at the bottom are arranged in tiers.
- Two detachable wooden planks are provided on one side of the box for transferring (Mixing) the beans.
- Beans are loaded on fermentation box and covered with banana leaves or gunny bags.
- Every alternate day, mix the beans after transferring to the next box after 24 hrs.
- Temperature of the fermentation mass will rise to 45-55°C after 48 hrs of fermentation.
- **Basket method of fermentation:** Bamboo or cane baskets are used for fermenting small quantity of beans.
- One or two layers of banana leaves are placed at the bottom with provision to drain the sweating.
- Fill the beans, cover with banana leaves and place a small weight and keep on raised surface to facilitate drainage.
- Finally, cover with thick gunny bags.

- Mix thoroughly on 3rd and 5th day and again cover with gunny bags.
- Fermentation will be completed in 6 days.
- Well fermented beans will be plumpy and filled with reddish brown exudates.
- Testa becomes loosened from the cotyledons.
- When cut open, cotyledons will have a brownish colored with lots of ridges and furrow appearance in the centre with a brownish ring in the periphery.
- During fermentation, sugars in the pulp of the bean is converted into alcohol.
- **Sun drying** to reduce moisture content around 6-8%.
- **Grading and storage**: Flat, slate, shrivelled, broken and other extraneous materials are removed and packed in fresh polythene lined (150-200 gauge) gunny bags.
- Bags are kept on raised platform of wooden planks.

7.14. Preparation of cocoa for market

- **Curing of fermentation**
  - This process involves maintenance of heat per milling, sweating to drain off at the same time to provide access of air.
- **Cacao products**
  - Fermented dried beans are cleaned, roasted and shelled.
  - Roasted product is known as cacao ribs which are used for the manufacture of products.
  - Cocao ribs are rich in fat and when they are subjected to hydraulic pressure, an oily fatty substance is obtained and that is known as cacao butter.
- **Chocolate liquor cacao mass**
  - When the cacao ribs are ground, the resulting product is known as the chocolate liquor or mass.

8

Betelvine (*Piper betel*) *Piperaceae*, 2n=78

8.1. Origin and distribution

- Originated in South and South East Asia.
- Cultivated mostly from Pakistan to Papua New Guinea.

8.2. Economic importance

- In India and Sri Lanka, sheaf of betel leaves are traditionally offered as a mark of respect and auspicious beginnings like greeting elders at wedding ceremonies, celebrating the New Year, and offering payment to Ayurvedic Physicians and Astrologers (To whom money and/or areca nut placed on top of the sheaf of leaves are offered in thanks for blessings).
- Consumed in Asia as betel quid or paan with arecanut and/or tobacco.
- Used in cooking usually raw for its peppery taste.
- Medicinal use over a 300 years history in China.

8.3. Varieties

- **Peninsular India**: Kapoori and Sanchi
- **North India**: Bangla and Deswari
- **West Bengal**: Meetha
- **Tamil Nadu**: Thulasi, Venmani, Arikodi, Kalkodi, Karilanchi, Karpuram, Chelanthi karpuram, Koottak kodinandan, Perum kodi, Amaravila and Pramuttan, Kallar kodi, Karpura kodi, Revesi, Karpuri, SGM 1, SGM (BV) -2, Vellaikodi, Pachaikodi, Anthiyur kodi, Kanyur kodi, Bangla and Konur (Salem).

8.4. Soil and climate

- Well drained fertile clay loams.
- Does not tolerate saline and alkaline conditions.
- Requires a cool humid with considerable humidity.

8.5. Propagation and planting

- Propagated by terminal stem cutting or setts of 30-45 cm long from the top portions of the vines.
- 1,00,000 setts/ha
- Planted at the base of the live supports, which are to be planted 4 to 5 months earlier.

Row spacing (cm)	Vines/hectare	
	Single vine	Double vine
20	50,000	1,00,000
30	30,000	60,000
45	22,500	45,000

8.6. Season

- November-December and January-February.

8.7. Field preparation and planting

- Beds of 2 m wide are formed to a convenient length.
- Provide drainage trenches of 0.5 m width by 0.5 m depth in between two adjoining beds.
- Plant seeds of agathi (*Sesbania grandiflora*) in long rows.
- 750 banana suckers are planted at bed edges; used for tying vines on live support and packing leaves.
- When agathi reaches 4 m height, they are topped off for maintaining the height.
- Crop is planted in two rows in beds of 180 cm width on agathi with a spacing of 45 cm between plants.

8.8. Fertilizer application

Time of application	N	P (kg/ha)	K
Basal dressing	37.5	100	50
Top dressing @ 3 split doses	112.5	0	0

8.9. Irrigation

- Irrigate the field immediately after planting and afterwards, once in a week.

8.10. After cultivation

- **Training of live support:** Before establishment of vines, side branches of agathi upto a height of 2 m are cut for early creeping of the vines.
- **Training of vines**: For fixing the vine at intervals of 15 to 20 cm along the support loosely with the help of banana fibre; done at every 15-20 days interval.
- **Lowering of vines**: Under normal cultivation, the vines grow to height of 3 m in one year period. When they reach this height, their vigour to produce normal size leaf are reduced and they need rejuvenation by lowering during March-April. After the vine is lowered, the tillers spring up from the nodes at the bends of the coiled vines at the ground level and produce many primary vines. Irrigation should be given after each lowering.

8.11. Harvest

- Harvesting every day.

8.12. Yield

- 75 to 100 lakh leaves/ha/year.

9

Oilpalm *(Elaeis guinensis)* Arecaceae, 2n=32

9.1. Origin and distribution

- Originated from West Africa.
- Cultivated in Malaysia, Indonesia, USA, Cameroon, Côte d'Ivoire, Ghana, Liberia, Nigeria, Sierra Leone, Togo, Angola and Congo.
- Commercial scale plantation in Malaysia was found in 1917.

9.2. Economic importance

- Produces edible palm oil as well as palm kernel oil.
- Major constituent of oil is mono unsaturated oleic acid; rich in carotenoids.
- Unrefined palm oil is a significant source of tocotrienol, part of vitamin E.
- Oil content is 17 to 27% for palm oil, and 4 to 10% for palm kernels.

9.3. Varieties

- Dura, Tenera, Pisifera
- **Tenera**: It is a hybrid between shell less Pisifera and thick shelled Dura. It is a thin shelled and cultivated across the Globe.
- **Dura**: Thick shell of 2-8 mm and not commercially cultivated variety.
- **Pisifera**: Shell less variety.

9.4. Climate

- Temperature is 21 to 32°C, annual rainfall - 200 cm and RH - 75-100%.
- Altitude - 450-900m above msl.

9.5. Soil

- Moist deep, loamy, alluvial soils, rich in humus with good water permeability and pH of 4-6

9.6. Nursery technique

- 3-9 months old seeds are soaked in water for 5 days (Change water daily) and dry for 24 hrs.
- Sow in 250 gauge and 23 x 13 cm size polythene bags filled with top soil, sand and FYM in equal proportions
- Apply fertilizer mixture containing 15:15: 6 of NPK at 8 g in 5 litres of water for 100 seedlings.
- Ten to sixteen months old seedlings are used for planting.

9.7. Season and planting

- May-June
- Pit size 60 cm^3.
- Planting can be done in triangular system with spacing of 9 x 9 x 9 m, accommodating 140 palms/ha.

9.8. Manures and fertilizers

- For five year old palm, apply 50 kg compost or FYM.
- **Fertilizer schedule:** N: P: K : Mg SO_4 (g/palm/year)
- 1st year = 400:200:400:125
- 2nd year = 800:400:800:250
- 3rd year onwards = 1200:600:1200:500 (2 equal split doses on July-August and December-January)

9.9. Irrigation

- Yielding palm of 3 years age and above: Minimum of 150 litres/day.
- Drip irrigation: 5 h of irrigation per day using 4 drippers per palm at 8 litres of water/dripper.

9.10. Weed management

- Glyphosplate and paraquat gave good control of weeds (0.367 kg a.i/ha), salapon 10 a. i/ha.

9.10. Mulching

- *Puraria phaseoloides, Calapogonium mucunoides, Mucuna* sp and *Centrosema pubescens* are useful cover crops.
- Empty bunches, male flowers, coconut husk, sugarcane trash can be used as mulch.

9.11. Maintenance of paths

- In young plantation, maintenance of paths is important for inspection and in later years for harvesting.

9.12. Insect pollination in oil palm

- Pollinating weevil (*Eldeidobius kamerunicus)* carries maximum pollen during the third day of anthesis.

9.13. Ablation

- Removal of all inflorescences during first three years to gain adequate stem girth, vigour and root system.
- Ablation is done at monthly intervals by pulling out the young inflorescence.

9.14. Leaf pruning

- Palms aged 4-7 years should retain 6-7 leaves/spiral (48-56 leaves).
- Those aged 8-14 years should retain 5-6 leaves/spiral (40-49 leaves)
- Palms above 15 years should have 4-5 leaves/spiral (32-40 leaves).

9.15. Harvesting

- Fruits mature in 180 days after pollination.
- Loosening of fruits from the tip of bunch is an indication for harvesting stages.
- First harvest: 3.5 to 4 years after planting.
- Young palms: Chisel harvesting
- 10th year onwards: Harvesting by hook.
- Harvesting at 10-12 days interval and during rainy season, 6-7 days interval.

9.16. Yield

- 25-30 tons of fresh fruit bunches/ha.

9.17. Processing

- Fruits are perishable and hence, should be processed with in 24 hours.
- If not, it develops free fatty acid (FFA).
- There are three stages for processing.
  - Extraction of crude palm oil
  - Refining, bleaching and deodorization (RBD) of oil
  - Fraction of palm oil to produce palmolein.

9.17.1. Technology

- Palm is extracted from mesocarp of the fruit using wet extraction method.
- Steps involves harvesting, sterilization, stripping, digestion, processing, clarification and purification.
- Fresh fruit bunches is to be steam sterilized at a temperature of 120°C at a pressure of 3 kg/cm^3 for 1 hr.
- After that, digestion of fruits to loosen the pericarp and to break up fruit pulp physically.
- This is done by a mechanical agitation at a temperature of 95°C for 30 minutes and 20-30% oil is released from oil and water mixture.
- The extraction of oil from the pulp is carried out by using hydraulic process for small scale and screw press for large scale operation.
- After processing, oil water mixture is subjected to clarification.
- Mixture with water is heated to 95°C and the oil separates at top and water with residues settles at the bottom.
- Oil layer is decanted and the sludge is removed from the bottom of settling tank
- Centrifuging can be adopted for quick separation of oil from the sledge.
- Raw oil is extracted under controlled condition.
- Oil is purified for removal of dirt and moisture by using special centrifuge.
- For making edible oil, degumming neutralization, bleaching and deodorization are followed.
- Final step is fractionalization to produce liquid palmolein.

9.19. Extraction of kernel oil

- Nuts are shelled in a rotary shaker and the light fraction removed in a steam is to be used as fuel.
- Separation of shells from the kernel is done by hydro cycler.
- Oil palm kernels is extracted by extractors.

10

Palmyrah (*Borassus flabellifer*) Arecaceae, 2n=36

10.1. Origin and distribution

- Native to the tropical Africa and grows in the Nepal, India, Bangladesh, Sri Lanka, Cambodia, Laos, Myanmar, Thailand, Vietnam, Malaysia, Indonesia and the Philippines.
- Palm is widespread in Tamil Nadu, Kerala, Andhra Pradesh, Telangana, Odisha, West Bengal, Bihar, Jharkhand, Karnataka and Maharashtra.
- In Tamil Nadu, it is present in Tuticorin, Tirunelveli, Thenkasi, Virudhunagar, Ramanathapuram Madurai, Theni, Salem, Namakkal, Kancheepuram, Chengalpattu, Tiruvallur, Villupuram, Kallakurichi and Cuddalore districts.
- Palm is dioecious in nature.
- Palmyrah was declared as "State Tree of Tamil Nadu" by the Government of Tamil Nadu.

10.2. Economic importance

- **Sweet sap padaneer:** Tamil Nadu and Kerala padaneer considered a primary food.
- Contains potassium, ascorbic acid, calcium
- Increases body weight
- Cure people who are breath troubled.
- Upgrades the digestion and cools the body
- Young nungu mixed with padaneer heals intestine and stomach ulcers.
- Cleanses intestine and cures constipation.
- Improves health of people with leprosy.
- Vitamin A deficiency beriberi, scurvy, etc. can be treated by means padaneer sipping.
- Red blood cells of people with anemia (hemoglobin) can be increased
- **Value-added products**: Maple sugar, panam karkandu, panaitten, nungu candy, soft drinks, kir, fruit juice, potato pie, sweets, kesari, rice pudding, idli, pakkota, cimpu halwa, ice etc.

- **Flowers, fruits and seeds**: Used to prevent swelling of the spleen obtained from burned ash spathes.
- **Nungu**:Contains sweet taste.
- Slender, easily digestable.
- Young nungu mixed with curd used as a thyir paccadi and served at wedding functions.
- Applying nungu water on the body cures the rashes.
- Nungu water pouring on the eyes of the cornea will reduce heat.
- Skin of the nungu grind is given to small children in villages to solve a stomach huff.
- **Palmyrah fruit (Panampalam):** Ripe fruits are used to treat leprosy.
- **Palm sugar:** Dissolved in water and extracted liquid filter provided for typhoid patients may have their body strength.
- Dissolved in water is remedy for constipation to children.
- Children eating palm sugar solution regain their weight.
- **Panam karkandu:** Mix with milk will increase the agility.
- Alleviate sore throat with healing and reduce the body heat.
- Measles and smallpox aggressors are cured and cooling the body.
- Pepper powder with panam karkandu cures dry cough and throat irritation.
- **Root:** Cure people infected with leprosy.
- Provide cooling.
- Used to make up stomach discomfort.
- **Leaves:** Used as cattle feed in times of drought.
- Fibre and various handicrafts are made.
- In villages, used for roofing of house.

10.3 Variety

- SVPR is released from Palmyrah Research Station, Srivilliputhur

10.4. Climate

- Tropical and sub-tropical regions.
- Temperature: 30-45°C
- Elevation up to 500-800 m above msl.

10.5. Season

- Monsoon season

10.6. Soil

- Grows in fallow land and all soil types.
- Tolerates acidity of the soil (pH 5.5)
- East coast palm grows in sand ridges and prevents soil erosion.

10.7. Propagation and planting

- Propagated through seeds.
- Planted in round basin of 45 cm in diameter.
- Seeds germinate and produce a tuber and this tuber is highly priced and edible and good for health.
- Tuber is botanically called apacole.

10.8. Manuring

Year	FYM (kg/palm)	FYM (t/ ha)
Second year	10	11.10
Third year	20	22.20
Fourth year	20	22.20
Fifth year	30	33.30
Sixth year	30	33.30
Seventh year	40	44.40
Eighth year	40	44.40
Ninth year	50	55.50
Tenth year	50	55.50
Eleventh year onwards	60	66.60

10.9. Intercropping

- Groundnut, sesame, cowpea and green lentils,
- Pomegranate, aonla, custard apple, guava, West Indian cherry, ber.
- Moringa is grown as a border crop.

10.10. Cleaning of trees

- Sixth year onwards cleaning is necessary.

10.11. Padaneer extraction and yield

- From male and female inflorescence, padaneer extract is derived.
- A palm takes three to five months to produce padaneer.
- Cut 30% leaves to increase padaneer producing hormones.
- Padaneer takes 90-130 days in blossoming season (April-July).
- Padaneer begins to secrete less in April, increasing in May-June, and slow down in July.
- Female palms offer padaneer up to 33-50%.

- 12 to 15 years of palm: 100-200 litres/tree/season in 4 months.

Year	Yield of padaneer/palm/year (litres)
12	60
13	70
14	80
15	100

- **Padaneer storage:** Add lime at 1.25 g/litre to improve taste and delay fermentation.
- Stored in mud pots or refrigerated conditions.

10.12. Sugar and jaggery preparation

- Padaneer is cleaned using filter.
- Boiled in aluminum containers at 110°C.
- Add crushed castor seeds to reduce excessive cooking.
- Cool in cooling tanks with thick viscosity to obtain brown sugar.
- 7-8 kg sugar and 8 kg jaggery are obtained from 100 kg of padaneer.

10.13. Tapping

- Extraction of sap from the inflorescence is referred as tapping.
- Extraction of sap from male and female palm somewhat dissimilar.
- **Tapping in male palm**
    - In male palm, the sap is extracted from the two weeks old inflorescence, which is called aripanai.
    - When the extraction of sap from the florescence is done in one month old inflorescence is known as vellupanai.
    - inflorescence is wrapped with palmyra leaves and a pot is placed on the inflorescence for extraction of the sap.
    - To extract the sweet sap (padaneer) commonly lime is added to the sap If lime is not added to the sap, ferments and the liquid is called toddy.
- **Tapping in female palm**
- Sweet sap (Padaneer) from the female tree is obtained for a longer period (April to December) compared to male tree (December to February).
- Young female inflorescence is tapped using a method called thattupalai.
- Tappers soften the inflorescence tissues by hitting the main axis with wooden/ iron rod and fork is used to press the region from which the tender fruits develop when the inflorescence are two or three months old.
- Another method called Kaivetty which is employed during July to November.

11

Pests of Plantation Crops

11.1. Tea

11.1.1. Tea mosquito bug: *Helopeltis theivora*

Symptoms

- Adults and nymphs suck sap from buds, young leaves and tender stems.
- Punctures appear as reddish brown spots.
- Leaves curl up, badly deformed and shoots dry up.

Management

- Spray clothianidin 50 WDG 120 g/ha or thiacloprid 21.7 SC 375 ml/ha or thiamethoxam 25 WG 100g/ha or bifenthrin 8 SC 500 ml/ha.

11.1.2. Shot hole borer: *Euwallacea fornicatus*

Symptoms

- Presence of round shot holes in primary branches.
- Mortality of buds and dieback symptoms in branches.
- Presence of circular or longitudinal tunnels inside stem.

Management

- Remove and destroy affected twigs.

11.1.3. Red spider mite: *Oligonychus coffeae*

Symptoms

- Infest upper surface of mature leaves.
- Starts along midrib and veins and spreads to entire upper surface of leaves.
- Affected leaves become bronzed, dry and crumpled.

Management

- Spray azadirachtin 5 W/W at 200 ml/ha or bifenthrin 8 SC at 500 ml/ha or dicofol 18.5 EC at 1250 ml/ha.

11.1.4. Scarlet mite: *Brevipalpus californicus*

Symptoms

- Mites congregate on under surface of mature leaves.
- Feeding leads to brown discolouration of leaves.

Management

- Spray fenazaquin 10 EC at 1000 ml/ha or hexythiazox 5.45 EC at 300-500 ml/ha or propargite 57 EC at 750-1250 ml/ha or dicofol 18.5 EC at 1250 ml/ha.

11.1.5. Purple mite: *Calacarus carinatus*

Symptoms

- Mites feed on undersurface of mature leaves.
- Assam type of tea - more susceptible to purple mite.
- Damaged leaves - coppery brown discolouration.

Management

- Spray dicofol 18.5 EC at 1250 ml/ha or ethion 50 EC at 500 ml/ha.

11.1.6. Pink mite or orange mite: *Acaphylla theae*

Symptoms

- Mites are found on the undersurface (Abaxial) of young leaves.
- Affected leaves turn pale and upward curling.
- In severe infestation, leaves become leathery and brown.
- Damage restricts to top 10-15 cm of tender leaves.
- Assam type of tea is susceptible.

Management

- Spary ethion at 2ml/ lit of water or azadirachtin 5 EC at 200 ml/ha or dicofol 18.5 EC at 1250 ml/ha or fenazaquin 10 EC at 1000 ml/ha.

11.1.7. Yellow mite: *Polyphagotarsonemus latus*

Symptoms

- Mite is seen on young leaves especially top two to three leaves and bud.
- Affected leaves become rough and brittle and corky lines.
- Downward curling.
- Internodes get shortened; shoots are stunted and deformed.

Management

- Spray dicofol 18 EC at 2 ml/lit or ethion 50 EC at 2 ml/lit.

11.1.8. Aphid: *Toxoptera aurantii*

Symptoms

- Leaves curl up and shoot growth is stunted.
- Honey dew secreted development of sooty mould.

Management

- Spray dimethoate 30 EC or chlorpyriphos 20 EC at 2 ml/litre or phosalone at 2 ml/lit.

11.2. Coffee

11.2.1. White stem borer: *Xylotrechus quadripes*

Symptoms

- Larvae enter into hardwood and make tunnels which extend even into roots.
- Tunnels are filled with excreta of grubs.
- Yellowing and wilting of leaves.
- Young plants (7 to 8 years old) attacked by borer may die in a year.

Management

- Prune the infested plants, uproot and burn.
- Remove loose scaly bark of main stem and thick primaries using coir glove or coconut husk and kill eggs
- Spray with neem kernel extract.
- Field release of predators, *Apenesia* sp or white muscardine fungus, *Beaveria bassiana.*

11.2.2. Coffee berry borer: *Hypothenemus hampei*

Symptoms

- Pin hole at tip of berries (Navel region).
- Female beetle bores into berries through navel region, make tunneling and feed inside content.
- Powdery substance pushed out through holes.

Management

- Dipping infested berries in boiling water for 2-3 minutes kills all stages inside.
- Drying of coffee beans prevents breeding of beetles based on moisture content level.
    - Arabica (10% moisture)
    - Robusta (11% moisture)

- Install attractant trap or multiple funnel trap or pitfall trap to collect and kill the beetles.
- Chlorpyriphos is also effective in controlling beetle.
- Field release of *Cephalonomia stephendarix* and *Phymastichus coffea.*
- Application of white muscardine fungus, *Beauveria bassiana.*

11.2.3. Shot hole borer: *Xylosandrus compactus*

Symptoms

- Infests green succulent branches.
- Initial symptom-drooping of leaves.
- Withered (Faster in young branches and delayed in older twigs) or dried branches.
- Attacked leaves fall prematurely.
- Terminal leaves wilt, droop and dry up.
- Severe infestation - loss of considerable number of productive branches.

Management

- Prune affected twigs 5.0 to 7.5 cm beyond the shot-hole and burn (September).
- Remove and destroy the unwanted/infested suckers during summer (Avoid breeding).
- Maintain thin shade and good drainage.

11.2.4. Mealy bugs: *Planococcus* spp.

Symptoms

- Both nymphs and adults suck sap from the leaves.
- Severe infestation - chlorotic leaves, aborted flower buds and small berries.
- Honey dew excretes cause development of sooty mould fungus (Affects photosynthesis).

Management

- Control ants by dusting quinalphos 1.5% or malathion 5% around base of coffee and shade trees.
- Spray with quinalphos 25 EC at 300 ml.
- In 2-4 year old crop, drenching with dimethoate 30 EC @ 3 ml per litre of water.
- Field release of predatory lady bird beetle, *Cryptolaemus montrouzieri* and parasitoid, *Leptomastix dactylopii.*

11.2.5. Coffee bean beetle: *Araecerus fasciculatus*

Symptoms

- Infested coffee beans show circular holes (Bigger than made by the coffee berry borer).
- Attacked fruits in the field shrink and become black.

Management

- Maintain optimum temperature, RH and moisture (Less than 8%) in storage.
- Impregnate gunny bags with malathion 50 EC at 10 ml + pyrethrum colloid at 2 g per litre of water.

11.2.6. Red borer: *Zeuzera coffeae*

Symptoms

- Caterpillar bores into stem or branches to feed on wood.
- Young plants or branches show wilting.
- A pellet like excreta of larva hangs out and accumulates at base of plant.

Management

- Cut and burn affected plants or twigs.
- Field release of white muscardine fungus, *B. bassiana* and braconid parasite, *Amyosoma zeuzerae.*

11.3. Coconut

11.3.1. Rhinoceros beetle: *Oryctes rhinoceros*

- Beetle bores into unopened fronds and spathes.
- Yield loss upto 10 to 15%.
- Fully opened shows characteristic triangular cuts.
- Central spindle appears cut or toppled.
- Fully opened fronds show diamond shaped cuttings.
- Holes with chewed fibre sticking out at the base of central spindle.

Management

- **Cultural method:** Remove and burn all dead coconut trees.
- Collect and destroy various bio-stages of beetle from the manure pits.
- **Mechanical method:** Beetle may be extracted from palm crown using GI hooks.
- Set up light traps following first rains in summer and monsoon to attract and kill adult beetles.

- **Chemical method:** Top most three leaf axils around the spindle may be filled with:
  - Sevidol 8 G at 25 g + fine sand at 200 g thrice in a year in April-May, September-October and December-January.
  - For seedlings, apply naphthalene balls at 10.5 g (Approximately three to four balls) covered with fine sand, once in 45 days.
- Place phorate 10 G at 5 g in perforated sachets in two inner most leaf axils for 2 times at 6 months intervals.
- Treat manure pits and other possible breeding sites with 0.01% carbaryl (50% WP) on w/w basis and repeated every six months.
- **Trap method:** Set up rhino lure pheromone trap at 5 traps/ha.
- Dispenser may be hanged in a plastic bucket having 2 litres of insecticide solution once in a week.
- Trapped beetles can be disposed off.
- **Biological method:** Application of green muscardine fungus, *Metarhizium anisopliae* at 5 x 10^{11} spores / m^3
- Spray 250ml *Metarizhium* culture + 750 ml water in manure pits.
- Field release of *Baculovirus oryctes* inoculated adult rhinoceros beetle at 15 beetles/ha reduces the leaf and crown damage.
- Soak castor cake at 1 kg in 5 litres of water in small mud pots and keep them in coconut gardens to attract and kill adults.
- Apply mixture of either neem seed powder + sand (1:2) at150 g per palm or neem seed kernel powder + sand (1:2) at150 g per palm in base of 3 inner most leaves in crown.

11.3.2. Red palm weevil: *Rhynchophorus ferrugineus*

Symptoms

- Hole can be seen on stem with chewed up fibres protruding out.
- Oozing out reddish brown liquid from hole.
- Grubs feed on soft tissues of stem or crown and often cause severe damage; completely eaten and become fall of rotting fibres.
- Top of young palms withers, while in older palms top portion of trunk bends and breaks.
- Gnawing sound produced by feeding grubs inside will be audible.
- In the advanced stage, crowns fall down or dry up later when palm is dead.

Management

- **Cultural method:** Remove and burn all wilting or damaged palms.

- Avoid the cutting of green leaves.
- **Chemical method:**Through top most hole, pour 1% carbaryl (20g/l) or 0.2% trichlorphon at 1 litre / palm using a funnel. Then, plug this hole with cement / tar. Repeat after one week.
- Fill crown and axils of top most three leaves with a mixture of fine sand and neem seed powder or neem seed kernel powder (2:1) once in three months.
- **Trap method: Coconut log traps:** Setting up of attractant traps (Mud pots) containing sugarcane molasses 2½ kg or toddy 2½ litres (or pineapple or sugarcane activated with yeast or molasses) + acetic acid 5 ml + yeast 5 g + longitudinally split tender coconut stem/logs of green petiole of leaves of 30 numbers in one acre to trap adult red palm weevils in large numbers. Incorporate any insecticide to each trap to kill the weevils trapped.
- **Pheromone trap:** Install pheromone trap at one trap per ha
- Step 1: Specialized buckets with 3 to 4 holes are made and the bucket is wound with coconut fibre/ jute sack, so that the pests can enter.
- Step 2: Lure (Ferrolure +) is suspended inside bucket and one litre of water is added along with 100g pineapple/ sugarcane, 2g yeast and 2g carbaryl in bucket.
- Step 3: Bait buckets are placed at sites in farm, where infestation is seen most.
- Step 4: After a week, water is checked for catch and refilled to prevent mosquitoes from breeding.

11.3.3. Coconut Eriophyid: *Aceria guerreronis*

Symptoms

- 2-3 months old buttons show pale yellow triangular patches below the perianth.
- Later, the patches become brown to black necrotic lesions with longitudinal fissures on husk.
- Oozing of gummy exudation from affected surface of nuts.
- Distortion and stunting of nuts leading to reduction in copra yield.
- Malformed nuts with cracks and hardened husk.

Management

- **Cultural method:** Collect and destroy all fallen buttons of affected palm.
- Grow intercrop (Sunhemp, four crops/year) and shelter belt with *Casuarina* all round.

- Apply urea 1.3 kg, super phosphate 2.0 kg and muriate of potash 3.5 kg/ palm/year.
- Soil application of borax 50 g + gypsum 1.0 kg + manganese sulphate 0.5 kg/palm/year
- TNAU micronutrient solution at 200 ml/tree.
- Neem cake application @ 5 kg per palm per year.
- **Chemical method:** Round 1: Azadirachtin 1% (5 ml in one lit. of water).
- Round 2: Neem oil + teepol (30 ml in one lit. of water).
- Triazophos 40 EC at 5 ml/lit or carbosulfan 25 EC at 2 ml/ lit in alternation with neem azal 1% at 5 ml/lit as spot application.
- Apply neem oil + garlic emulsion (2%)
  - o Slice bar soap and dissolve in 500 ml lukewarm water. Grind 200 g of garlic and take extract in 300 ml of water. Pour 500 ml soap solution in 200 ml neem oil slowly and stir vigorously to get a good emulsion. Mix the garlic extract in neem oil + soap emulsion. Dilute this 1 litre stock solution by adding 9 litres of water to get 10 litres of 2% neem oil + garlic emulsion.
- **Biological method:** Entomo fungal pathogen, *Hirsutella thompsonii* and *Verticillium lecanii* manages mites effectively.

11.3.4. Leaf eating caterpillar/ black headed caterpillar: *Opisina arenosella*

Symptoms

- Dried up patches on leaflets of lower leaves, three or four youngest leaves at center remain green.
- Galleries of silk and frass on underside of leaflets.
- Whole plantations present a scorched appearance.

Management

- **Cultural method:** Affected leaves may be cut and burnt.
- **Chemical method:** Spray under surface of fronds with dichlorvos 100EC at 0.02% or malathion 50 EC at 0.05% (1 ml/lit.) or quinalphos at 0.05% or phosalone at 0.05%.
- **Biological method:** Larval parasitoids, Bethylid, *Goniozus nephantidis*, *Elasmus nephantidis* (Brown species) and *Brachymeria nosatoi* are the most effective at 3000/ha release.
- Release pupal (Eulophid) and (Chalcid) parasitoids and predators periodically from January onwords.

11.3.5. Slug caterpillar: *Parasa lepida* and *Contheyla rotunda*

Symptoms

- Young larvae feed on lower epidermis of leaf.
- As they mature, whole leaf blade is eaten leaving the midribs.
- In heavy infestation, defoliation occurs.

Management

- **Chemical method:** Spray carbaryl 50 WP at 2 g/litre or dichlorvos 76 WSC at 2 ml / lit.
- **Mechanical method:** Set up 5 light traps per hectare to trap and collect adult moths.

11.3.6. Coconut skipper: *Gangara thyrsis; Suastus gremius*

Symptoms

- One half of leaflets are cut and rolled into a case which dries.

Management

- Spray dichlorvos 76 WSC at 2 ml / lit.

11.3.7. White grub/ cockchafer beetle: *Leucopholis coneophora*

Symptoms

- In nursery, grubs feed tender roots and tunnel into bole of collar region resulting in drying and yellowing.
- In older plantations, yellowing of leaves, premature nut shedding, delayed flowering, retardation in growth and yield decline.

Management

- **Cultural method:** Repeated ploughing once a week for 4 to 5 times after first rain in summer.
- Plant neem twigs with leaves after rain to attract and kill adult beetles
- **Mechanical method:** Set up light trap at 1/ha
- **Chemical method:** Soil application with malathion 5 D at 25 kg/ ha at the time of planting.
- Treat soil with phorate 10G @ 100 g/palm or drench with chlorpyrifos at 0.04% first during April-May and second in September
- **Physical method:** Collect and destroy adult beetles attracted to trees like neem, *Ailanthus* and *Acacia* on receipt of monsoon showers (In evening).

11.3.8. Termite: *Odontotermes obesus*

Symptoms

- Damages transplanted seedlings particularly in the earlier stage results in wilting of seedlings.
- Base of trunks plastered with runways made of soil and fibres.

Management

- **Cultural method:** Locate termite mounds in or near the coconut nursery or garden and destroy.
- Removal of organic matter in nursery and covering germinating nuts with a layer of river sand.
- **Chemical method:** Spray copper sulphate 1% or cashew nut shell oil 80% or chlorpyriphos at 3ml/lit of water
- Spray neem oil 5% or NSKE 20% to preserve plaited coconut leaves from termite attack.
- Apply calcium at base of trunk for control of termite attack.
- Swabbing with neem oil 5% once on base and up to 2 m height of trunk.

11.4. Arecanut

11.4.1. Spindle bug: *Carvalhoia arecae*

Symptoms

- Sap sucking bug - damage unopened spindle leaf
- Inhabit inner most leaf axils, usually below spindle.
- Severe infestation - blackish brown linear lesions on spindle leaf
- Stunted growth and twisted

Management

- Spray dimethoate at 0.05%
- Filling inner most leaf axils with phorate 10% G (10g/palm)

11.4.2. Cholam or white mite: *Oligonychus indicus*

Symptoms

- Adults and nymphs feed on lower surface of leaves.

Management

- Removal of heavily infested and dried leaves and burning
- Spraying with dicofol (2 ml/litre) or rogor (1.5 ml/litre)

11.4.3. Root grub: *Leucopholis burmeisteri*

Symptoms

- Grubs feed on growing roots
- Infested palms show a sickly appearance
- Yellowing of leaves

Management

- Addition of organic amendments and antifeedants (Neem and pongamia oil cake)
- Application of phorate (Thimet 10G) at 15g per palm.

11.4.4. Inflorescence caterpillar: *Tirathaba mundella*

Symptoms

- Caterpillars feed on inflorescence (Tender female flowers) and rachillae
- Webbing and feeding inflorescence
- Spathe opening is delayed
- Yellowing of spadices
- Presence of small holes with frass and drying on the spathe

Management

- Infected spadices may be forced open and sprayed with malathion 0.05%

12

Diseaes of Plantation Crops

12.1. Tea

12.1.1. Algal leaf spot: *Cephaleuros virescens*

Symptoms

- Leaves develop lesions that are roughly circular, raised and purple to reddish brown.

Management

- Removal of infected portions and spraying of Bordeaux mixture
- Improving soil nutrient status by application of nitrogen, phosphorus and potassium

12.1.2. Brown blight, grey blight: *Colletotrichum* sp., *Pestalotiopsis* sp.

Symptoms

- Small, oval, pale yellow green spots appear on young leaves, surrounded by a narrow, yellow zone.
- As the spots grow and turn brown or grey, concentric rings with scattered tiny black dots become visible leading to defoliation.

Management

- Spray copper oxy chloride or Bordeaux mixture 0.1% during winter and summer seasons.

12.1.3. Blister blight: *Exobasidium vexans*

Symptoms

- Small, pinhole size spots on young leaves of less than a month old.
- As the leaves develop, spots become transparent, larger, and light brown.
- After 7 days, lower leaf surface develops blisters with dark green, water soaked zones surrounding blisters.
- Following release of fungal spores, blister becomes white and velvetty.
- Subsequently, blister turns brown and young infected stems become bent, distorted and break off or die

Management

- Spraying of Bordeaux mixture or copper oxy chloride 0.1%
- A mixture of 210g of copper oxy chloride + 210g of nickel chloride/ha sprayed at 5 days interval from June to September and October to November
- Spray Tridemorph at 340 and 560 ml/ha.

12.1.4. Horse hair blight: *Marasmius crinisequi*

Symptoms

- Black fungal threads resembling horse hair are attached to upper branches and twigs by small brown discs.
- Fungus infects twigs from discs and produces volatile substances that cause rapid leaf drop.

Management

- Prune out infected or dead branches from the plant canopy

12.1.5. Twig dieback, stem canker: *Macrophoma theicola*

Symptoms

- Browning and drooping of affected leaves.
- Entire branch can die from tip downward.
- Dying branches often have cankers - shallow, slowly spreading lesions surrounded by a thick area of bark.

Management

- Remove diseased twigs by cutting several inches below cankered areas and disinfecting them.

12.1.6. Black root rot: *Rosellina arcuata*

Symptoms

- At surface of soil, mycelium surrounds stem and kills bark for 7.5-10.0 cm length.
- A swollen ring of tissue is formed round the stem above the dead patch.

Management

- Removal and destruction of infected plant.
- Dig a drench around infected bush to provide sunlight.

12.2. Coffee

12.2.1. Rust: *Hemileia vastatrix*

Symptoms

- In earlier stage, 2-3 mm size yellow spots develop on lower surface of leaves.
- Rust lesions develop as orange yellow spots on lower surface of leaves.
- Upper surface of leaves become yellow to brown.
- At later stages, orange yellow spots become black spots and then, necrosis.
- Plants remain stunted with reduced yield.

Management

- Grow S 238 and S 395 resistant varieties.
- Spraying of Bordeaux mixture 0.5% in May, August-September and October.

12.2.2. Brown leaf spot: *Cercospora coffeicola*

Symptoms

- Brown spots with grey centre and brown margin surrounded by yellow hallow.
- On berries, dark blotches are developed, shrink and drop.

Management

- Spraying with Bordeaux mixture 0.5%.

12.2.3. Damping off / Collar rot: *Rhizoctonia solani*

Symptoms

- In post emergence damping off, collar region near soil level is infected leading to death of seedlings.

Management

- Soil drenching with copper oxy chloride 0.25%

12.2.4. Die back or Anthranose: *Collectorichum coffeasum*

Symptoms

- On leaves circular to greyish spots of 2-3 mm in dia.
- On berries, small dark coloured sunken spots are formed.
- Die back also occurs.

Management

- Spraying mancozeb 0.25%

12.2.5. Black rot or koleroga: *Corticium salmonicolor*

Symptoms

- Dark brown or black decaying leaves, twigs and berries from which black rot is derived.
- Sclerotia can be seen all over the affected areas.

Management

- Spraying with Bordeaux mixture 1%

12.2.6. Berry blotch: *Cercospora coffeicola*

Symptoms

- Necrotic spots on exposed surface of green berries enlarge and cover the major portion.
- Fruit skin shrivels and sticks fast.

Management

- Spray 1% Bordeaux mixture during June and late August and maintain medium shade overhead.

12.3. Coconut

12.3.1. Tanjore wilt/ basal stem end rot/ Ganoderma wilt: *Ganoderma lucidem* and *Ganoderma applanatum*

Symptoms

- Yellowing and drooping of outer whorl of leaves.
- Exudation of reddish brown liquid through cracks at base of trunk and oozing spread upwards.
- Bark turns brittle and often gets peeled off in flakes, leaving open cracks and crevices.
- Internal tissues are discoloured, disintegrated and emitting a bad smell.
- Bracket formation at base of trunk during rainy season.
- Ultimately, palm dies off.

Management

- **Cultural method**
- Remove and destroy all affected palms.
- Green manure crops must be raised and ploughed *in situ* before flowering.
- **Biological method**
- Apply *Bacillus subtilis* (*Pf*-1) at 200g/palm + *Trichoderma asperellum* at 200g/palm/year.

- Apply 200g Phosphobacteria and 200 g Azotobacter mixed with 50 kg of FYM/palm.
- Apply FYM 50 kg + neem cake 5 kg once in 6 months along with fertilizers.
- **Chemical method**
- Bleeding patches in stem may be chiselled and protected with tridemorph (5% calxin) and subsequently with hot coal tar.
- Aureofungin-sol 2 g +1 g copper sulphate in 100ml water applied as root feeding.
- Root feeding with tridemorph 2ml or hexaconazole 1 ml with 100 ml of water (3 times at 3 months interval).
- Drench 40 litres of 1% Bordeaux mixture around trunk in a radius of 1.5m.

12.3.2. Bud rot: *Phytophthora palmivora*

Symptoms

- Seedlings show yellowing of younger leaves.
- Basal tissues of leaf rot quickly and can be easily pulled out from the crown.
- In the later stages, spindle withers and drops down.
- Tender leaf base and soft tissues of crown rot into a slimy mass of decayed material emitting foul smell.
- Ultimately, entire crown falls down and the palm dies
- In adult palms, colour change of the spear, which becomes pale and breaks at the base and hangs down.
- Rotting slowly progresses downwards, finally affecting meristem and killing palms accompanied by drooping of successive leaves

Management

- **Cultural method**
- Provide adequate drainage in gardens.
- Adopt proper spacing and avoid over crowding in bud rot prone gardens.
- **Chemical method**
- Drenching the crown with copper oxy chloride 0.25%.
- Apply Bordeaux paste and protect it from rain till normal shoot emerges.
- Spray 0.25% copper oxy chloride or 1% Bordeaux mixture on crown of neighbouring palms.
- For palms sensitive (Dwarf palms) to copper fungicides, tie 2 g mancozeb in small, perforated sachets to top of leaf axil which is released when raining to leaf base, thus protecting the palm.

- Remove infected tissues from crown and dress with Bordeaux paste or spray with 1% Bordeaux mixture as pre-monsoon spray (May and September).

12.3.3. Leaf blight (LB): *Lasiodiplodia theobromae*

Symptoms

- Affected leaflets start drying from tip downwards and exhibit a charred or burnt appearance.
- Dark grey to brown lesions with wavy to undulated margins appear from apex of nuts.
- Decay of endosperm.
- Affected nuts are desiccated, shrunken, deformed and dropped prematurely and resulting in nut yield loss up to 10 to 25%.

Management

- **Biological method**
- Application of 200g *Bacillus subtilis* along with 50 kg of FYM+ 5 kg of neem cake/ palm/year.
- **Chemical method**
- Spray 1.0% Bordeaux mixture or 0.25% copper oxychloride (2 times at 45 days interval during summer months).
- Root feeding of carbendazim 2 g or hexaconazole/ Tridemorph 2 ml + 100 ml water (3 times at 3 months interval).

12.3.4. Stem bleeding disease: *Thielaviopsis paradoxa*

Symptoms

- Exudation of a dark reddish brown liquid from longitudinal cracks in bark and wounds on stem trickling down for a distance of several inches to several feet.
- Liquid oozing out, dries up and turns black.
- Tissues below lesions become rotten and turn yellow first and later black.
- In advanced cases, inner portions of affected trunks are hollow due to decay of inner tissues.
- Outer whorl of the leaves turns yellow, dry and shed prematurely.
- Production of bunches is affected adversely.
- Nut fall is also noticed.

Management

- **Cultural method**
- Destroy the chiseled materials by burning.
- Along with 50 kg FYM, apply 5 kg neem cake with *Trichoderma asperellum* at 200g/palm/year to the basin during September.
- **Chemical method**
- Chisel out completely affected tissues and paint wound with tridemorph 5% or Bordeaux paste. Apply coal tar after 1-2 days on treated portion. Burn off chiselled pieces.
- Root feed with tridemorph 5ml in 100 ml water, thrice a year during April-May, September-October and January-February.

12.3.5. Leaf rot disease: *Colletotrichum gloeosporioides, Exserohilum rostratum* and *Fusarium* spp.

Symptoms

- Appearance of water soaked brown lesions in the spear leaves of root wilt affected palms.
- Spots enlarge and coalesce resulting in extensive rotting.
- As the leaf unfurls, rotten portions of lamina dry and get blown off in wind, giving a "fan" shape to leaves.

Management

- Pour fungicide solution of hexaconazole 2ml or mancozeb 3g in 300ml water per palm to base of spindle leaf, 2-3 rounds of spraying.
- Spray crowns and leaves with 1% Bordeaux mixture or 0.5% copper oxy chloride or 0.4% mancozeb in January, April-May and September.

12.4. Arecanut

12.4.1. Foot rot or Anabe roga: *Ganoderma lucidum*

Symptoms

- Leaf lets in outer wholes of leaves become yellow and spreads to whole leaf and leaves drooping down.
- Later, inner whole leaves also become yellow.
- Subsequently, all leaves droop, dry up and fall off, leaving the stem alone.
- Then, stem also become brittle and easily broken by heavy wind.
- Base of stem shows brown discoloration and oozing of dark fluid.
- Bracket shaped fructifications of fungus called 'anabe' appears at base of the trunk.
- Roots become discoloured, brittle and dried.

Management

- Digging trenches to avoid root contact from diseased to healthy plants.
- Drench 1% Bordeaux mixture at frequent intervals.

12.4.2. Mahali / kolerogo/ fruit rot: *Phytophthora arecae*

Symptoms

- Rotting and excessive shedding of immature nuts.
- Water soaked spots initially develop at base of nut.
- Nuts show large vacuoles and dark brown radial strands.
- Top of affected trees dries up resulting in withering of leaves and bunches.
- Affected nuts fall off and show the white mycelial growth of the fungus.

Management

- Spray with Bordeaux mixture 1% before onset of monsoon and second spray 40-45 days.

12.4.3. Inflorescence die back and button shedding: *Colletotrichum gloeosporioides*

Symptoms

- Disease appears on rachilla of male flowers and then, in main rachis as brownish patches which soon spread from tip downwards covering entire rachis causing wilting.
- Female flowers of the infected rachis shed and whole inflorescence show 'die back' symptom.

Management

- Spraying with copper oxy chloride 0.25% or mancozeb 0.2%

12.5. Cocoa

12.5.1. Seedling blight: *Phytophthora palmivora*

Symptoms

- On leaves, small water soaked lesions appear which later coalesce in to blighting of leaves.
- On stem, water soaked lesions develop initially and later turn to black causing seedling death.

Management

- Spray 1% Bordeaux mixture or 0.2% copper oxy chloride just before onset of monsoon and thereafter, at frequent intervals.

12.5.2. Black pod rot: *Phytophthora palmivora, P. megakarya, P. citrophthora and P. capsici*

Symptoms

- Chocolate brown spot spreads rapidly and occupies the entire surface of the pod.
- Affected pods turn brown to black.
- Internal tissues and beans become discoloured.

Management

- Spray 1% Bordeaux mixture with onset of monsoon and also frequent intervals.
- *Bacillus subtilis* liquid formulations @ 0.5% as soil and foliar spray (3 times per year- June, October and February) found effective against cocoa pod rot.

12.5.3. Stem canker: *Phytophthora palmivora*

Symptoms

- Greyish brown water soaked lesion on the outer bark.
- A reddish brown liquid oozes out from these lesions, which later dries up to form rusty deposits.

Management

- Excision of diseased bark followed by dressing with Bordeaux mixture or copper oxy chloride paste.

12.6. Rubber

12.6.1. Birds eye spot: *Drechslera heveae (Helminthosporium heveae)*

Symptoms

- Circular spots with grey centre with brown border occur in large number in nursery or on young plants.

Management

- Spray with 1% Bordeaux mixture

12.6.2. Corynespora leaf spot: *Corynespora cossicola*

Symptoms

- Circular spots with brown or papery centre and dark brown ring appear on young leaves.
- Yellow hallo develops around the ring.
- Affected leaves dries up.

Management:

- Spray with 1% Bordeaux mixture

12.6.3. Pink disease: *Corticium salminicolor*

Symptoms

- Fungal growth encircles stem, penetrates bark and cortical tissues which eventually decay.
- Bark splits and peels off.

Management

- Affected parts should be pasted with Bordeaux paste or painting tar.

12.6.4. Powdery mildew: *Oidium heveae*

Symptoms

- White powdery fungal growth appears on young and matured leaves.
- Infected leaves curl, crinkle, role inwards and fall off leaving petiole attached to tree giving a broom stick appearance.
- Infected flowers and tender fruits shed.

Management

- Spray sulphur 3-5 times at 15 days interval.
- Carbendazim 0.1% or tridemorph 1.5% dusting alternated with sulphur dust 70%

12.6.5. Brown root disease: *Fomes noxius*

Symptoms

- Affected roots are encrusted with mass of soil and small stones held by network of mycelium.
- Affected tap root becomes rotted and entire tree is killed.

Management

- Apply lime at 2.5t/ha
- In the area of dead tree, apply lime at 25 kg/ dead tree.

12.6.6. Colletotrichum leaf disease: *Colletotrichum acutatum, C. gloeosporioides*

Symptoms

- Spots are small, brown in colour and are surrounded by a yellow hallo.
- Infected leaves often crinkle and become distorted before shedding.

Management

- Spraying with Bordeaux mixture 1%, copper oxy chloride 0.125%, mancozeb 0.2% or carbendazim 0.05% at 10-15 days intervals.

12.7. Oil palm

12.7.1. Anthracnose: *Botryodiplodia palmarum*

Symptoms

- Regular or irregular brown to black leaf blotches surrounded by yellow haloes, which develop along the margin, centre or tip of the leaves.

Management

- Spraying mancozeb at 200 g/100 litres of water.

Section IV

Facts and Figures of Spices, Condiments and Plantation Crops

1

Common Name, Scientific Name, Family, Chromosome Number, Origin, Inflorescence and Fruit Type of Spices and Plantation Crops

Common name	Scientific name	Family	Chromosome number	Origin	Inflorescence	Fruit type
Spices						
Black pepper	*Piper nigrum*	Piperaceae	2n = 52	Western Ghats and Eastern Himalayas	Spike	Drupe
Cardamom	*Elettaria cardamomum*	Zingiberaceae	2n = 48, 52	Western Ghats in Southern India.	Panicle	Capsule
Clove	*Syzygium aromaticum*	Myrtaceae	2n = 22	Moluccas Islands of Indonesia.	Cymose	Berry
Cinnamon	*Cinnamomum zeylanicum*	Lauraceae	2n= 24	Central hills of Sri Lanka and Malabar Coast	Terminal or Axillary panicle	Berry
Nutmeg	*Myristica fragrans*	Myristicaceae	2n = 42, 44	Moluccas of Spice Islands of Indonesia	Axillary umbellate	Rounded fleshy berry
Turmeric	*Curcuma longa*	Zingiberaceae	2n = 63	South and South East Asia	Spike	Capsule
Ginger	*Zingiber officinale*	Zingiberaceae	2n = 22	Tropical South East Asia	Spike	Capsule
Garlic	*Allium sativum*	Alliaceae	2n = 16	Central Asia, Middle Asia and North eastern Iran	Umbel	Capsule
Paprika	*Capsicum annuum* var. *longum*	Solanaceae	2n = 24	North America, in particular Central Mexico.	-	Berry
Tamarind	*Tamarindus indica*	Caesalpiniaceae	2n = 24	Tropical Africa	Terminal raceme	Pod (Indehiscent legume)
Vanilla	*Vanilla planifolia*	Orchidaceae	2n = 28	Atlantic Coast from Mexico to Brazil.	Raceme	Capsule
Allspice	Pimenta dioica	Myrtaceae	2n = 22	West Indies (Jamaica), Central and South America	Umbel	Berry
Coriander	*Coriandrum sativum*	Apiaceae	2n = 22	Southern Europe and	Compound	Schizocarp

				Northern Africa to Southwestern Asia.	umbel	
Fenugreek	*Trigonella foenum graecum*	Fabaceae	2n = 16	South Eastern Europe, West Asia and Mediterranean region.	Compound umbel	Pod
Fennel	*Foeniculum vulgare*	Apiaceae	2n = 22	Southern Europe, Mediterranean region	Compound umbel	Schizocarp
Cumin	*Cuminum cyminum*	Apiaceae	2n = 14	Egypt, Syria, Middle East and stretching East to India	Umbel	Fusiform or ovoid achene
Indian mustard	*Brassica juncea*	Brassicaceae	2n = 36	China	Raceme	Siliqua
Plantation crops						
Tea	*Camellia sinensis*	Theaceae/ Camelliaceae	2n = 30	China in South East Asia	Solitary or cluster	Capsule
Coffee	*Coffea arabica*	Rubiaceae	2n = 44	Ethiopia	Axillary cluster	Drupe
	Coffea canephora		2n = 22			
Coconut	*Cocos nucifera*	Arecaceae	2n = 32	South Asia, Malaysia, Pacific	Spadix	Drupe
Arecanut	*Areca catechu*	Arecaceae	2n = 32	Indian ocean, Malaysia or Philippines	Spadix	Berry
Rubber	*Hevea brasiliensis*	Euphorbiaceae	2n = 36	Tropical rainforest of Amazon basin of Brazil	Panicles of small cyme	Capsule
Cashewnut	*Anacardium occidentale*	Anacardiaceae	2n = 42	America and Brazil.	Indeterminate panicle	Accessory fruit (Pseudocarp)
Cocoa	*Theobroma cacao*	Malvaceae	2n = 20	Tropical America, Amazon Basin	Axillary cluster	Pod
Betelvine	*Piper betel*	Piperaceae	2n = 78	South and South East Asia	Cylindrical spike.	Fleshy drupe
Oilpalm	*Elaeis guineensis*	Arecaceae	2n = 32	West Africa.	Compound spadix	Drupe
Palmhrah	*Borassus flabellifer*	Arecaceae	2n = 36	Tropical Africa	Axillary spadices	Drupe

2

An Abstract on Production Technology of Spices and Plantation Crops

An Abstract on Production Technology of Spices

Common name	Varieties	Propagation	Season	Planting	Fertilizer	Special practices	Harvest season and yield
Pepper	Panniyur 1,2,3,4,5,6,7&8, Karimunda, Sreekara, Subhakara, Panchami, Pournami, IISR Thevam, IISR Malabar Excel, IISR Girimunda, IISR Sakthi, PLD-2, Vadakkan tropical cultivar	Aerial shoots, runner shoots, fruit bearing lateral branches	June-December	Pit size: 50 x 50 x 50 cm Pit spacing: 2 to 3 m in either direction	100:40:140g of NPK/vine	To increase berry size: NAA @ 400 ppm	November-March, Hand picking, 2 to 3 kg dry berry /vine/ year
Cardamom	Malabar type: Mudigree 1, Mudigree 2, PV 1, ICRI 1, ICRI 3, TKD 4, IISR Suvarna, IISR Vijetha, IISR Avinash. Mysore type: ICRI 2, Vazhukka, PV 2, Njallani (Green gold).	Seeds: 600g/ha Suckers: 6800 plants/ha	June-December	Large type: 2.5 x 2.0 m Smaller type: 2.0 x 1.5 m Sucker: 1.8 x 0.6 m	75:75:150kg NPK/ha	Provide moderate shade of 50-60%	Harvest once in a month, 200-250 kg/ha
Turmeric	CO 1, CO 2, BSR 1, Swarna, Sugandham, BSR 2, Ranga, Rashmi, Rajendra Sonia, Krishna, Suroma, Allepy Finger Turmeric (AFT), IISR Prabha, IISR Prathiba, IISR Alleppey Supreme, IISR Kedaram, Sona, Varuna, IISR Pragati, Suvarna	Finger rhizome: 2500 kg/ha (25-30g weight)	May-June	45 x 30 cm	150:60:108 kg NPK/ha	Provide minimum of 50% shade	Fresh rhizome: 25-30 t/ha Cured rhizome: 5-6 t/ha
Ginger	IISR Varada, IISR Mahima,	1000-1200 kg	May-June	Irrigated:	75:50:50 kg	Seed treatment:	Harvest at 8-9

	IISR Rejatha, Suprabha, Suruchi, Suravi, Subhada, Himagiri, Athira, Karthika, Aswathy	bit rhizome/ha. Hilly areas: 2000 to 2500 kg/ha.		40 x 20 cm- Ridge and furrows Rainfed: 20 x 20-25 x 25 cm as Raised beds	NPK/ha	Mancozeb (Copper oxychloride) 3g/lit	months after planting, 12-15 t/ha
Garlic	Agrifound White (G-41) Yamuna Safed (G-1) Yamuna Safed-2 (G-50) Yamuna Safed-3 (G-282) Yamuna Safed-4 (G-323) Yamuna Safed-5 (G-189) Godavari, Yamuna Shweta, Phule Baswant, GG 4 Ooty 1, Ooty 2 ,VL Garlic 1, VL Lahsun 2, Agrifound, Parvati Agrifound, Parvati-2 (G-408), Bhima Purple, Bhima Omkar	500-600 kg cloves/ha having 8-10 mm diameter	March-April September-November	Form ridges & furrows at 30 cm spacing Cloves are planted at 15 x 10 cm spacing	40:75:75 kg/ha NPK	Controlling rubberisation: Avoid excessive application of N Use ammonium sulphate or urea Spray 1500 ppm of CCC on 30 DAP	8-12 t/ha
Clove	Penang, Amboyna, Zanzibar	Seeds	June-December	Pit size: 30 cm^3 Spacing: 6 x 6 m	20:20:60 g NPK/plant- One year old plants. 300: 300:960 g NPK/plant 7 years old bearing tree	Provide shade for seedlings	Harvest after 6th year, 2-3 kg dried buds/tree
Nut meg	Vishwashree, Konkan, Sugandha, Konkanswad, Keralashree	Seeds/ approach grafting Budded plants	June - December	Pit size: 60 cm^3 Spacing: 8 x 8 m	20:20:60 g NPK/plant - One year old plants	Provide shade	Bearing: 6-7 years after planting. Aril commonly

		Patch budding			400:300:1200 g NPK/plant-7 years old bearing tree		known as mace Fruits: 1000-2000 kg/tree. Dried nuts: 5-7 kg/tree Dried mace: 0.5-0.7 kg/tree.
Cinnamom	YCD 1, PPI 1, Nithyasree Navasree, Konkan Tej, Susadhini	Seed/semi hard wood cuttings	June-December	Pit size: 60 cm^3 Spacing: 2 x 2 m	20:18:25 g NPK/tree - 1st year 200:180:250 g NPK/tree 10th year	Coppicing - cut at ground level for side shoots	Harvest at 4 to 5 years after planting. Bark harvest: May-November. 100g dried bark/ tree/ year. Cinnamon oil: Leaves 35 kg leaf oil/ha/ year
Tamarind	PKM 1, Urigam Hasanur Tumkur Prathisthan DTS1, Yogeshwari	Seeds/ softwood/ grafting Air layering IBA 4000 ppm for rooting	June-December	Spacing: 8-10 m x 8-10 m Pit size: $1m^3$	200:150:250 g NPK/tree/ year	Early training to form high head, Pruning to remove dried, diseased and criss cross branches	Harvest at 4 to 6 years March-April 150-200 kg/tree/ year
All spices		Seeds / Air layering	January	6 x 6 m	20:180:50 g NPK/tree - 1st year 300:250:750 g NPK/tree - 15 years	Male: female ratio for planting:1:10	Harvest seeds after 6 years. 50-60 kg dry berries/year/ tree
Coriander	CO1, CO2, CO3, Gujarat	Seeds:	June – July	20 x 15 cm	Basal	Thinning 30 days	For leaf – pull out

	Corinader-1, Gujarat Coriander-2, Rajendra Swati, RCR-41, Swati, Sadhana, Sudha (LCC-128), Sindhu, APHU Dhania-1 (LCC-170), Suguna (LCC-236), Suruchi (LCC-234), Susthira (LCC-219)	10-12 kg/ha (Irrigated), 20-25 kg/ha (Rainfed)	October – November		10:40:20 kg NPK/ha Top dressing 10 kg N/ha	after sowing Spraying CCC @ 250 ppm induces drought tolerant in rainfed crop.	after 30-40 days Rainfed - 300-400 kg/ha. Irrigated - 500-600 kg/ha. Leaf - 6000-7000 kg/ha
Cumin	RZ 19, RZ 209, GC 1, GC 2, MC-43	Seed: 12–15 kg/ha	November - December	20-30 x 0-15 cm	30 kg N/ha 15 kg P/ha		Average yield of 0.5 t/ha Improved varieties yield up to 0.7-0.8 t/ha.
Fenugreek	CO 1, CO 2, Rajendra Kanti, Pusa Early Bunching, Lam Sel.1, Hisar Sonali, Hisar Suvarna, Hisar Madhavi Hisar Muktha, RMt 305, Guj Methi 1, RMt 143, Pant Ragini, AM-01-35	Seed: 20 kg/ha	June – July October – November	20 x 15 cm	Basal: 30:25:40 kg NPK/ha Top dressing: 20 kg of N	-	Harvest at 20-25 days for greens and 90-100 days for grains, 4-5 t/ha - greens 500-700 kg/ha - grain
Fennel	CO 1, UF 32, PF 35, Gujarat Fennel 1, Ajmer Fennel 1, Ajmer Fennel 2 (AF-2), Hisar Swarup, HF-143, Rajendra Saurbha	Seed: Direct sowing: 9 to 12 kg/ha Transplanting: 3-4 kg/ha	Hills: May-April Plains: October-November	60 x 30 cm	50:10 kg NPK/ha	-	Harvest at 7-8 months after sowing, 4-5 harvest at 10-15 days interval. 500-1200 kg/ha
Vanilla	Madagascar, Mexican, Indonesian, Tahitian, Indian, Tonga	Stem cuttings of 60-120 cm	In May-June, plant Standards: Glyricidia Erythrina	Plains - 2 to 2.5 x 1.2-1.5 m Hills: 1.5 x 1.5 m Pit size:	40-60:20-30: 60-100 g NPK/vine/year	Training the vines to height of 1.2-1.5 m Pinching of top 7.5-10 cm vine	Harvest at 6-9 months after flowering. Green to pale yellow mature

			Jatropha, Plumeria, Casuarina 6 months after planting Standard (August-November) plant vanilla	30 cm³		@ 6-8 months before flowering Artificial pollination and 1kg cured	beans. 300-600 kg cured bean/ha/year. 6 kg green pod beans/vine.
Paprika	Arka abir, KTPL-19, Byadagi Kaddi, Byadagi Dabba, Paprika King	Seeds: 600 g/ha Transplant 35-40 days old seedlings	June-July	Hills: 45 x 45 cm Plains: 60 x 45 cm	120:100:120kg NPK/ha	-	Harvest at 70-80 days of planting 3-6 harvests 25-35 t/ha
Indian mustard	RH 0749, RH 0406, T 59 (Varuna), Pusa Agrani, Pusa Mustard 21, Pusa Mustard 24, NPJ 112, Pusa Mustard 26, Pusa Mustard 28 NRCDR 2, NRCHB 506 Hybrid, NRCDR 601, NRCHB 101, DRMRIJ-31 (Giriraj), DRMR 150-35, DRMR 1165-40, NRCYS-05-02 (Yellow Sarson)	Seed rate: 6.5-7.5 kg/ha	Rabi season-September to October, February to March. Rainy season-July to November in South India	30-45 x 10-15 cm	60-90:60:40 kg NPK/ha	-	Rapeseed: 1.4-2.0 t/ha Mustard: 2.0-2.5 t/ha

An Abstract on Production Technology of Plantation Crops

Crop	Varieties	Propagation	Season	Planting	Fertilizer	Special practices	Harvest season and yield
Tea	Pandian Sundaram Golconda	Semi hard wood cuttings with one leaf and internodes	May – June September-October	Single hedge: 1.2 x 0.75 m 10800	1st year: 180:270 kg NK/ha	**Centering**: To induce more laterals at 3-5 months after	Plucking from 3 years old onwards. Rush period: Harvesting
	Jayaram Evergreen Brookland BSS 1 to BSS 5	with a slanting cut at the bottom. April-May August-September		plants/ha Double hedge: 1.35 x 0.75 x 0.75 m 13200 plants/ha	2nd year: 240:360 kg NK/ha 3rd year: 300:350 kg NK/ha 4th year: 300:300 kg NK/ha 80 - 100 kg/ha as rock phosphate once in a year	planting **Tipping**: 1st at 35 cm height. 2nd at 60 cm from ground level **Pruning:** To maintain convenient height and vegetative growth **Skiffing:** Removal of only top 5-8 cm new growth to obtain a uniform level (More than 65 cm).	of 2-3 leaves with bud at 7-10 days interval. Lean period: 2 leaves with bud at 10 - 15 days interval. Green leaves: 10 t/ha
Coffee	Arabica: Cauvery, Chandragiri, Sanroman, Sln. 795, 7, 9, 10 Robusta: Sln. 274, 270, Sln. 3	Seeds	June-December	Arabica: 1.5 - 2 m Robusta: 2.5 m either way Dwarf variety Sanroman: 1 x 1 m	**Arabica** 1st year: 45:30:45 kg NPK/ha 2nd & 3rd year: 60:45:60 kg NPK/ha 4th year: 80:60:80 kg	Centering and desuckering at 5-6 years after planting Cover digging: Weeds & debris are turned and buried in soil	Fly picking: Small scale picking (October - February) Main picking: Well formed and ripened berries in December. Stripping: Harvesting all berries Gleaming: Picking of

					NPK/ha One ton/ above: 160: 120:160 kg NPK/ha **Robusta:** Less than 1 t/ha- 80:60: 80 kg NPK/ha 1 t/ha and above - 120: 90:120 kg NPK/ha		drop down berries 750-1000 kg dry parchment/ha
Cashew	VRI 1 VRI 2 VRI 3 VRI 4 VRI (CW) H1	Softwood and epicotyl grafting Air layering	June-December	Normal planting: 200 plants/ha 7 m either way High density Planting: 5 x 4 m 500 plants/ha	100:125: 250 g NPK/tree	Training and pruning: Dried twigs and branches are removed every year	Harvest ready from 3rd year onwards. Peak months: March-May, 3 to 4 kg/tree/ year 375 kg of nuts and 1250 kg of cashew apple/ha.
Cocoa	Criollo Forestero Trinitario Amalinado CCRP 1 to 7, TNAUCC1 to TNAUCC 10, VTLCC-1 Vittal Cocoa Clone 1 VTLCH-1 VTLCH-2 VTLCH-3	Seeds/patch budding/ softwood grafting	June-July, September - October	Pit size: 50 cm^3 Spacing: 3 x 3 m	100:40:140 g/tree for trees of 3 years of age and above	Pruning: Unproductive dead and diseased shoots are removed	Bearing 3rd year onwards Economic yield from 6th year onwards. 500 to 1000 kg of dry beans/ha or 1-2 kg of pods/tree/year; 50-70 pods/tree/year.

	VTLCH-4 VTLCH-5 (Nethra Centura)						
Rubber	Tjir 1, PB 86, BD 5, BD 10, PR 17, GT 1, RRII 105, RRIM 600, PB 28/59, PB 217, PB 235, RRIM 703, RRII 5, PCK-1, 2, PB 260	Green budding, brown budding and crown budding, modified forkert method.	June-July	Budded plants: Hills: 6.7 x 3.4 m Plain: 4.9 x 4.9 m Seedling: Hills: 6.1 x 3 m Plain: 4.6 x 4.6 m	10:10:4 g NPK/plant	Rain guarding: Fix polythene rain guard to the trunk above tapping panel Tapping stage: 7 years	375 kg/ha from seedling tree 800-1000 kg/ha from budded plants
Coconut	Hybrid: VHC 1, 2, 3 Tall: VPM 3, 5, ALR 1, 2, 3, WCT Dwarf: COD, CYD, CGD, MYD	Seedings	June-July December - January	Tall varieties: 7.5 x 7.5 m Dwarf varieties: 6.5 x 6.5 m Hybrids: 8.5 x 8.5 m Pit size: 3 cubic feet	1.23:2:1.9 kg NPK/tree from 5th year onwards	TNAU coconut tonic: Root feeding at 200 ml/palm once in 6 months. Intercrop: Below 7 years: Annual crop. 7-20 years: Green manures and fodder crop. > 20 years: Groundnut, banana, cocoa, pepper, nutmeg, vanilla.	Hybrid nut yield: 100 nos/tree/year Tall: 60-80 nos/tree/year Dwarf: Tender coconut 70-90 nos/tree/year
Arecanut	Mangala Sumangala Subamangala Mohitnagar Srimangala Samruthi Thirthahalli dwaft Shatamangala	Seeds	June - December	Pit size: 90 cm^3, 2.75 m either way	100:40:140 g NPK/tree	Provide shade during summer months. Growing banana or other crops in advance may also provide shade.	Bearing after 5 years of planting, 1250 kg/ha

Oil palm	Dura Tenera Pisifera	Seeds, Four – five leafed seedlings	May - June	Pit size: 60 cm^3 9 x 9 x 9 m 140 palms/ha.	3rd year onwards: 1200:600: 1200 g NPK/ palm/year 2nd year: 800: 400:800g NPK /palm/year	Ablation: Removal of all inflorescence during initial 3 years	3.5 to 4 years after planting: 1st harvest 25-30 t of fresh fruit bunch/ha
Palmyrah	SVPR 1	Seeds	June – August	3 x 3 m	Sheep penning is practiced	Tapping: Extraction of sap from inflorescence	150 litres padaneer/ tree/year. 24 kg jaggery/tree/year. Jaggery recovery per litre of padaneer: 180-250 g
Betel vine	Thulasi, Venmani, Arikodi, Kalkodi, Karilanchi, Karpuram, Chelanthikarpuram, Koottakkodinanda, Perumkodi, Amaravila, Pramuttan, Kallarkodi, Karpurakodi, Revesi, Karpuri, SGM 1, SGM (BV) 2, Vellaikodi, Pachaikodi, Anthiyur kodi, Kanyur kodi, Bangla, Konur	Terminal stem cuttings or setts of 30 - 45 cm long	November-December January-February	Two rows on bed size of 180 cm width on agathi plants and 45 cm between plants	150:100:50 kg NPK/ha/year	Training: Fixing the vine at intervals of 15 to 20 cm along the standards loosely with the help of banana fibre at every 15 - 20 days interval.	75 to 100 lakh leaves/ ha/year

3

Plant Part Used, Important Chemical Content and Essential OilS in Spices and Plantation Crops

Spices and Condiments	Plant part used	Important chemical content	Essential oil %
Black pepper	Fruit, seed	Piperine	1-3
Cardamom	Fruit, seed	Cineol	-
Ginger	Rhizome	Zingiberene	1.9-2.2
Turmeric	Rhizome	Curcumine	0.24
Cinnamon	Bark	Cinamaldehyde	5.2
Nutmeg	Seed, aril	Terbein	-
Clove	Unopened flower bud	Eugenol	-
All Spice	Fruit, seed	Phenol	-
Capsicum	Fruit	Solanine	0.1 -2.6
Curry leaf	Leaves	Koeniging	2.6
Asafoetida	Oleoresin gum	Ferumine	10-17
Saffron	Stigma	Cicrocrocin	0.5-1.0
Sweet flag	Rhizome	Acorin	-
Dill	Fruit	Carvone	1.5-4.0
Coriander	Leaves , seed	Linalool	0.1-1
Cumin	Fruit	Cuminol	2.5-3.5
Fennel	Fruit	Limone	0.7-1.2
Aniseed	Fruit	Anithol	-
Bishop' weed	Fruit	Thymol	-
Vanilla	Fruit	Vanillin	-
Fenugreek	Seed	Diosgenin	-
Kokum	Fruit	B- Hydroxy acetic acid	-

Unit IX
Production Technology of Medicinal and Aromatic Crops

Section I: Production Technology of Medicinal Plants

1

Senna (*Cassia angustifolia*; syn: *Senna angustifolia*),Caesalpiniaceae, 2n=28

1.1. Origin and distribution

- Native of Yemen, South Arabia, Sudan and Sinai.
- Introduced into Tamil Nadu in eighteenth century grown as an annual crop (5 to 7 months).
- ***C. angustifolia*** **(Tirunelveli senna)**
- Indigenous to Somalia, Southern Arabia, part of Sindh (Pakistan) and Kutch area of Gujarat.
- Cultivated in Gujarat, Rajasthan and Tamil Nadu (Tirunelveli, Ramanathapuram, Madurai, Salem, Tiruchirapalli)
- ***C. acutifolia*** **(Alexandrian senna)**
- Indigenous to Sudan and Sinai.
- ***Cassia abovata***
- Common all over India.
- ***C. italica***
- Grows in west Rajasthan and Saurashtra (Gujarat)
- New source of the senna drug
- Both possess good anthraquinone content
- Germany, Hungary, Japan, Netherlands and USA are main markets.

1.2. Economic importance

- Exported as leaves, pods and sennoside concentrates
- Manufacture of calcium sennoside granules, tablets and syrups
- Medicine for cathartic especially useful in habitual constipation.

- Increases peristaltic movement of colon.
- Laxative principles are two glycosides, *viz.*, sennoside A and B.
- Also contains sennosides C and D, beta-sterol (0.33%) and flavanols-kaemferol, kaempferin and isorhamnetin.
- Used as decoctions, powders, confections and many other household preparations.

1.3. Varieties

- **KKM Sel 1:** Selection from Thenkalam local contains 2.34% sennoside released by Tamil Nadu Agricultural University, Coimbatore. High yielding recording 38.5% higher leaf yield and 69.88% higher pod yield than local. This is highly suited for all soils of Tirunelveli and Tuticorin districts and ideal for rainfed cultivation.
- **ALFT 2:** Released by Gujarat Agricultural University, suitable exclusively as a leaf crop.
- **Sona:** Released by CIMAP, Lucknow, high yielding variety.

1.4. Climate

- Requires bright sunshine and dry weather.
- Minimum and maximum temperature between 10-42°C.
- Heavy rains and cloudy weather are harmful.
- An average rainfall of 25-40 cm, distributed from June to October.

1.5. Soil

- Red loam, alluvial loam, sandy loam or laterite soils with rich clayey rice fields and pH of 7-8.5.
- Very sensitive to water logging.

1.6. Season and growing condition

- Grown under rainfed and irrigated conditions.
- Grown in early summer (February-March) or June-July.
- September-October in South India.

1.7. Propagation and planting

- Seed rate - About 15-20 kg/ha
- For quick germination, pound seeds lightly with coarse sand in a mortar and soaking overnight.

1.8. Spacing

- 45 x 30 cm

1.9. Irrigation

- About 5-8 light irrigations and heavy irrigations are injurious.

1.10. Weed management

- Weeding cum hoeing is done at 25-30, 75-80 and 110 days

1.11. Manures and fertilizers

- FYM at 10-15 t/ha
- Basal: N: P: K at 40:40:40 kg/ha.
- 40 DAS: 40 kg N/ha

1.12. Crop rotation and intercropping

- Crop rotation - Paddy - Senna
- Intercropping with cotton, sesame, chilli, brinjal, okra and tomato.

1.13. Harvest

- First picking: 60 DAS.
- Second picking: 90-100 DAS
- Third picking: 130-150 DAS

1.14. Yield

Yield	Leaves (kg/ha)	Pods (kg/ha)
Rainfed	1000	150-200
Irrigated	2000	400

- Young leaves (2.0-2.5 %) and pods (2.5-3.0 %) contain more sennosides than the mature ones
- Bluish green, matured leaves fetch better price in market.

1.15. Postharvest handling and storage

- Harvested leaves spread in a thin layer under open sun for 6-10 h to reduce moisture content.
- Further drying is done in well ventilated drying sheds.
- Takes 3-5 days to dry produce in sheds by frequently turning them all over.
- Reduce to 7-8% moisture and maintain light green to greenish yellow colour.

1.17. Grading

- **Grade A / first grade**
    - Extra-large and bold leaves and pods having yellowish green colour
- **Grade B / second grade**
    - Produce having brownish leaves and pods
- **Grade C / third grade**
    - Smaller and broken leaves and pods

2

Periwinkle (*Catharanthus roseus*), Apocynaceae, 2n=16

2.1. Origin and distribution

- Native of Madagascar and spread to India, Indonesia, Indo-China, Philippines, South Africa, Israel, USA and other parts of the world.
- In India, Tamil Nadu, Karnataka, Andhra Pradesh, Madhya Pradesh, Gujarat and Assam are the major states that cultivate periwinkle.
- USA - world's largest user as raw material.
- Germany, Italy, Netherlands and the UK are interested in roots.

2.2. Economic importance

- Economic parts: Leaves, stem and roots
- Used as an abortificent, purgative, antidiabetic, diuretic, hemorrhagic antimalarial, antidysentric and against skin diseases
- Alkaloids ajmalicine (Raubasin), serpentine and reserpine present in roots have hypotensive and antispasmodic properties.
- Vincristine and vinblastine alkaloids from leaves - cancer therapy.
- Vincristine sulphate, (ONCOVIN), used against acute leukemia and vinblastine sulphate (VELBE) to cure Hodgkin's disease.
- Alkaloids leurosidine, leurosovine and rovidine - Anticancer properties

2.3. Varieties

- Three variants in periwinkle
- Rose purple flowers - Higher alkaloid content
- White flowers - Nirmal and Dhawal by CIMAP, Lucknow
- White flowers with a rose purple spot in centre.

2.4. Climate

- Tropical and sub-tropical climate.
- Elevation - 1300 m above msl.

- Rainfall - 100 cm or more
- Suitable for rainfed crop.

2.5. Soil

- Hardy and wide variety of soils, except alkaline or water logged.
- Deep sandy loam to loamy of medium fertility

2.6. Season

- June-July or September-October.

2.7. Propagation

- Seeds or vegetatively through cuttings.
- Directly sown in field at 2.5 kg/ha
- Nursery preparation and transplanting
- Raised nursery beds in March-April in rows
- 500g/ha

Vegetative propagation

- Soft wood cuttings from lateral shoots are better than either hard or semi hard wood cuttings.
- 10-15 cm length with 5-6 nodes results in 90% rooting.
- Soaking overnight in NAA at 25 or 50 ppm improves rooting upto 96%.

2.8. Planting

- After 2 months, transplanting at 45 x 30 cm.

2.9. Irrigation

- 4-5 irrigations

2.10. Weed management

- Weeding at 60 and 120 DAS.
- Mulching with cut grass or rice straw minimize weed
- Pre-emergence - fluchloralin at 0.75 kg/ha or alachlor at 1.0 kg/ha.
- Mixture of 2,4-D and gramaxone at 25 kg/ha before sowing

2.11. Manures and fertilizers

- FYM at 10-20 t/ha
- Basal - 25:50:70 kg NPK/ha/year
- Top dressing - 50 kg N on 60 DAT.

2.11. Harvest

- First leaf stripping - 180 DAS and Second - 270 DAS
- Third - when whole plant is harvested.
- **Roots:** At 12 months of sowing, cut at 7.5 cm above ground level and dry the stem, leaves and seeds.
- **Seeds:** Collect from matured pods at 2 to 3 months before harvest of whole plant.
- Aerial part between 7.5 and 25 cm above ground level for marketing.

2.12. Yield

Yield	Leaves (t/ha)	Stem (t/ha)	Roots (t/ha)
Irrigated	4	1.5	1.5
Rainfed	2	0.75	0.75

- Total alkaloid content in leaf - 0.15 to 1.34%
- Vinblastine - 0.002%
- Vincristine - 0.005%

3

Coleus (*Coleus forskholii,* syn: *Coleus barbatus*), Lamiaceae, 2n=30

3.1. Origin and distribution

- Native of South East Asia
- Distributed all over tropical and sub-tropical regions of India, Pakistan, Sri Lanka, tropical east Africa, Brazil.
- In India, sub-tropical Himalayan regions from Kumaon to Nepal.
- Cultivated in parts of Rajasthan, Maharashtra, Karnataka and Tamil Nadu.

3.2. Economic importance

- Dried roots - rich source of forskholin (Coleonol) (0.10 to 0.80%)
- Used for treating hypertension, glaucoma, asthma, congestive heart failures and certain types of cancer.
- Tuberous roots, resembles carrot in shape and brown in colour - commercial part.
- Useful against cholesterol and activating the cardio vascular system
- Used in cosmetics.

3.3. Varieties

- **TNAU - medicinal coleus CO 1**
- **A selection, K 8:** 0.5% forskolin, higher tuber yield.
- **Aisiri:** Forskolin 0.7% from UAS, Bengaluru.
- **Manganiperu:** Cultivated in and around Belgaum districts of Karnataka and Tamil Nadu.
- **Garmai:** Cultivated in Gujarat.

3.4. Climate

- Tropical crop
- Grown on barren hills

- Altitude - 2400 m
- Humid climate with RH of 83 to 95%
- Temperature - 10-25°C.
- Annual rainfall - 100-160 cm
- Also in less humid and warmer regions of South India like Coimbatore, Salem, Thiruvannamalai districts as irrigated crop.

3.5. Soil

- Porous and well drained soils with marginal fertility and pH of 5.5-7.0
- Red, sandy loam soils are ideal.

3.6. Season

- June-July

3.7. Propagation

- Terminal cuttings (10 cm) or rooted cuttings.
- 10-12 cm long with 3-4 pairs of leaves.

3.8. Spacing

- 60 x 45 cm spacing (37,030 plants/ha)
- Low fertile soils - 60 x 30 cm (55, 500 plants/ha)
- Drip irrigation system - broad row ridges 90×60 cm and planting at 60 x 45 cm

3.9. Irrigation

- First - immediately after transplanting.
- Initially irrigated once in 3 days and later weekly irrigation

3.10. Weed management

- 2 or 3 weedings and after second weeding, earthing up is given
- As roots are shallow, deep digging is avoided.

3.11. Manures and fertilizers

- FYM – 15 t/ha
- 40:60:50 kg NPK/ha, half dose of N and full dose of P and K as basal and remaining N is applied as top dressing on 30th day.
- 10 kg $ZnSO_4$/ha

3.13. Harvest

- Harvest at 5-6 months after planting.

3.14. Yield

- Fresh tubers: 15-20 t/ha
- Dry tubers: 2000-2200 kg/ha

3.15. Post-harvest handling

- Harvested tubers are cut into small pieces and dried under shade up to 8% moisture.

4

Ashwagandha (*Withania somnifera*) Solanaceae, 2n=48

4.1. Origin and distribution

- It is called as Indian ginseng.
- Native of Africa, India and Mediterranean region.
- Wild in grazing grounds in Mandsaur and forest lands in Bastar district of Madhya Pradesh, all over foothills of Punjab and Himachal Pradesh and Western Uttar Pradesh, in Himalayas.
- Also wild in Mediterranean regions in North Africa.

4.2. Economic importance

- Winter cherry and Indian Ginseng
- Economic part: Root
- Types of alkaloids - Withanine and somniferine (0.13-0.31%)
- Used in Ayurvedic and Unanic preparations.
- Withaferin A - antibiotic and antitumor properties.
- Used for curing inflammation of tubercular glands and roots for curing skin diseases, bronchitis and ulcers.
- Used as aphrodisiac, remunerative tonic, diuretic, hypnotic, sedative and restorative
- Useful in rheumatism, cough debility from old age, dropsy and general weakness.
- New alkaloids with a sominiferol A, B and C.

4.3. Varieties

- High root and alkaloid yielding variety Poshita and Rakshita: Released from CIMAP, Lucknow
- Jawahar - 20: Highest dry root yield.
- Jawahar Aswagandha B4 - Hybrid.

- WSR: CSIR-Regional Research Laboratory, Jammu.
- Nagori: Local variety with starchy roots

4.4. Soil and climate

- Sandy loam soil and slightly alkaline soil with pH of 7.5 to 8
- Altitude: 600-1200 m above msl.
- Semi-tropical areas receiving low rainfall.
- Dry weather is suitable.
- Optimum temperature: 20°C to 35°C

4.5. Season

- Late rainy season (August-September) and harvesting in May.

4.6. Propagation and planting

- For broadcasting, 10-12 kg/ha and for transplanting, 5 kg/ha
- 6 weeks old seedlings are transplanted.

4.7. Spacing

- 60 x 60 cm

4.8. Irrigation

- Light irrigation after transplantation
- No need, if rainfall occurs since dry land crop.
- Avoid excessive rainfall/ water

4.9. Weed management

- Hand weeding at 30 DAS

4.10. Harvest

- Ready for harvest in 150 to 180 DAS
- **Maturity indices:** Drying of leaves and berries turn red.

4.11. Yield

- Dried roots: 400-1200 kg/ha
- Seeds: 200-500 kg/ha.

4.12. Post-harvest handling and storage

- Entire plant: Uproot and separate roots from aerial parts by cutting 1-2 cm above the crown.
- Root is cut transversely into small pieces (7 to 10 cm) or dried in sun.

4.14. Grading

- Dried roots are hand sorted into 4 grades
- **Grade A:** Root pieces 7-10 cm long, diameter 1 to 1.5 cm. roots brittle, solid and pure white from inside.
- **Grade B:** Root pieces 5-6 cm long, diameter 1 cm, roots brittle, solid and white from inside.
- **Grade C:** Root pieces solid, 3-4 cm long, diameter <1 cm.
- **Lower grade D:** Small pieces of root, somewhat hollow, yellowish from inside and <1 cm in diameter.

5

Glory Lily (*Gloriosa superba*) Colchicaceae, 2n=22

5.1. Origin and distribution

- Native of tropical Asia and Africa.
- Tropical India, from North-west Himalayas to Assam and Deccan plateau.
- In Karnataka, it is grown along Western Ghats.
- Important species in India - *Gloriosa superba* and *G. rothschildiana.*
- Tamil Nadu leads in production of glory lily in India
- State flower of Tamil Nadu

5.2. Economic importance

- Economic part: Seed
- Rhizomes are used as a tonic, antiperiodic, antihelminthic and also against snake bites and scorpion stings.
- Used for promoting labour pains and an abortifacient.
- Useful in colic, chronic ulcers, piles and gonorrhoea.
- Used against parasitic skin diseases and as a cataplasm in urological pains.
- Leaf paste applied to forehead and neck cures asthma in children.
- Leaf juice is used against head lice.
- Alkaloids like colchicines (0.5-0.7%), colchicoside and gloriosine.
- Colchicines are used in treatment of gout and also used as polyploidizing agents in polyploid breeding.
- Colchicines - 0.15 to 0.30% in tubers and 0.7 to 0.9% in seeds.

5.3. Varieties

- Singaleri selection: 0.25% colchicine

5.4. Climate

- Tropical plant and warm humid regions.
- Elevation – 2120 m above msl.

- Annual rainfall - 370 cm
- Temperature - 15-20°C

5.5. Soil

- Sandy loam soils, acidic pH 6-7 with good drainage
- In Southern India, red and black loamy soils are preferable.

5.6. Season

- June-July

5.7. Propagation and planting

- Underground V-shaped biforked rhizomes or seeds.
- 2.5 to 3.0 tons tubers/ha.

5.8. Spacing

- 30 to 45 cm.
- It requires staking for its proper growth.
- More care should be taken for growing tips, since it does not sprout again during the season.

5.9. Irrigation

- Avoid excess watering
- Immediately after planting and subsequently at 5 days interval.

5.10. Drip and fertigation system

- 150:100:300 kg NPK/ha through water soluble fertilizers

5.11. Weed management

- Frequent weeding

5.12. Fertilizer requirement

- 120:50:75 kg NPK/ha

5.13 After cultivation

- Artificial pollination is done between 8-11 am to increase the yield

5.14. Harvest

- Pod harvesting at 160-180 DAP.
- Capsule turns light green from dark green
- Fruit skin shrunken and light in weight
- Pod gives crinkling sound.

5.15. Yield

- Seeds: 200-250 kg/ha

6

Sarpagandha *(Rauvolfia serpentina)*, Apocynaceae, 2n=22

6.1. Origin and distribution

- Indigenous to moist deciduous forest of south East Asia (Myanmar, Bangladesh, Sri Lanka, Malaysia and Andaman Islands)
- In India, Uttar Pradesh, Bihar, North Bengal, Parts of Western Ghats and Assam are the major cultivators.

6.2. Economic importance

- Economic part – Root.
- Used in treatment of snake bite, stings, nervous disorders etc.,
- Root bark - 40-60% alkaloids helps in reducing high blood pressure and sedative or tranquilizing agent.
- Fresh roots - Emit acrid aroma very bitter.
- Roots - High alkaloid concentration.
- Reserpine - Used for hypertension in allopathic medicine.
- Alkaloids-Nescinamine, doserpidine, serpentine, ajnaline, ajnatinine, rauvolfinine etc.

6.3. Varieties

- Local collection from wild
- RS1: Developed by Jawaharlal Nehru Krishi Viswa Vidyalaya; Alkaloid 1.64 - 2.94% at 18 months, yield of air dried root gives upto 2.5 t/ha

6.4. Climate

- Hot humid climate; 500 m altitude.
- Rainfall: 250-500 cm.
- North East region
- Temperature: 10-30°C

6.5. Soil

- Sandy alluvial loam to red lateritic loam
- Humus rich sandy loam soil with pH 4.5-6.5
- Properly drained soil

6.6. Season

- June-July

6.7. Propagation

- Seed – Best method
- Vegetative means like root cutting, root stumps, and stem cutting.

6.7.1. Seed propagation

- April - May-Suitable for nursery sowing.
- Germination - 15-20 DAS
- 6 kg/ha

6.7.2. By stem cutting

- Hard wood stem cutting - 15-20 cm length planted during June in nursery beds
- Treat with rooting hormone - indole acetic acid at 30 ppm for 12 h and root develops within 15 days.

6.7.3. By root cutting

- Lateral secondary rootlets of 2.5- 5 cm length, 0.25 cm diameter are planted in March-June
- 100 kg of root cuttings/ha.

6.7.4. By root stump

- 5 cm root
- Transplanted in May-July.

6.7.5. Planting

- Transplanting of seedlings of 40-50 days old with 4-6 leaves

6.8. Spacing

- 30×30 cm

6.9. Irrigation

- Rainfed crop.
- 4 irrigations in summer and 2 in winter at one month intervals

6.10. Weed management

- Weeding, cleaning, and hoeing - twice during rains and after rains.

6.11. Manures and fertilizers

- FYM at 20-25 t/ha.
- Basal dose - 30:60:30 kg NPK/ha
- Top dressing: 10 kg/ha of N each at 50 and 170 DAP

6.12. Intercrop

- Soybean, onion or soybean and garlic

6.13. Defloration

- Flowering and fruiting – 180 DAP, depress shoot and root growth
- Recommended for better root growth and yield.

6.14. Harvest

- 2-3 years after planting.
- Roots - 40 cm deep
- Winter months - Ideal
- Roots - Cleaned, washed and 12-15 cm pieces for drying and storage.
- Dry root - 10-12% moisture stored in polythene lined gunny bags

6.15. Yield

- 2-2.5 t/ha at 30 months.
- 2.2 t/ha air dried from 2 years old crop.
- 3.3 t/ha from 3 years old crop.

7

Dioscorea (*Dioscorea floribunda*), Dioscoreaceae, 2n=36

7.1. Origin and distribution

- Native of South East Asia, Central America.
- Mexico - largest producer of diosgenin.
- In India, it is grown in North eastern states and Goa
- *Dioscorea alata* and *Dioscorea esculenta* - edible tubers.
- *Dioscorea floribunda, Dioscorea villosa, Dioscorea composita* - widely grown
- Major producing countries - Mexico, Guatamala, Costa Rica, India and China.

7.2. Economic importance

- Diosgenin, a steroidal sapogenin from rhizomes.
- Major base chemical for steroid hormones including sex hormones, cortisones, other corticosteroids.
- Active ingredient in oral contraceptive pills.
- Other important sapogenins - yamogenin, botogenin and kryptogenin.
- Minor sapogenins - pannogenin and tigogenin.

7.3. Varieties

- **FBC 1**
  - Composite strain of *D. floribunda*
  - Diosgenin content - 3 to 3.5%.
  - Suitable for Bangalore and Coorg (Karnataka), Goa, Assam, Meghalaya, Tamil Nadu, Maharashtra and Andaman.
  - Tuber yield is 20-25 t/yr (1 kg/plant) or 60 t for a 2 year old crop (2.5 kg/plant).
- **Arka Upkar** - IIHR variety

  - High yielding clone
  - Very vigorous with a stout, robust vine, bearing broad dark green leaves
  - Tuber branches - thick, broad and deep
  - Higher diosgenin content - 3.5 to 4.0 %.
  - Tuber yield exceeds 6t (in a 2-year-old crop)
- **Pusa-1**
  - Selection from germplasm
  - Tuber yield -1.5 kg/vine after 18 months

7.4. Climate

- *D. floribunda* and *D. composita* - tropical
- *D. prazeri* - temperate
- Grows well in temperate regions of Kashmir and Himachal Pradesh
- Elevation - 1500 m above msl.

7.5. Soil

- Well drained clay loamy soil
- Medium loam deep soils rich in organic matter with pH of 5.5 to 6.5
- For red soils, *D. floribunda* and *D. composita* are suitable
- Avoid highly acidic and highly alkaline soils

7.6. Season

- June-July

7.7. Propagation

- Tuber pieces, single node stem cuttings or seed.
- Commercial - tuber pieces.

7.7.1. Propagation from tuber pieces

- Cutting tubers into pieces of 50-70 g weight.
- Three types of tuber pieces.
  - Crowns - produce new shoots at 30 DAP, commercial one
  - Medians and tips - 100 DAP.
- Dipping tuber for 5 min. in 0.3% benlate followed by dusting 0.3% benlate in talcum powder.

7.7.2. Single leaf node cuttings

- For rapid multiplication
- Single leaf with petiole of 0.8 cm from non-flowering plants.
- Quick dip with 500 ppm of IBA and plant in mist chamber in sand beds.
- Within 8-10 weeks, transferred to plastic bags
- Transplanting after six months

7.7.3. Propagation by seed

- Fresh seed is sown in polybag at depth of 0.5 to 1.0 cm in February.
- Transplanting: June-July.

7.8. Spacing

- Deep furrows of 60 × 30 to 45 cm

7.9. Provision of support

- Trained over pandal system or trellis.
- 2 m high stone pillars spaced 9 m apart
- Galvanized wires (Nos. 6 and 8) on boundaries and interconnecting wires of 12 gauge thickness.
- Supported on coir ropes tied to wires and pegged to ground by wooden stakes.

7.10. Irrigation

- 4 to 5 days interval in summer
- 7 to 10 days in winter

7.11. Weed management

- Periodic hand weeding

7.12. Manures and fertilizers

- Basal dose - FYM at 20 to 25 t/ha
- 300:150:150 kg NPK/ha

7.13. Intercropping

- Legumes like cowpea, horse gram, cluster bean and French bean are cultivated.

7.14. Harvest

- Tubers at 25 to 30 cm depth.
- Done by manual labour in February-March.

7.15. Yield

Particulars	First year	Second year
Fresh tubers	15 to 20 t/ha	40 to 50 t/ha
Diosgenin content	2.5-3.0%	3.0-3.5%

- Tubers dried under sun to 6-7% moisture.

8

Aloe *(Aloe vera,* Syn: *A. barbadensis)* Liliaceae, 2n=14

8.1. Origin and distribution

- Indigenous to Eastern and Southern Africa, the Canary Islands and Spain.
- Spread to Mediterranean basin, West Indies, India, China.
- Also cultivated in India. It is the miracle plant.

8.2. Economic importance

- Source of drug aloe; aloe gel is a mixture of glucoside aloin, main constituent of atom is barbation.
- Commercially important species - *A. barbadensis, A. ferox, A. africana* and *A. spicata*
- Yellow bitter juice from leaves consists of aloin and gel.
- Other products - dehydrated aloe powder, concentrates.
- Contains cathartic anthrax - glycosides 4.5 to 25% of aloin.
- Used as active ingredients in laxative and antiobesity preparation, as moisturizer, emollient or wound healer.
- Used in sunscreens, x-ray /burns, dermatitis, cutaneous leishmaniasis.
- Extracts used as flavouring alcoholic and non-alcoholic beverages and in candy to impart bitter.

8.3. Varieties

- Two strains: Sweet and bitter
- Gurguva aloe: *Aloe vera (Aloe barbadensis)*
- Cape aloe: *Aloe ferox*
- Sagotrin aloe: *Aloe perryi*
- Cultivated types - Zanzibar, Uganda, Natal and Hyderabad aloe

8.4. Climate

- Warm, humid or dry climate

- Rainfall: 35-40 to 150-200 cm
- Temperature: 25-40°C

8.5. Soil

- Sandy coastal to loamy soils with pH of 7.5 to 8.5.
- Avoid water logged conditions and problematic soils

8.6. Season

- June-July and September-October

8.7. Propagation

- Root suckers or rhizome cuttings (5-6 cm length with 2-3 nodes).
- 10000 suckers/ha and 15-18 cm long root suckers

8.8. Spacing

- 60 x 30 cm or 60 x 45 cm.

8.9. Irrigation

- Rainfed crop
- 4 to 5 irrigations per year

8.10. Weed management

- Weeding twice a year

8.11. Manures and fertilizers

- FYM at 10t/ha
- Mixture of 150 kg/ha of NPK

8.12. Harvest

- 8 months after planting.
- Manually or tractor drawn disc harrow or cultivator
- Broken rhizome left in soil for new sprouts
- Commercial yield from second year and up to 5 years.
- Extraction with in 6 hr of harvest

8.13. Yield

- 10 to 15 t/ha

9

Phyllanthus *(Phyllanthus amarus)* Euphorbiaceae, 2n=26

9.1. Origin and distribution

- Originates in Tropical America
- Distributed in all tropical and sub-tropical regions

9.2. Economic importance

- Economic part - Whole plant
- Phyllanthin (0.4-0.5%) and hypophyllanthin
- Antioxidant with free radical scavenging properties due to presence of high levels of super oxide dismutase.
- Effective in treatment of peptic ulcer, dyspepsia, diabetes etc.
- Treatment of Hepatitis B and jaundice

9.3. Varieties

- Navyakrit (CIMAP) - High herbage and active constituents

9.4. Climate

- Tropical and sub-tropical rainfed crop

9.5. Soil

- Well drained sandy loam/ black soil with pH of 6.5 -7.5

9.6. Season

- June-July

9.7. Propagation

- 1 kg/ha seed
- 8 lakh seedlings/ha
- Germination - 7 DAS and maintain up to 20 days
- Soaking seeds in water for 20-30 min.

- Seed treatment - GA_3 at 200 ppm for 6 hr
- 3-4 week old seedlings are transplanted.

9.8. Spacing

- 10 × 15 cm

9.9. Irrigation

- Necessary irrigation

9.10. Weed management

- Maintain weed free field

9.11. Manures and fertilizers

- FYM at 10-20 t/ha.
- 50:25:25 kg NPK/ha

9.12. Harvest

- September - October
- September - High phyllanthin
- Ready for harvest at 80-100 DAP

9.13. Yield

- Fresh herbage - 17.5 t/ha
- Dry herbage - 1.75 t/ha

10

Creat or Green Chireta *(Andrographis paniculata)*, Acanthaceae, 2n=50

10.1. Origin and distribution

- Nativc of India and tropical South East Asia.
- Distributed throughout plains of India from Uttar Pradesh to Assam, Madhya Pradesh, Tamil Nadu and Kerala.

10.2. Economic importance

- King of bitters.
- Economic part – Dried leaves and tender shoots (Whole plant)
- Leaves contain three bitter principles: Deoxy andrographolide, Andrographolide and Neo andrographolide.
- Roots contain flavonoids like andrographin, panicothin, aplegenin etc.
- Andrographidin content of A, B, C, D, E and F varies from 0.01 to 0.15.
- Leaf - 2.5% chemical constituents
- Acrid, cooling, laxative, antipyretic, anti periodic, anti inflammatory, expectorant, sudorific, digestive and stomachic.
- Useful in burning sensation, chronic fever, malaria and intermittent fever, inflammation, cough, bronchitis, skin diseases, intestinal worm, dyspepsia, flatulence, colic, diarrhoea, dysentery and hemorrhoids
- Used as a substitute for Chirayita (*Swertia chirayita*).

10.3. Varieties

- Local lines

10.4. Climate

- Well in tropical and sub-tropical regions all over India
- Hardy species
- Hot and humid climate
- If grown during cooler climate, gives more bitter ingredients
- Ratoon crop in North India

10.5. Soil

- Medium fertile sandy loam to clay loamy soil

10.6. Season

- October-November and January-May

10.7. Propagation and planting

- Seed - easy and economical
- Vegetative method - cuttings.

10.7.1. Raising of nursery

- Seeds soaked in water for 24 h and sown in September or May
- 650-750 g/ha
- Transplanting - Six weeks old seedlings of 10-25 cm long
- Direct sown crop by broadcasting - 1.5 kg/ha

10.8. Spacing

- 45 × 30 to 60 × 45 cm; 30 x 15 cm, 15 x 15 cm.

10.9. Irrigation

- 4-6 light irrigations

10.10. Weed management

- 2 to 3 weedings at 20 and 60 DAT

10.11. Manures and fertilizers

- FYM at 20 t/ha.
- 75:75:50 kg NPK/ha in split doses at planting and 40 DAP

10.12. Harvest

- Harvest at 120 DAS
- Uprooting of plants at bloom stage and dried in sun for 2 days

10.13. Yield

- 3.5-4 t/ha

11

Medicinal Solanum (*Solanum khasianum*), Solanaceae, 2n=24

11.1. Origin and distribution

- Originates in India
- Sikkim, West Bengal, Odisha, Upper Gangetic Plains and in Nilgiris, ascending to altitude of 1600 m.
- North-east, North-west, Southern as well as Central India, and extends into Myanmar and China.

11.2. Economic importance

- Economic part – Whole plant
- Raw materials for production of steroidal drugs, steroid bearing solanum
- Yields a glyco alkaloid, solasodine, a nitrogen analogue of diosgenine.
- Anti-inflammatory, anabolic and anti-fertility properties
- Cures toothache, laxative, treatment of piles.
- Useful in bronchitis, asthma, fever relieving and thirst.

11.3. Varieties

- Galaxo strain, BARC strain, Pusa-1, RRL 20-2 and RRL-G, SL-6 - Completely spineless.
- Arka Sanjeevini and Arka Mahima - diploid and tetraploid varieties developed by IIHR, Bangalore.

11.4. Climate

- Elevation upto 2000m.
- Moderate climatic conditions.
- Warm season crop
- 21-27°C temperature is optimum.

11.5. Soil

- Well drained loamy soil (Red lateritic) with pH of 7 to 8
- Red lateritic soil with medium fertility.
- Not perform well in very clayey soils.

11.6. Season

- June to September or October

11.7. Propagation

- Seeds at 1.25 kg/ha.
- Germination: 7-10 DAS
- Transplanting: 4-5 weeks after sowing, 10-12 cm height of seedlings with 6 leaves

11.7.1. Nursery bed size

- 1 to 1.5 cm deep furrows, 10 cm apart

11.8. Spacing

- 50 x 50 cm, 75 x 75 cm and 90 x 120 cm

11.9. Irrigation

- Once in a week at first month
- Later - Once at fortnight interval

11.10. Weed management

- First - 20 to 30 DAP
- Second - 2 to 3 MAP

11.11. Manures and fertilizers

- FYM at 25 t/ha
- Basal dose: 50:40:80 kg NPK /ha
- Top dressing: 25 kg N at 2 MAP

11.12. Harvest

- High alkaloid in fruits - 50-60 DAP.
- Alkaloid - 60% in seeds and 40% in pericarp.

- 6 MAP - ready for harvesting.
- Change in fruit colour from green to yellow with green streaks
- Berries dried at 10% moisture

11.13. Yield

- Fresh berries: 10 t/ha
- Dried berries: 2-2.5 t/ha
- Solasodine content: 2.0 to 2.5%

12

Isabagol (Blond Psyllium) (*Plantago ovata*), Plantaginaceae, 2n=8

12.1. Origin and distribution

- Native of Persia and neighbouring tracts of Western and Southern Asia
- In India, North Gujarat, Rajasthan, Haryana and Bihar are the major states.
- Husk and peel - exported to USA, Germany, UK and France

12.2. Economic importance

- Psyllium from Greek meaning flea, in reference to colour, size and shape.
- Economic part - Seed coat, husk used in indigenous medicine. Seeds contain glucoside and tannin.
- Works as an anti-diarrhoea drug.
- Seed - Pentose, a sugar is isolated.
- Beneficial in chronic dysenteries of amoebic and bacillary origin.
- Also cooling and demulcent effect
- Used in Ayurvedic, Unani and Allopathic medicines.
- Husk yields colloidal mucilage consists of xylose, arabinose and galacturonic acid.

12.3. Varieties

- Gujarat Isabgol 1: Yield potential is 800-900 kg/ha
- Gujarat Isabgol 2: Yield potential is 1000 kg/ha
- TS -1- 10
- EC 124345
- Niharika
- Haryana Isabgol - 1-5
- Jawahar Isabgol - 4

12.4. Climate

- Warm temperate regions.
- Cool and dry sunny weather.
- In India, it is grown in winter season.

12.5. Soil

- Light well drained sandy, silty loam to rich loamy soil with pH of 4.7-7.7.

12.6. Season

- November - December to March – April

12.7. Propagation

- Seeds at 4-8 kg/ha

12.7.1. Bed size

- Flat bed of convenient size
- 1.0 m x 3.0 m or 2.5 m

12.8. Spacing

- 15cm apart in beds (30-45 cm)

12.9. Irrigation

- Immediately after sowing
- 8 to 10 irrigations

12.10. Weed management

- 20-25 DAS - First weeding
- 2-3 weedings within 2 months of sowing.

12.11. Manures and fertilizers

- FYM at 15-20 t/ha.
- 50:25:30 kg NPK/ha

12.12. Harvest

- Maturity period: 110-130 DAS.
- Crop turns yellowish and spike turns brownish.
- Husk: seed ratio - 25:75 by weight.

12.13. Yield

- 700-1000 kg/ha.

13

Opium Poppy (*Papaver somniferum*) Papaveraceae, 2n=22

13.1. Origin and distribution

- Native to Turkey Western Mediterranean region.
- Cultivation in India - Confined to states of Madhya Pradesh, Rajasthan and Uttar Pradesh

13.2. Economic importance

- Opium and codeine - important medicines used for analgesic and hypnotic effects.
- A semi-synthetic derivative of drug from morphine - Heroin.
- Seeds contain linoleic acid which lowers blood cholesterol.
- License from Government is mandate for cultivation

13.3. Varieties

- Heavy black soils: Talia, Ranghatak and Dhola Chota Gothia
- MOP-3 from JNKVV, Mandsour.
- MOP-16 from Jawaharlal Nehru Krishi Vishwa Vidyalaya, Mandsour.
- Shama by CIMAP, Lucknow in 1983.
  - Main alkaloids - Morphine (14.51-16.75%), codeine (2.05-3.24%), the baine (1.84-2.16%), papaverine (0.82%) and narcotine (5.89-6.32%)
- Shweta by CIMAP, Lucknow
- BROP 1 (Botanical Research Opium Poppy-l) (NBRI-3)
  - Synthetic variety at National Botanical Research Institute, Lucknow by crossing selections from Kali Dandi, Suyapankhi and Safaid Dandi.
- Kirtiman (NOP-4)
  - Narendra Dev University of Agriculture and Technology, Kumarganj, Faizabad, through selection from local races.
  - Morphine content 12%.
- Chetak (U.O.285)

    - Rajasthan Agricultural University, Udaipur.
    - 12% morphine.
    - Trishna - National Bureau of Plant Genetic Resources, New Delhi, 14.78% morphine.
- Jawahar Aphim 16 (JA-16)
    - Pureline selection of 10 local landraces developed at Jawaharlal Nehru Krishi Vishwa Vidyalaya College of Agriculture, Mandsaur (Madhya Pradesh).
    - 45-54 kg/ha of latex, 8-10 q/ha of seeds
    - 12% morphine.
- Recently released three varieties
    - NBRI-3 - Opium
    - Sujatha - Opium-free poppy for oil and seed
    - Shubhra - High morphine and seed yield

13.4. Climate

- Temperate climate - Long cold season (20°C).
- Winter in subtropical regions.
- Temperature: 30-35°C during reproductive period.
- Cool climate favours higher yield.
- In February-March, latex flow is high.

13.5. Soil

- Well drained, highly fertile, light black or loam soil with pH of 6.0-7.5.

13.6. Season

- October-November

13.7. Propagation and planting

- Seed at 7-8 kg/ha for broadcasting
- 4-5 kg/ha for line sowing.
- Germination - 5 to 10 DAS
- Thinning - When plants at 5-6 cm high, 3-4 leaves within 3-4 weeks after sowing.

13.8. Spacing

- 30 x 30 cm

13.9. Irrigation

- Light irrigation - Given immediately after sowing
- Second - 7 DAS
- Third irrigation - 12 to 15 days interval subsequently
- 12-15 irrigations

13.10. Weed management

- Necessary weeding

13.11. Manures and fertilizers

- FYM at 1-20 t/ha
- 80:50:30 kg NPK/ha.

13.12. Lancing and latex collection

- Flowering - 95-115 DAS.
- Capsules mature after 15-20 DAF.
- Lancing of capsules (Industrial maturity) - Compactness and change in colour from greenish to light green ring in capsule
- Lancing - Done with knife having three or four equispaced pointed ends which does not penetrate more than 1-2 mm in capsule early morning at 2 days interval.
- Avoid - Too deep or too shallow incision
- Length of incision - 1/3 or less than full length of capsule.
- Lancing instrument is known as Nastar or Naka.

13.13. Harvest

- Crop left for drying for 20 to 25 days at last lancing of capsules which stops which exudation of latex.
- Harvested capsules are dried in open yard and seeds are collected by beating with wooden rod.

13.14. Yield

- Raw opium: 2-30 kg/ha.
- Seeds: 400-500 kg/ha.

14

Safed Musli (*Chlorophytum borivilianum*), Liliaceae, 2n=16

14.1. Origin and distribution

- Originated from Africa and distributed throughout warmer regions of world
- Naturally grown in hilly areas of Gujarat, Rajasthan and Madhya Pradesh

14.2. Economic importance

- Economic part - Root
- Roots of safed musli contains 2-15% saponin having medicinal property of enhancing vitality and immunity to human beings.
- Also helps in correcting gynecological disorders.
- Dried tubers - Used as curative for pre-natal and post-natal illness, arthritis, restorative and a health tonic
- Known as divya aushadhi and ayurvedic anori.

14.3. Varieties

- RC-2, RC-16, RC-36, RC-20, RC-23, RC-37 and CT-1: These varieties are maintained and collected by RAU, Udaipur. They are found to give good yield and high saponin content.
- MDB-13 and MDB-14: Developed by Maa Danteshwari Herbal Research Centre, Chikalputi. This variety gives high yield, disease/fungal resistance, maximum saponin and alkaloid content.
- Jawahar Safed Musli 405 and Rajvijay Safed Musli 414: Released by Rajmata Vijayaraje Scindia Krishi Vishwa Vidyalaya, Mandsaur, Madhya Pradesh and are good for cultivation.

14.4. Climate

- Growing in thick forests in its natural form.
- Warm and humid climate.

- Also grown in hot and subtropical climate.
- Rainfall: 500 to 1000 mm

14.5. Soil

- Loamy to sandy loam soil.
- Well drained soils with rich mineral content.
- Avoid hard and acidic soils

14.6. Season

- Kharif crop: June-August

14.7. Propagation

- Fleshy roots, rhizome fingers or seeds or tissue culture plantlets.
- 80,000 fingers 1-1.2 t/ha; 250-300 kg of fleshy roots/ha.

14.8. Spacing

- 30 × 15 cm

14.9. Irrigation

- Rainy season crop
- Avoid water logging
- Irrigation at 2 weeks interval; 6 to 8 irrigations

14.10. Weed management

- Frequent weeding - 3 to 4 times in first 60 DAS

14.11. Inter cropping

- Crop is raised in the plantation of teak, neem, mango, papaya and banana.

14.12. Manures and fertilizers

- FYM at 20-25 t/ha, 20-40 kg P_2O_5/ha.

14.13. Harvest

- Three to four months after planting
- Yellowing, dry and fall off leaves
- Skin of mature tubers turn dark brown.

14.14. Yield

- Fresh roots: 2 tons/ha
- Dry safed musli: 350 to 400 kg/ha

15

Stevia (*Stevia rebaudiana*) Asteraceae 2n=22

15.1. Origin and distribution

- Also known as Cheeni tulsi or Mou tulsi.
- Originated from Seirra Amambay in North eastern Paraguay and adjacent areas in Bolivia and Brazil.
- Widely distributed in USA and Brazil.

15.2. Economic importance

- Economic part - leaves
- Candy leaf
- Major constituents-Stevioside (3 to 10%), rebaudiside-A (13%), rebaudiside-B and C
- Produces sweet steviol glycosides in leaves
- Source for the production of natural sweetener (food), as a source of chlorophyll, phytosterols - Medicine.
- Sweetener converted into gibberellins by fermentation - Agrochemicals
- Vegetative residue used as animal feed and stalks used as source of cellulose-Cellulose industry.
- Regulating blood sugar, preventing hypertension, treatment of skin disorders, and prevention of tooth decay.
- Best alternate source for diabetes sufferer.

15.3. Varieties

- S.R.B 123 - Glycoside 9-12%.
- S.R.B 512 - Suitable for Northern India and glycoside 9 to 12%.
- S.R.B 128 - Highest stevioside and suitable for Northern India.

15.4. Climate

- Semi humid, subtropical plant

- Warm temperature, 31°C.
- Long spring and summer days favour leaf growth.
- Short days trigger blossoming.
- Prefers partial shade during summer sunshine.
- Rainfall - 140 cm/year.

15.5. Soil

- Well drained, fertile sandy loam or loam soil, high in organic matter with pH of 6-7.
- Avoid water logging.

15.6. Season

- March-May

15.7. Propagation

- Stem cuttings
- 2-4 inches (15cm) long, two leaf buds above ground with 2 or 3 small leaves.
- Transplanting - 25 to 30 days of rooting
- Paclobutrazol at 100 ppm - Induces root initiation.
- Raised bed - Most economical, 15 cm in height and 60 cm in width.

15.8. Spacing

- 40 x 45 cm

15.9. Irrigation

- Light irrigation frequently
- Summer - 3 to 5 days interval

15.10. Weed management

- Regular weeding every 2 months

15.11. Manures and fertilizers

- FYM at 50 t/ha
- 27.5:112.5:112.5 kg NPK/ha.

15.12. Harvest

- 4 months after planting and subsequently once in 3 months
- Plants should have 40 to 60 cm height

- Cut the branches with pruning shears before stripping the leaves.
- Better to cut plants leaving about 10 cm stem from ground.
- Facilitate new flushes to emerge

15.13. Drying

- Hang the plants upside down and dried in shade.
- 24 to 48 h drying at 40°C to 50°C.
- Home dehydrator also used
- After adequate drying, leaves are stripped from stems by thresher/separator

15.14. Yield

- Green herb: 15 t/ha
- Dried leaves: 3 to 3.5 t/ha

16

Horse Eye or Velvet Bean (*Mucuna pruriens*), Fabaceae, 2n=22

16.1. Origin and distribution

- Native of Asia
- Widely distributed in tropical regions of Asia and Pan tropics
- Most part of India up to 1000m elevation includes Andaman and Nicobar Islands.

16.2. Economic importance

- Economic part - Seed, roots.
- Black seed variety contains L - DOPA CL - 34 dihydroxy phenylalanine non-protein amino acid.
- Used as tonic, aphrodisiac and treatment of Parkinson's disease
- Used in rheumatic ailments.
- Alkaloid - Mucumine and mucunadine
- Roots are pungent stimulant, diuretic, purgative and tonic.
- Raised as green manure and fodder

16.3. Varieties

- Four colour varieties: Speckled, White, Yellow and Black
- Arka Dhanvantari, Arka Daksha and Arka Shukla

16.4. Climate

- Sub-tropical to tropical climate and withstand drought.
- Temperature - 15°C to 38°C
- Coastal humid to dry humid climate
- Rainfall - 650 to 2500 mm

16.5. Soil

- Sandy loam soil with good drainage and pH of 6 to 7.5

16.6. Season

- June-July
- Spring and summer

16.7. Seed rate

- Seeds: 20-40 kg/ha
- Germination - 8 to 10 DAS

16.8. Spacing

- 100 × 20 to 80 cm; 60 x 45 cm.
- Being a twiner, if requires support with sticks of 150-180 cm height or tied to an overhead trellis.

16.9. Irrigation

- Fortnight irrigation - Dry season
- One irrigation/month - Winter

16.10. Weed management

- Frequent weeding

16.11. Manures and fertilizers

- FYM at 10 to 20 t/ha
- 100:80:40 kg NPK/ha.

16.12. Harvest

- Matures at140 DAS
- Pods turns grayish brown

16.13. Drying

- Dried in sunlight for 4 to 7 days to reduce 7 to 8% moisture

16.14. Yield

- Seeds: 2.5 to 3.0 t/ha

17

Vallarai (Indian Pennywort) (*Centella asiatica* L.), Apiaceae, 2n=18

17.1. Origin and distribution

- Native of India, China, Indonesia, Madagascar and Africa.

17.2. Economic importance

- Economic part - Whole plant mainly leaves and roots.
- Cures leprosy, skin diseases, diarrhoea, anti-tumor and anti-ulcer activity.
- Fresh herb - Aromatic odour when crushed.
- Nauseous bitter taste
- Kitchen herb - Chutneys, pickles, refreshing drinks
- Possess insecticidal properties
- Used as a nerve tonic for improving memory and in insanity.

17.3. Varieties

- Vallabh medha: Single plant clonal selection. It is bigger in size in respect to all the morphological characters and leaf area is more than 4.5 times than the local cultivar.
- Arka Prabha (CA-13)
- Arka Divya (CA-1)

17.4. Climate

- Tropical and sub-tropical regions.
- Needs shady and moist place.

17.5. Soil

- Moist and marshy land, often grown around water bodies
- Prefers acidic soils
- Clayey soil with good organic matter

17.6. Season

- October-November

17.7. Propagation and planting

- Stem cuttings - Rooted nodes with leaves
- 1,00,000 plants/ha
- Either directly planted in main field or in nursery.

17.8. Spacing

- 30 x 30 cm

17.9. Irrigation

- Every 4 to 6 days interval

17.10. Weed management

- During initial stages weeding is done at 15 to 20 days interval

17.11. Manures and fertilizers

- FYM at 5 t/ha
- 100:60:60 kg NPK/ha.

17.12. Harvest

- 90 DAP
- Intermittent harvesting at fortnight intervals
- Few stolons left for regeneration

17.13. Yield

- Fresh herb: 5.5 t/ha for single harvest
- Dry herb: 2 t/ha
- Asiaticoside: 20 kg/ha

18

Brahmi (*Bacopa monnieri*) Scrophulariaceae, 2n=64

18.1. Origin and distribution

- Native of wetlands of Southern and Eastern India, Australia, Europe, Africa, Asia, and North and South America
- Cultivated throughout tropical and subtropical India.

18.2. Economic importance

- Economic part - Whole plant
- Chemical constituents: Bacosides
- Used as a nerve tonic and for epilepsy and insanity.
- Used as diuretic and for treating rheumatisim, asthma and hoarseness.
- Good potency in controlling cough, fever and diabetes.
- Inherent potential of enhancing memory and vitality,
- Celestrial drugs (Divyaausadhi), when consumed with milk for six months.
- Useful against painful joints, swelling in joints, peripheral neuritis, constipation and burning urination.
- Also used in convulsions, mental retardation, chest congestion and laryngitis.

18.3. Varieties

- **Subodhak**: A selection from wild collections providing a dry herb yield of 4.7 t/ha harvest (with 1.6% bacoside A).
- **Pragyashakthi**: A selection from Odisha with dry herb yield of 6.5 t/ha harvest (with 1.8% bacoside A).
- **CIM - Jagriti**: Has a potential of producing 85 kg/hectare of bacoside A from a dry herb yield of 4 t/ha.

18.4. Climate

- Growing along the canals and water bodies and in marshy area throughout India.
- Grows under sub-tropical and tropical climate.

- Elevation - 1300 m
- 33° - 40°C temperature and 60 to 65% RH.

18.5. Soil

- Grows well in poorly drained soils.
- Acidic nature is preferred

18.6. Season

- Transplanted in the month of July-August

18.7. Propagation

- Soft herbaceous cuttings.
- Plants cut into small divisions of 5-6 m long with few leaves and nodes and planted in sunken beds.
- 62,500 cuttings/ha

18.8. Spacing

- 10 x 10 cm

18.9. Irrigation

- Flood irrigation - quickly just after planting.
- Subsequently at 7 to 8 days interval

18.10. Weed management

- Hand weeding - Every 15-20 days interval

18.11. Manures and fertilizers

- FYM at 5 t/ha
- 100:60:60 kg NPK/ha.

18.12. Harvest

- Ready for harvest on 75-90 DAT
- October - November
- Ratoon crop also taken
- Upper portions of stem 4-5 cm from base are removed and rest left for regeneration.

18.13. Yield

- Fresh herbage: 30 t/ha
- Dry herbage: 6 t/ha
- Ratoon dry herbage: 4 t/ha

19

Black Nightshade (*Solanum nigrum*) Solanaceae, 2n=24

19.1. Origin and distribution

- Native to Eurasia and introduced in America, Australia and South Africa.
- Grown throughout dry parts of India

19.2. Economic importance

- Whole plant - Useful for rheumatalgia, swellings, cough, asthma, bronchitis, wounds, ulcers, flatulence, dyspepsia, hepatomegaly, otalgia, hiccough, opthalmopathy, vomiting
- Used in cardiopathy, leprosy, skin diseases, fever, splenomegaly, emollient
- As an antiseptic, anti-inflammatory, expectorant, anodyne, vulnerary, digestive, laxative, diuretic, cardiotonic, depurative, diaphoretic, febrifuge, rejuvenating, sedative, alterant and tonic.

19.3. Climate

- Well in tropical and subtropical climate
- 2000 m altitude.

19.4. Soil

- Raised on a variety of soils good in organic matter.
- Usually grows in moist habitat in wastelands or as weed.

19.5. Season

- April-May

19.6. Seeds and planting

- Seeds.
- Raised in the nursery and transplanted after 30-45 DAS
- Seedlings of 8-10 cm height is used for transplanting.
- Rainy season - Planting on ridges

- Summer - Furrows are formed for planting.
- Transplanted seedlings are given temporary shade for 2-4 days during summer.

19.7. Spacing

- Spacing between plants - 30-90 cm
- Between rows - Convenient
- 60×60 cm - Preferred
- Need staking to avoid lodging due to heavy bearing

19.8. Irrigation

- 3 to 4 days interval in summer and alternate days during fruiting

19.9. Weed management

- Necessary weeding

19.10. Manures and fertilizers

- FYM at 20-25 t/ha
- 75:40:40 kg NPK/ha.

19.11. Harvest

- Harvest at 4 to 6 months after planting
- Plants are collected and dried in shade

19.12. Yield

- Fresh herbage: 12 to 20 t/ha

20

Sappan Wood (*Caesalpinia sappan*), Caesapinaceae, 2n=24

20.1. Origin and distribution

- Native of Indo-Malayan region
- Grown in all southern states and extends northwards to Odisha and Bengal.
- In India, it is cultivated in gardens and nurseries as live fence in Tamil Nadu, Kerala, Karnataka, Andhra Pradesh and West Bengal.

20.2. Economic importance

- Economic part – Wood, leaves, pods and seeds.
- Sappan wood or East Indian red wood - Multipurpose tree, yielding valuable natural dyes with medicinal properties.
- Source of commercial redwood or Brazil wood.
- Used for inlaying work, cabinet making, violin bows and for walking sticks
- Stem produces gum, tannin or dye stuff
- Heart wood yields valuable red crystalline dye used on cotton, silk and wool fabrics.
- Leaves contain pleasant smelling volatile oil.
- Medicine: An astringent used in mild cases of dysentery and diarrhoea.
- Also given internally for certain skin ailments.
- Given as tonic to women after confinement and to relieve vomiting of blood.
- One of ingredients in mixture prescribed for malaria.
- Dried heart wood is widely used in oriental medicine, particularly against inflammation.
- Seeds serve as a sedative.

20.3. Climate

- Xerophyte in nature and grows in tropical and sub-tropical regions with hot and dry climate.

- Tolerates precipitation up to 700-4300 mm
- Mean annual temperature of 24-28°C
- Withstands drought

20.4. Soil

- Grows well in all kinds of soil especially red loamy soil
- Lushly growth in red soil
- Avoid too wet soil conditions

20.5. Season

- June-July

20.6. Seeds and planting

- Seed: 600g/ha.
- Renewed by coppicing.
- Mature pods burst open and scatter in dry season which remain dormant until rainy season.
- Pods gathered, pounded and put into cold water.
- Seed germination - Enhanced by dipping and wrapped in cotton cloth in boiling water 5 sec.
- 30-45 days old seedlings are transplanted in pits of 45 cm^3.

20.7. Spacing

- 4×4 m

20.8 Irrigation

- Weekly or fortnight interval

20.9. Weed management

- Once in 4 months

20.10. Manures and fertilizers

- FYM at 5 kg/tree and 100:50:50 kg NPK/ha

20.11. Harvest

- For dyewood, harvested every 6-8 years and trunk diameter should be 5-6 cm.
- Trees are cut at 1 m above ground to allow sprouts from stump.
- Harvest main branches along with stump

- Seeds are harvested at second year of planting and heartwood at 6 to12 years.
- Harvested wood is to be chipped into pieces and dye extracted by boiling with water.
- While extracting, few paddy grains thrown into boiling liquid to check completion of extraction process.
- If husk scales off, boiling is sufficient.

20.12. Yield

- Inner pulp: 80 kg/tree.
- Pods: 2-2.5 t/ha
- Seeds: 200-250 kg/ha.

21

Gymnema (*Gymnema sylvestre*), Apocynaceae, 2n=22

21.1. Origin and distribution

- Native to tropical Asia and China
- Widely growing in tropics of Africa, Asia and Malaysia.
- In India, it is grown in forests of Karnataka, Tamil Nadu and Bihar.
- Due to raising demand in South East Asia, plant is endangered particularly in Tamil Nadu.

21.2. Economic importance

- Madhunashini - Important medicinal climber acclaimed for its anti-diabetic properties.
- Preparations - Profound actions on the modulating taste suppressing sweet taste sensations.
- Pronounced effects on blood sugar, among the diabetics.
- Play a role in treatment of Type 1 diabetics, who typically need daily injections of insulin.
- Taken orally lowers blood glucose levels and improves blood fat and cholesterol profiles.
- Used as a remedy for rheumatism, cough, ulcer and pain in eyes.
- Also useful in inflammations, dyspepsia, constipation, jaundice etc.,
- Roots as remedy for snake bite.
- Anti-diabetic property - Presence of mixture of triterpines and saponins in leaves.
- Designated as gymnemic acids A, B, C and D which have the gymnemagenin and gymnestrogenins.

21.3. Types

- **Small leaf type**: Leaves are oval measuring 1.0-3.5 cm length and 1.5-2.5 cm in width, very soft, found in dry regions.

- **Broad and pubescent type**: Leaves are oval measuring 3-6 cm in length and 3.5-5.0 cm in width; dark green compared to small leaf type and also are pubescent.

21.4. Climate

- Prefers tropical and sub-tropical type of climate.
- Areas with high or medium well distributed rainfall

21.5. Soil

- Red sandy loam or medium deep black soil; avoid water logging

21.6. Season

- June - July

21.7. Propagation and planting

21.7.1. Propagation by seeds

- 2-3 kg/ha.
- Soaked overnight in water and next day seeds are sown in seed pan containing soil mixed with sand.
- Seeds germinate in 15 DAS.
- Transplanting - 40-50 DAS

21.7.2. Propagation by cuttings

- Matured stem cuttings of 15 cm length; treated with 500 ppm IBA for 18 h and 1% carbendazim.
- Root development – 90 DAP in the main field.
- 45cm^3 sized pit at 2.5 m between rows and 1.75 m between plants (within row).
- Pits dug open 15 days earlier to planting and filled with green leaves and top soil and 2 kg of well rotten manure per pit.

21.7.3. Training

- Since a climber, it trained on trellis of "Y" shape iron structure.
- Two main stems are trained on opposite directions.
- Also trained to wire fence which serve double purpose.

21.8. Irrigation

- Immediately after planting and later, once in 5-6 days

21.9. Weed management

- Hand weeding at periodic intervals

21.10. Manures and fertilizers

- FYM at 5 kg/tree and 95:45:35 kg NPK/ha

21.11. Harvest

- Ready for harvest - second year after planting.
- Leaves - economic part
- Leaves harvested along with flowers either by hand or cut with sickle/knife.
- Dried under shade for 7-8 days and direct sunlight is avoided

21.12. Yield

- Dried leaves: 5-6 kg/plant at 3-4 years old plant
- Dried leaves: 10-15 t/ha.
- Cultivated for 10-15 years

22

Tulsi (*Ocimum sanctum* L.) Lamiaceae, 2n=32

22.1. Origin and distribution

- Queen of herbs
- Native to India
- Wide distribution covering entire Indian sub-continent up to 1800m in Himalayas and Andaman and Nicobar Islands.
- *Ocimum tenuiflorum (Ocimum sanctum)* - Holy basil
- *Ocimum basilicum* (2n=48) - Culinary or ornamental basil

22.2. Economic importance

- Have a religious attachment and grown in shrines and homes as an aromatic perennial shrub.
- Part of routine worship and scientific background as antimicrobial and antiviral properties and purifies the air.
- Commercial production due to valuable aroma chemicals extracted from essential oil of tulsi.
- Leaves - bright yellow, volatile oil useful against insects and bacteria.
- Principle constituents - eugenol, eugenol methyl ether and carvacrol.
- Inhibits the *in vitro* growth of *Mycobacterium tuberculosis* and *Micrococcus pyogenes* var. *aureus*.
- Fragrant leaves and flowers are used in form of tincture, tea or decoction
- Used in treating coughs, bronchitis, skin diseases, and diarrhoea.
- Prophylactic against epidemics including cholera, influenza and malaria.
- Tulsi seeds taken, mixed in water, juice or cow's milk, are used as antioxidant, nourishing, mucilaginous and demulcent.
- Used in treating low energy, ulcers, vomiting and diarrhoea or as an overall tonic.

- Powder of dried root taken in milk, ghee or as a decoction to treat malarial fever as an analgesic application to the bites and string of insects and also to increase sexual stamina and prevent premature ejaculation.
- Herb improves resistance to stress and has a normalizing influence on blood pressure and blood sugar imbalances.
- Prove prophylactic against negative effects of environmental toxins, including cancer.
- Also richly endowed with bioavailable antioxidants, vitamins A and C and calcium.
- Marked insecticidal activity against mosquitoes.

22.3. Varieties

- There are no named varieties under this crop and only the types
  - Green type (Sri tulsi)
  - Purple type (Krishna tulsi).

22.4. Climate

- Wide adaptability and grown in tropical and sub-tropical climates.
- Also grown in temperate climates
- Rainfall: 50-120 cm.
- Altitude: 2000 m above msl
- Long days with high temperature

22.5. Soil

- Grows in moist soil
- Avoid highly saline, alkaline or water logged conditions.
- Sandy loam soil with good organic matter.
- pH: 6.5-8.5.

22.6. Season

- October-December.

22.7. Propagation and planting

22.7.1. Propagation through seeds

- 350-450 g/ha sown in nursery beds.
- Seed beds - 4.5 x 1.0 x 0.2 m size.
- Germinate in 8-12 DAS
- Transplanting - 6 weeks old seedlings at 4-5 leaf stage.

22.7.2. Vegetative propagation

- Terminal cuttings of 8-10 nodes and 10-15 cm length
- Rooting in 4-6 weeks

22.8. Spacing

- 40 × 40 cm, 40 × 50 cm or 50 × 30 cm

22.9. Irrigation

- Necessary irrigation

22.10. Weed management

- First weeding - one month after planting
- Second after 30 days and after each harvest

22.11. Manures and fertilizers

- FYM at 15 t/ha
- 120:60:60 kg NPK/ha

22.12. Harvest

- First harvest after 90 DAP and subsequently at every 75 days interval at full bloom stage by cutting plants at 15 cm from ground level

22.13. Yield

- Fresh herbage: 10 t/ha/year.
- Essential oil: 0.1 to 0.23%; 10-20 kg/ha
- Irrigated tulsi: Herbage yield up to 20 t/ha and oil yield up to 40 kg/ha

23

Noni (*Morinda citrifolia*) Rubiaceae, 2n=44

23.1. Origin and distribution

- Native from South Eastern Asia (Indonesia) to Australia
- Indian mulberry

23.2. Economic importance

- Widely used for health restorative properties, over years popularity diminished due to unpleasant smell from ripened fruit.
- Helps in reducing high blood pressure and cures heart disease and stroke due to presence of scopoletin which dilate blood vessels resulting in lower blood pressure.
- Stimulates body's production of nitric oxide, a chemical which allows the blood vessel to dilate more easily and more elastic.
- Anti-inflammatory and histamine inhibiting effects, excellent for promotion of smooth joint movement.
- Helps to enhance functioning of pancreas and immune system.
- Helps in curing allergy and inflammation which causes for asthma.
- Helps in the treatment of malaria, general febrifuge and analgesic.
- Stem bark used for curing jaundice.
- Seed oil is used in preparation of scalp insecticide.
- Leaf or fruit poultices help in curing tuberculosis, sprains, deep bruising and rheumatism.
- Fruits are used as an appetite and brain stimulant.
- Bark contains red pigment and roots contain yellow pigment used in making dyes; used to color clothing and fabrics.

23.3. Climate

- Withstands 20-35°C

- Mean annual rainfall: 250-4000 mm.
- Ability to survive in harsh environments.

23.4. Soil

- Wide range of soils
- Tolerates water logging
- Prefers free, well drained soils and grows in wide range of acidity levels.

23.5. Season

- Summer is suitable
- Flowers and fruits round the year.

23.6. Propagation and planting

23.6.1. Propagation through seed

- Pretreatment - Scarification
- Germination - After 20 to 120 DAS
- Transplanting - 2 to 12 months after germination

23.6.2. Propagation from stem cuttings

- 20-40 cm cutting
- Rooting takes place in 3 weeks
- Transplanting in 6-9 weeks.

23.7. Spacing

- Plant spacing - 4 x 4 m.
- 3 × 3 m (High density or dwarf varieties)

23.8. Fertilizer requirement

- 50:225:50 kg NPK/ha

23.9. Harvest

- Fruits turning white or fully ripe i.e., turned soft, translucent and characteristic odour.
- Starts yielding at age of three years and regular yield from 5th year.

23.10. Yield

- Fruit: 80 t/ha or more

24

Tippili (Long Pepper) (*Piper longum* L.), Piperaceae, 2n=52

24.1. Origin and distribution

- Originated in hills of South Western India, Indo-Malaya region.

24.2. Economic importance

- Used to improve appetite and digestion and treat stomachache, heart burn, indigestion, intestinal gas, diarrhoea and cholera.
- Fruits are stomachic and carminative.
- Used for lung problems including Asthma, bronchitis and cough.

24.3. Varieties

- Vellanikara -1 (Visvum tippili) and Yercaud (Pl) - 9.

24.4. Climate

- Elevation between 100 and 1000 m.
- Shade intensity: 20-25%.
- Lower elevations with high humidity
- Temperature: 30-32°C; RH 60%
- Annual rainfall - 150 cm.

24.5. Soil

- Well drained red loamy soils rich in organic matter

24.6. Season

- June-July and September-October

24.7. Propagation and planting

- Rooted cuttings/suckers (2/pit).
- Planting in coconut and arecanut plantations in multitier cropping pattern.
- Pits of 15 cm depth and rooted cuttings are planted at centre of pit.

24.8. Spacing

- Hills: 2 x 2 m
- Plains: 60 x 60 cm

24.9. Irrigation

- Once a week in pure condition.

24.10. Weed management

- Weeding followed by earthing up
- Mulching with dried leaves.

24.10. Manures and fertilizers

- Plains: FYM 20-25 t/ha; 50:20:70 kg NPK/ha.
- Hills: 10 t/ha

24.11. Intercrop

- Plains: Cultivation under coconut or arecanut plantations.
- Hilly areas: Planted as intercrop in between banana or casuarina.
- To prevent from excess temperature during late afternoon, shade trees like banana are planted.
- Annually trim shade trees to ensure adequate shade.
- Additional shade in arecanut gardens is provided by banana

24.12. Harvest

- When fruits attain dark green
- Maintained up to three years.

24.13. Yield

- First year: 750 kg dried fruits/ha
- Second year: 1500 kg dried fruits/ha
- Roots: 500 kg/ha in the third year.

24.14. Grading

- Dried thicker parts of the stem and roots are called piplamool.
- Three grades:
    - Grade I - Thick roots and underground stem.
    - Grade II and grade III - Consists of either roots or stems or fragments.

25

Digitalis (Wooly Foxglove) (*Digitalis lanata*), Scrophulariaceae, 2n=56

25.1. Origin and distribution

- Native to Western and South Western Europe, Western and Central Asia and Erstwhile USSR countries.

25.2. Economic importance

- Dried leaves constitute the drug, digitoxin and gitoxin.
- Used to treat congestive heart failure and heart rhythm problems.
- Increase blood flow throughout the body and reduce swelling in hands and ankles.

25.3. Varieties

- Strain E.C. 115996 - Higher foliage yield with high glycoside content

25.4. Climate

- Requires a cool and mild climate (20-30°C)
- Well in hill slopes at elevation of above 1250 m above msl

25.5. Soil

- Well drained sandy soil rich in organic matter and slightly acidic (pH 5.5-6.5).

25.6. Season

- May – June for direct sowing and February for nursery sowing.

25.7. Seed rate

- Seeds: 5-8 kg/ha for direct sowing; 1-2 kg/ha for transplanting.

25.8. Spacing

- 45 x 30 cm.

25.9. Weed management

- 1-2 weedings

25.10. Manures and fertilizers

- FYM at 10 t/ha; 170:30:30 kg NPK/ha.

25.11. Harvest

- Harvesting after 2 years.
- First year crop: 2-3 harvests; 8-10 cm long leaves are collected between July and August.
- Second harvest: 2 harvests; one and half months later, dry leaves by hot air at 60°C.

25.12. Yield

- Dried leaves: 2-3 t/ha
- Active principle in dried leaves: 0.44 to 0.71%.

26

Pyrethrum (*Chrysanthemum cinerariaefolium* Vis), Asteraceae, 2n=18

26.1. Origin

- Native to Dalmatia (Yugoslavia), Herzegovina and Montenegro.

26.1. Economic importance

- Natural insecticide from dried flower heads.
- Seed or achene contains Pyrethrin I, II, Cinerin I, II, Jasmolin I, II collectively called Pyrethrins.

26.2. Varieties and types

- KKL-1 released from TNAU.
- Hansa and Jhelum released from CIMAP, Lucknow.
- Sel-2 - developed at the Kodaikanal Centre of CIMAP.
- Two types
  - Dalmatian pyrethrum: *C. cinerariaefolium*, grown on the Easte Coast of the Adriatic Sea.
  - Japanese pyrethrum also derived from *C. cinerariaefolium*.
- Concentration of active principles (pyrethrin) in the Dalmatian and Japanese type of pyrethrum flowers ranges from 0.38 to 0.58 and 0.58 to 1.21%, respectively.

26.3. Climate

- Mild cool climate, short winter and cool summer.
- Annual rainfall of 100 cm or more.
- Elevation: 1500-2400 m.

26.4. Soil

- Well drained sandy soils; red laterite loams and light and medium loamy soils are also suitable.

26.5. Season

- April-May and October-November.

26.6. Seeds and planting

- Seeds: 2-3 kg/ha.
- Seedlings in nursery beds and transplanted after 6-8 weeks.

26.7. Spacing

- 45 x 60 cm.

26.8. Fertilizer requirement

- 40-60 kg N, 40-50 kg/ha P and 50-80 kg K /ha.

26.9. Harvest

- Flowers: Fortnightly intervals from third month of planting
- Dried immediately under sun or in electric oven having not more than 50°C

26.10. Yield

- Dried flowers: 500 kg/ha.

Section II
Production Technology of Aromatic Crops

1

Palmarosa (*Cymbopogon martini*) Poaceae, 2n=40

1.1. Origin and distribution

- Native to South East Asia especially India
- In India, it is cultivated in MP, Maharashtra and AP

1.2. Economic importance

- Oil of palmarosa - rusa or rusha or rosha
- Oil obtained from *C. martini var. sofia* - ginger grass oil.
- Essential oil – Geraniol- East Indian Geranium.
- Motia grass - fine yellow stem with dark green leaves and height 1.80-2.40 m.
- Sofia grass - purple stem, shorter (90-120 cm)
- Used as base for fine perfumery
- Useful in imparting rose like aroma to a wide variety of soaps, tobacco products etc.

1.3. Varieties

- RRL(B)-71, RRL-82 (Jamrosa)
- High yielding synthetic variety - Trishna
- Motia, Sofia, Tripta - Popular varieties
- PRC-1, IW 31245, IW 3629, IW 3244, OPD-1, OPD-2

Varieties	Characteristics
IW-31245	Profuse tillering. 20-25 t/ha of herbage under low irrigation. Yield-150 bags in 3 cuts during 16-18 months. Oil contains 90% total geraniol. It has superior rosaceous green odour of palmarosa.
Trishna	It possesses straw yellow clumps, long compact inflorescence and has profuse tillering. Oil contains 93% geraniol.
Jamrosa	It has vigorous growth, high yield of 35-45 t/ha of herbage, it yields between 0.3-0.4% of oil in the herb which is rich in geraniol content (54-83%) and geranil acetate (17.89%). Oil yield was estimated between 130-150kg/ha in the second year.

1.4. Climate

- Warm tropical, hardy plant and grows in varying altitude right from sea level.
- Rainfall: 75 cm to 150 cm.
- Temperature: 10-36°C.
- Avoid water stagnation.

1.5. Soil

- Well drained loamy soils of neutral to alkaline reaction (Tolerates up to pH of 9.5), ideal pH 6 -7.
- Poor sandy to heavy fertile soils of arid tracts and saline soil conditions are suitable.
- Also in marginal and sub-marginal lands.

1.6. Season

- June-July

1.7. Propagation and planting

1.7.1. Propagation through seeds

- 2.5 kg/ha.
- Raised bed nursery of 15-30 cm apart
- Transplanting - 3 to 4 weeks old seedling.

1.7.2. Propagation through slips

- 28,000 slips/ha.
- By root cuttings for higher yield and quality oil.

1.8. Spacing

- 60 × 60 cm

1.9. Irrigation

- Life irrigation - Third day
- 7-15 days interval

1.10. Weed management

- 1-2 weeding at early stage
- Earthing up after each harvest.
- Apply diuron 1.15 kg a.i; oxyfluorfen 1.5 kg a.i/ha.

1.11. Manures and fertilizers

- FYM at 10t/ha
- 60:50:40 kg/ha (20:50:40 kg NPK/ha basal and top dressing of 40 kg N in three split doses after each harvest) and $ZnSO_4$ at 25 kg/ha.

1.13. Harvest

- First harvest: After six months from transplanting.
- Subsequent harvests - 3 to 4 months interval up to 5 years.
- Cutting upper third of stem along with leaves.
- Shade drying of leaves for 2-3 days
- Steam distillation is better than other type of oil extraction.

1.14. Yield

- Herbage: 20-30 t/ha/yr with oil of 50-60 kg/yr.
- Maintain up to 8 to 10 years; oil yield declining from fifth year.
- Oil yield: First year - 20 kg/ha.
- Second year- 60 kg/ha.
- Third and fourth year - 70 kg/ha.

Plant part	Essential oil (fresh weight)	Geranial (%)
Whole plant	0.53	72.4 to 86.5
Inflorescence	0.83	72.5 to 84.5
Leaves	0.58	92.9 to 94.6
Stalk	0.01	

2

Lemon Grass (*Cymbopogon flexuosus*) Poaceae, 2n=20

2.1. Origin and distribution

- Native to India and Sri Lanka.
- Widely cultivated in tropics and subtropics.

2.2. Economic importance

- Cochin grass or Malabar grass or East Indian or True Lemon grass; perennial and multi cut aromatic grass.
- Lemon like odour due to presence of 70-90% citral, a cyclic monoterpene.
- Other compounds are geraniol, citronellal etc.
- A good source of natural citral used as a basic raw material for synthesis of β-ionone used for synthesis of number of useful aromatic compounds and Vitamin A.
- Used as main substitute for Cod liver oil.
- Used in perfumery for grades of soaps, detergents, cosmetics and flavour agent for soft drinks.
- Used in Ayurvedic preparation like Balm increasing.

2.3. Varieties

- OD-19 (Sugandhi), OD-408, OD-440, RRL-39, SD 68, Pragathi, Praman, CKP-25 (*C. khasianus* x *C. pendulus)*, Krishna, Nima, Jor Lab L-2 and Cauvery.

Variety	Description
Sugandhi (OD 19)	It is adapted to a wide range of soil and climatic condition. A red stemmed variety with plant height of 1 to 1.75 m and profuse tillering. Oil yield ranges from 80 to 100 kg/ha with 85-88% of total citral produced under rainfed conditions (with life saving irrigation).
Pragati	It is a tall growing variety with dark purple leaf sheath suitable for North Indian Plains and Tarai belt of subtropical and tropical climate. Average oil content is 0.63% with 75-82% citral.
Praman	Evolved through clonal selection and belongs to species *C. pendulus*. It is a medium sized variety with erect leaves and profuse tillering. Oil yield is high with 82% citral.
Jama Rosa	Very hardy with vigorous growth. Variety yields about 35 tonnes of herbage per ha containing 0.4% oil (FWB). Variety yields upto 300 kg oil in 4-5 cuts in 16-18 months growing period.
RRL 16	Average herbage yield of this variety is 15 to 20 t/ha/year giving 100 to 110 kg oil. Oil content varies from 0.6 to 0.8% (fresh weight basis) with 80% citral.
CKP 25	A hybrid between *C. khasianum* × *C. pendulus.* Gives 60 t/ha herbage in North Indian plains under irrigation. Oil contains 82.85% citral.
Other varieties	OD-408 is white stemmed selection from OD-19 and is an improvement in yield in terms of oil and citral content. Kaveri needs high soil moisture to produce luxuriant growth and is evolved for river valley tracts.

2.4. Climate

- Warm humid climate with plenty of sunshine.
- Well under tropical and subtropical conditions.
- Elevation: 1000-1200 m.
- Rainfall: 1800 to 3000 mm.

2.5. Soil

- Wide variety of soils from rich loams to poor laterite and pH of 6
- Also suitable to Jhum fallow, hill slopes and flood free degraded land
- Best suited to well drained sandy loam; avoid water logged conditions

2.6. Season

- Transplanting during June to July and October to November

2.7. Propagation and planting

- Vegetatively by well grown clumps or slips.
- Seeds: 3-4 kg/ha.
- Transplanting - 25-30 days after sowing.

2.8. Spacing

- Seedlings: 30 x 30 cm with a plant density of 1,11,000/ha.
- 55,600 slips/ha at 60 x 45 cm and 37, 000 slips/ha at 90 x 60 cm.

2.9. Irrigation

- Life irrigation – third day
- 7-15 days interval

2.10. Weed management

- Hand hoeing and weeding at 25-30 days after planting
- Second at 40-60 days after planting
- Mulching - Distillation waste (Spent grass) as organic mulch at 3 t/ha in between rows

2.11. Manures and fertilizers

- FYM: 20-25 t/ha
- 150:60:60 kg NPK/ha/year

2.12. Harvest

- First harvest at 90 DAP
- Cutting grass 10 cm above the ground level.
- First year: 3 cuttings/yr
- Subsequently: 5 cuttings/yr
- Oil is extracted by water or steam distillation
- Oil recovery - 0.3-0.5%.

2.13. Yield

- Herbage: 20 to 30 t/ha
- Oil: First year - 25 kg/ha; second year onwards: 350-400 kg/ha.

2.14. Distillation

- Oil extracted from wilted herb by steam distillation in stainless steel unit.
- Factors influencing oil production are storage of plant material, treatment (Wilting and cutting into pieces) of material and method of distillation.
- Major source of loss is by oxidation and resinification of essential oil.
- Essential oils are present in oil glands, oil sacks and glandular hairs of plant.
- Before distillation, wilted plant cut into small pieces to expose directly.

- Wilted leaves are steam distilled about 3 h.
- Dipping chopped lemongrass in 1-2% sodium chloride solution for 24 h before distillation

2.15. Utilization of spent material

- Residue thrown out after extraction of oil is spent grass.
- Cattle relish it when it is hot or converted into silage by adding a dilute solution of molasses.
- Used as fuel for distillation after drying and also cheap packing material for fragile objects.

3

Citronella (*Cymbopogon winterianus* Jowitt), Poaceae, 2n=20

3.1. Origin and distribution

- Originated from Sri Lanka.
- Cultivated throughout tropics.
- Presently major producers - China, Vietnam, Indonesia, Argentina, Taiwan, Sri Lanka, Brazil, Madagascar and India.

3.2. Economic importance

- Two types of citronella oil: Java citronella oil obtained from *Cymbopogon winterianus* which is superior.
- Another is Ceylon citronella oil which is obtained from *Cymbopogon nardus* inferior type oil.
- Citronella oil - essential oil containing citronellal, geranial and hydroxy citronellal and high value perfumery bases obtained on steam distillation of citronella grass.
- Used for perfuming soaps, detergents, cosmetics, agarbattis and for making mosquito repellent creams.
- Also used to isolate citronellal, geranial in turn converted into citronellal, hydroxy citronellal, synthetic menthol and esters of geranial and citronellal.
- Used for making high grade blended perfumes.

3.3. Varieties

- Java-2, Jerlab-2, Java citronella, Java II, Ceylon citronella, BIO-13, Mandakini and Manjusha
- Jor Lab C-2, Jalpallavi, Manjari.
- Jorhat- C2, Java-2, and CIMAP- Bio-13 are the high yielding varieties of this crop for Southern and Eastern India.
- Manjusha and Mandakini are the varieties released by the CIMAP, Lucknow for the North Indian plains.

- An elite mutant clone of Manjusha M3-8 named Manjari which has been found to possess 50-90% more oil, high citronellol and low elemol content on an average, over the other varieties. Manjari is an erect growing herb with yellowish green leaves and a dark purple stem. It is profuse tillering and rapid growing ability hence, produces a high herb yield.

3.4. Climate

- Thrives in tropical and sub-tropical conditions.
- Grows well in Assam where abundant moisture, sunshine and high humidity (70% and above) are available.
- Temperature - 20°C to 38°C
- Elevation: 1000-1500 m above msl.
- Rainfall: 2000-3000 mm.
- Grows best in sandy loam soils with good organic matter and pH of 6 to 8
- Avoid heavy clayey soils tend to water log and light sandy soils
- Does not tolerate water stagnation.

3.6. Season

- May-July

3.7. Propagation and planting

- Propagated by clumps or slips; 28,000 slips/ha
- Vigorously growing plant and 6 months to one year old plants are selected.
- Fibrous roots and leaves trimmed off at 25-30 cm height before planting.

3.8. Spacing

- Slips are spaced at 60 × 60 cm or 60 × 45 cm

3.9. Irrigation

- Life irrigation - Third day
- 7-15 days interval

3.10. Weed management

- 1 to 2 weeding at early stage
- Earthing up after each harvest

3.11. Manures and fertilizers

- FYM: 10 t/ha
- 100:40:40 kg/ha

3.12. Harvest

- First harvest: 6 months after planting
- Second and subsequent harvests: 2 to 3 months interval
- Cut bush by leaving 10-15 cm above ground level

3.13. Yield

- Herbage: 20-30 t/ha/yr
- Oil content - 0.2 to 0.3%
- Oil yield
  - First year: 50-100 kg/ha
  - Second year: 100-150 kg/ha
  - Fifth year: 250-300 kg/ha

4

Vetiver (*Vetiveria zizanioides*) Poaceae, 2n=20

4.1. Origin and distribution

- Indigenous to Indian sub-continent.
- Native of Tropical Asia and introduced into tropics of both hemispheres.

4.2. Economic importance

- Economic part-roots up to 30-40 cm, cooling properties
- Major constituents - vetivone, vetiverols and vetivernate; obtained by steam distillation of roots
- One of finest oriental perfumes with persistent fragrance.
- Used in manufacture of soaps, cosmetics, perfumery, agarbattis, soft drinks, pan masala etc.
- Acts as excellent fixatives for volatile compounds.

4.3. Varieties

- Two types - North Indian and South Indian.
    - Flowering or seeding found in North Indian.
    - Non-flowering or non-seeding found in South Indian.
- North Indian type yields superior quality oil, but its rooting finds to be shallow, especially in damp ground.
- South Indian types are the cultivated types with a thicker stem, less branching roots, and wider leaves. It is the non-seeding type. Among South Indian types, Pusa Hybrid-7, Hybrid-8, CIMAPKS-2, Sugandha, KH-8 and KH-40, are varieties available for commercial cultivation.
- Cultivars Dharini, Gulabi, Nilamber and Kesari were released by CIMAP, Lucknow and were developed by repeated selection of germplasm collections from different parts of India.
- Other varieties are ODV-3 and KS-1 (Bharatpur type).

4.4. Climate

- Moist, sunny with no prolonged hot or dry period.
- Grows well in wastelands, sandy, marginal lands.
- Extremely hardy and tolerate submergence and withstand drought situation.
- 3-4 cycles of cultivation of vetiver improves sandy soils

4.5. Soil

- Well drained sandy loamy and red laterite soil with pH of 8.5-10.
- Grows in saline and alkaline soils.
- 21-44°C and 25-35°C are most suitable
- Rainfall - 100 to 200 cm
- Temperature - 20° to 30°C

4.6. Season

- June-July
- March-April with irrigation facility.

4.7. Propagation material

- Tillers and slips/seeds.
- Tillers - longer time for growing; 1,50,000 to 2,25,000 slips with 2 to 3 slips per pit
- Slips - better planting material
- Plants cut at 25-30 cm above ground

4.7.1. Planting

- System 1: Conical ridges, 30-38 cm high, 48 cm apart and slips are planted 23 cm apart.
- System 2: 30 high, 68 cm wide and 45 cm apart and slips planted in row of 22.5 cm apart.
- System 3: 45 cm high, 60 cm wide, 30cm apart and planting at 30 × 30 cm

4.8. Spacing

- 45 × 30 cm is also optimum.

4.9. Irrigation

- Necessary irrigation as and when needed.

4.10. Weed management

- 2-3 weeding at 1 month interval during initial period of growth.
- Second year - one weeding before harvest
- Atrazine at 0.5 kg a.i/ha (50 % a.i) as pre-emergence for weed free condition of first 75-90 days

4.11. Manures and fertilizers

- FYM at 25-30 t/ha
- 60:22.5:22.5 kg NPK/ha

4.12. Shoot cutting

- Cutting of aerial portion with commencement of winter (November) increases tillering and done twice.
- First shoot cutting - 30 cm height in planting year
- Second cutting - 15-20 cm height before harvesting

4.13. Inter cropping

- Greengram, blackgram, cluster bean in kharif and toria and coriander in rabi season
- Cowpea, daincha in row spaces and incorporates after 50-60 DAS as green manure.

4.14. Harvest

- 15-18 months after planting.
- Best quality oil: December to February.
- Digging: Stem portion cut at height of 15-20 cm and clumps uprooted.
- Thicker roots of length 10-35 cm produce more oil.

4.15. Washing and drying

- Immediately after harvest, roots washed in clean running water to remove adhering earth
- Cleaned roots spread on drying grounds for 1-2 days.
- Dried roots are sent to distillery or storage shed for 60-70 days where allowed to mature.

4.17. Yield

- Dry root: 3-4.5 t/ha.
- 1.4-1.8 t/ha under North India.
- Oil: 15 to 16 kg/ha

5

Geranium (*Pelargonium graveolens* (L.) Hert.), Geraniaceae, 2n=22

5.1. Origin and distribution

- Native of the Cape Province in South Africa.
- Grown in India, South Africa, Belgium, China, Congo, Egypt, France, Madagascar, Morocco, Reunion Islands, Spain, former USSR countries.
- Commercially cultivated mainly in Nilgiris and Kodaikanal hills of Tamil Nadu, and in and around Bangalore in Karnataka.

5.2. Economic importance

- Very profound and strong rose like odour.
- Also known as rose geranium.
- Chicf constitucnt of oil - gcraniol and citroncllol.
- Oil also contains α-pinene, β-pinene, α-terpinene, myrcene, α-phellandrene, limonene, cis-ocimene, trans ocimene, p-cymene, terpinolene, cis-roseoxide, trans-rose oxide, methone, trans-linalool, iso-menthone, caryophyllene, geranyl acetate, nerol, geranyl formate, geranayl butyrate and geraniol.
- Widely used in scenting soaps and for isolation of rhodinal forms high grade perfumes.

5.3. Varieties

- Three types: Zonal geranium, ivy-leaved geranium and scented leaved geranium.
- In India, two types namely algerian or tunisian and reunion or bourbon.
- **Algerian or Tunisian**: Scented geranium, is slender with flowers of a dark pink colour. It is being grown in the Nilgiris and is unsuitable for wet conditions. This variety yields 50-60% more oil with a more delicate odour than that of the reunion type.
- **Reunion or Bourbon**: Grown in the Nilgiris and Anamalai hills, the plant is sturdier with light-pink flowers and more suitable for wet conditions. Oil

content is higher during the summer months from April to June. The terminal portion with 6 to 12 leaves contains more oil than the middle and basal portions.

- **KKL-1**: In the evaluation trial of the algerian and reunion types, PG-7 and PG-20, respectively at the Horticultural Research Station, Kodaikanal, the clone PG-7 recorded 0.3% essential oil which has been released under the name KKL-1.
- **Sel-8**: Indian Institute of Horticultural Research (IIHR), Bangalore, has found Sel – 8, a reunion type, as the highest yielder under Bangalore conditions and has recommended it for cultivation.
- Hemanti, Bipuli and Kunti are the other varieties released by the CIMAP, Lucknow, for cultivation in the plains of North India.
- Kelkar and Ooty are the other varieties available in this crop.

5.4. Climate

- Grown in temperate, subtropical and tropical climates at altitudes from 1000 to 2200 m above msl.
- Thrives best in subtropical climates with temperature of 5° to 23°C.
- Higher growth and yield at 15-18°C.
- Annual rainfall: 100-150 cm.
- Avoid water logging.

5.5. Soil

- Shallow rooted crop.
- Well drained porous soil.
- Well in red lateritic soils with pH of 5.5-8.0
- Calcium rich porous soil is the best.

5.6. Season

- June-July

5.7. Propagation and planting

- Propagated by cuttings: 30,000 cuttings/ha.
- Terminal cuttings of 20 cm long with 8 nodes.
- Treating with IBA or IAA for 6 min. at 200 ppm.
- Cuttings planted at spacing of 8-10 cm.
- Before planting, cuttings are dipped in 0.1% benlate solution.

5.8. Spacing

- 60 × 60 cm, 60 x 45 cm.

5.9. Irrigation

- Immediately after planting.
- Alternate days for 10-15 days and reduced to twice a week.

5.10. Weed management

- Pre-emergence application of TOK-E 25 at 4 l/ha.
- Weeds are removed periodically.

5.11. Manures and fertilizers

- FYM at 25 t/ha and 150:150:100 kg NPK/ha.
- $ZnSO_4$ at 20 kg/ha and boron at 10 kg/ha.

5.12. Crop rotation and intercropping

- Cowpea or black gram.

5.13. Pruning

- Branches cut and leave 15-20 cm once in 4-5 years.

5.14. Harvest

- 7 to 8 months after transplanting.
- 3-4 harvests in a year.
- Leaves turn light green and change from lemon like odour.
- Tender tip portion with 6-12 nodes.
- Withered in shade for 2-3 years.

5.15. Yield

- Herbage: 20-25 t/ha
- Oil yield: 15-20 kg/ha

6

Davana (*Artemisia pallens* Wall) Asteraceae, 2n=16

6.1. Origin and distribution

- Grows wild in temperate Himalayas.
- Common in Kashmir Valley, Shimla and Nainital Hills.
- Commercially cultivated in Karnataka, Maharashtra, Kerala, Tamil Nadu and Andhra Pradesh

6.2. Economic importance

- Sweet worm wood or annual worm wood
- Chemical constituents - Artimisinin
- Commonly used in garlands, bouquets and religious offerings in most part of the country.
- Leaves and flowers contain essential oil valued for its exquisite and delicate aroma and is used in high grade perfumes and cosmetics.
- Oil of davana contains hydrocarbons (20%), esters (65%) and oxygenated compounds (15%).
- Esters - major constituents responsible for characteristic smell of Davana.
- Essential oil - brown, viscous liquid with rich, fruity odour, a sesquiterpene ketone known as artemone.
- Used for flavouring cakes, pastries, tobacco and beverages.

6.3. Varieties

- Jeevanrasha, Suraksha and Asha

6.4. Climate

- Grown during winter season
- Light showers with moderate winter conditions
- High temperature and heavy rains at the time of flowering affect plant growth and reduces oil content

6.5. Soil

- Various types of soils from sandy loam to medium black.
- Fertile, well drained, sandy loam soil rich in organic matter

6.6. Season

- November-December

6.7. Seeds and planting

- Seeds: 1.5 kg/ha
- Transplanting seedlings having 10 cm height on 6-8 weeks from DAS.

6.8. Spacing

- 15 × 7.5 cm

6.9. Irrigation

- Immediately after planting
- 10-12 days interval

6.10. Weed management

- 2-3 hand weedings

6.11. Manures and fertilizers

- FYM at 6 t/ha
- 120:40:40 kg NPK/ha
- Spray GA_3 at 200 ppm after five weeks of transplanting to increase the herb and essential oil yield.

6.13. Harvest

- Cutting plants from base.
- Ready for harvest at 120-125 DAS
- Dried under shade for 2 days

6.14. Yield

- Fresh herbage: 12-13 t/ha, oil content 0.29%.
- Oil: 12-13 kg/ha

7

Mint (*Mentha* spp.), Lamiaceae, 2n = 36

7.1. Origin and distribution

- Originated in the Mediterranean basin and spread to rest of world by both natural and artificial means.
- USA - major producer of pepper mint and spear mint
- Bergamot mint - native of Europe and naturalized in eastern USA
- Grows successfully in temperate and subtropical region of India.
- India - Japanese mint - Terai region, extending to parts of Indo Gangetic plains and parts of Punjab in North-Western India.

7.2. Economic importance

7.2.1. Japanese mint or corn mint or field mint (*M. arvensis)*, 2n: 36, 72.

- Native of China and Brazil.
- Primary source of menthol.
- Fresh leaves - 0.4-6.0 % oil.
- Main constituents of oil are menthol (65-75%), menthane (7-10%) and menthyl acetate (12-15%) and terpenes (Pinene, limonene and camphene).
- Menthol - used in flavouring of a large number of pharmaceutical and oral preparations like tooth pastes, dental creams, confectionery, beverages and other items like tobacco, cigarettes and paan masala.
- Excellent carminative and gastric stimulant.
- Acts as a mild analgesic.

7.2.2. Pepper mint (*M. piperita*), 2n:36

- Fresh herb - 0.4 to 0.6%.
- Menthol content - 35-50%.
- Constituents - menthyl acetate (14-15%), menthone (9-25%), menthofuran and terpenes like pinene and limonene.

7.2.3. Bergamot mint (*M. citrate*), 2n:96

- Linalool and linalyl acetate - main constituents
- Used directly in perfumes.
- Cosmetic preparations like scents, soaps, after shave lotions

7.2.4. Spear mint (*M. spicata*), 2n:48

- Principal constituent - carvone (57.71%)
- Used mostly as a flavouring in tooth pastes and as food flavouring in pickles and spices, chewing gum and confectionery, soaps and sauces.

7.3. Varieties

7.3.1. Japanese mint

- Himalaya (MAS-I), MA-2, Kalka (Hybrid-77), Shivalik, EC-41911, Kosi

7.3.2. Pepper mint

- Kukrail

7.3.3. Bergamot mint

- Kiran

7.3.4. Spear mint

- MSS-l, MSS-5, Punjab spearmint-l and Neera.

7.4. Climate

- Japanese mint grows in all tropical and sub-tropical areas
- Temperature: 20-25°C promotes vegetative growth; at 30°C, oil content is increased.
- Pepper mint and spear mint - grown in humid and temperate areas.
- Bergamot mint - both in temperate and sub-tropical areas.
- Rainfall: 100-150 cm

7.5. Soil

- Medium to fertile deep soil, rich in humus with pH of 6 to 7.5
- Good water holding capacity; avoid water logging

7.6. Season

- June-July
- Plains: Winter month
- Temperate climate: Autumn or spring

7.7. Propagation and planting

- Mints propagated through creeping stolons or suckers or terminal cuttings.
- Pepper mint and bergamot mint: Runners
- 400 kg stolons/ha; cut into small pieces (7-10 cm) and planted in shallow furrows of 7-10 cm deep

7.8. Spacing

- 40 × 40 cm

7.9. Irrigation

- Twice in a week

7.10. Weed management

- Weeding and hoeing at regular intervals upto 75 DAP
- Combining organic mulch with combination of 0.5 kg/ha of oxyfluorfen herbicide.
- Application of pendimethalin 1 kg/ha or dalapon 2 kg/ha or gramoxone 2.5 kg/ha as post-emergence.

7.11. Manures and fertilizers

- FYM at 25 to 30 t/ha.
- 75:75:50 kg NPK/ha

7.12. Crop rotation and inter cropping

- Mint - maize - potato.
- Mint - early paddy and potato.
- Mint-late paddy - sweet pea.
- Mint - maize -rape seed/mustard.
- Mint - maize - potato or mint - paddy rotation.

7.13. Harvest

- Japanese mint: 100-120 DAP
- Lower leaves turn yellow.
- Cutting green herb at 2-3 cm above ground.
- Second harvest: 80 days after first harvest
- Third: After 80 days from second harvest.

7.14. Yield

- Fresh herb: 25-30 t/ha.
- Oil: 100-150 kg/ha.

7.15. Storage of oil

- Light and golden coloured, motile liquid

8

Rosemary (*Rosmarinus officinalis* L.) Lamiaceae, 2n=20

8.1. Origin and distribution

- Native of Mediterranean regions of Europe, Asia Minor and North Africa.
- Grown in Spain, Italy, France, Algeria, Morocco and Portugal for its essential oil.
- Cultivated to a limited extent in Nilgiris in South India.

8.2. Economic importance

- Economic part - leaves and flowering tops
- Oil is valued for its use in culinary, medicine, perfumery and cosmetic industries.
- Excellent fixative material and oil also contributes a strong fresh odour
- Also serves to mask unpleasant smell.
- Rosemary oil - antimicrobial activity against certain gram positive and gram negative organisms.
- Also used in formulations of compounded oils for flavouring meat, sauces, condiments and other food products.
- Leaves used in cooking.
- Distilled water from flowers used for soothing eye wash.
- Oil-1, 8 cineole (20-50%), borneol (20%), camphor, linalool, α-pinene, camphene, β–pinene, sabinene, myrcene, α-phellandrene, α-terpinene, limonene, α-terpinene, p-cymene, terpinolene, thujone, copaene, terpinen-4-ol, caryophyllene, methyl chavicol, α-terpineol, thymol and carvacrol.

8.3. Types and varieties

- French rosemary: Produces white flowers and oil is superior in quality
- Italian rosemary: Flowers are purple.
- Ooty-1

8.4. Climate

- Prefers a low humidity, warm winters and mild summers
- Altitude - 2500 m above msl
- Temperature less than 30°C

8.5. Soil

- Very hardy plant and found growing on rocky terrains in temperate parts of world.
- In India, light loamy soils and sandy loam soils with pH of 5.5-7.0 are suitable
- Laterite soil is the best.

8.6. Season

- September-November

8.7. Propagation

8.7.1. Vegetative propagation

- Stem cuttings.
- 10-15 cm in length
- After 6-8 weeks, transplanting into main field.

8.7.2. Seed propagation

- Seeds: 0.2 to 2.5 g/sq.m
- Seedlings of 8 to 10 weeks are transplanted

8.8. Spacing

- 45 x 45 cm.
- 50,000 seedlings/ha

8.9. Irrigation

- Irrigated twice a week

8.10. Weed management

- Weed free condition should be maintained.

8.11. Manures and fertilizers

- 20 t/ha of FYM
- 100:40:40 kg NPK/ha

8.12. Pruning

- After 2 to 3 years, pruning is done to promote formation of numerous shoots

8.13. Harvest

- 8 months after planting
- Only 2 harvests are obtained.
- In subsequent years, 3 to 4 harvests at 100 to 120 days intervals

8.14. Yield

- 12 to 15 t/ha/year of fresh herbage yielding 85 to 100 kg of oil
- 2.5 t/ha of dry leaves

9

Thyme (*Thymus vulgaris*) Lamiaceae, 2n=30

9.1. Origin and distribution

- Originated in Southern Mediterranean.
- Distributed in India, Asia, Australia, Canada, Europe, France, Germany, Greece, Italy, Morocco, Spain, USA

9.2. Economic importance

- Commonly called as common thyme or garden thyme.
- The active principle alkaloid is thymol.
- Constituents - triterpinoid saponins, flavones, ursolic acid (1.5% in tops), coffeic acid, tannins and resins.
- Oil of thyme - colourless, yellow or red liquid with a characteristic pleasant odour, and a pungent taste.
- Leaves and flowers - food flavourants and seasoning for various food items, especially fish and meat preparations and for garnishing.
- Leaves - possess laxative, stomachic and tonic properties, good for kidney and eye and are blood purifiers.
- Also insect repellent property.
- Used in perfumery and in liquor distillery.
- Thyme of commerce; Thymi Herba in pharmacy.
- Possess antiseptic, anti-helminthes, expectorant, carminative, diuretic, alexiteric, and sedative properties.
- Used in mouth washes and gargles.
- Treatment of whooping cough and bronchitis, oil is used in soaps, perfumes and for flavouring food products such as a meat, sausages, sauces and canned food.

9.3. Varieties

- Creeping Thyme, Lemon Thyme, Golden scented Thyme
- Garden Thyme, St. Louis and Common Thyme

9.4. Climate

- Elevation: 1525-4000 m.
- Warm climate
- Grown both in hills and on plains.
- Hilly regions - best

9.5. Soil

- Prefers light fertile and calcareous soil with pH of 5.5-8

9.6. Season

- May -June

9.7. Propagation and planting

- Seeds and vegetatively by cuttings or by layering of side shoots.
- Nursery in April.
- Late summer - transplanting

9.8. Spacing

- 90 × 30-45 cm - seeds.
- 60 × 30-45 cm - cuttings.

9.9. Irrigation

- Frequent irrigations during dry period.

9.10. Weed management

- Inter culture operations and weeding at regular intervals.

9.11. Fertilizer requirement

- 75:40:40 kg NPK/ha.

9.12. Harvest

- Leaves and flowers for culinary and medicinal purposes - five months after sowing/planting.
- Shoots of 15 cm cut from plants, dried in shade or dryer immediately after harvest and stored in airtight containers.
- Dried leaves are curled, brownish green.

9.13. Yield

- Dry herb: 1.1-2.2 t/ha.
- Oil distilled from fresh flowering and from tops by steam distillation.
- Herb - 2% essential oil and total oil recovery by distillation - 21 kg/ha.

10

Patchouli (*Pogostemon patchouli*) Lamiaceae, 2n=34

10.1. Origin and distribution

- Native of Philippines
- Grows wild in Malaysia, Indonesia and Singapore.
- Natural habitat extends across Paruguay, Penang, East and West Indies, sub-tropical Himalayas and Deccan peninsula.
- Introduced into India by Tata Oil Mills
- Cultivated in Bangalore and coastal areas of South India, Bengal, Odisha and Assam.

10.2. Economic importance

- Oil of patchouli is called patchoulol. Oil contains benzaldehyde, eugenol and cinnamic aldehyde etc.
- Patchouli oil - used as flavouring ingredient in major food products, including alcoholic and non-alcoholic beverages, frozen dairy desserts, candy, baked goods, gelatin, meat and meat products.
- One of most important essential oils of perfumery industry, because oil blends well with other essential oils like vetiver, sandalwood, geranium, lavender, cedar wood derivatives and clove oil.
- Oil of patchouli - used as ingredient in insect repellent preparations and antibacterial properties

10.3. Varieties

- Johore, Singapore and Indonesia - commonly cultivated types. Of these, Johore yields the best quality oil in terms of chemical composition and odour value whereas, the other two have high herbage and oil yield potential, but their oils are of inferior quality
- CIM – Shresta: Recently released variety of patchouli

10.4. Climate

- Partially shaded.
- Prefers warm, humid climate and varied climatic conditions.
- Rainfall: 150-300 cm/year.
- Temperature - 24-28°C and RH 75%
- Altitude of 800 to 1050 m above msl.

10.5. Soil

- Hardy and grows on wide range of soils.
- Well drained fertile soils.
- Deep, loamy soil rich in humus and nutrient with pH of 5.5-6.2.

10.6. Season

- August-October

10.7. Propagation and planting

10.7. 1. Nursery

- Treated with suitable nematicide furadan (3% a.i) 20 kg/ha.
- Stem cuttings from 9 months old branches, 10-12 cm length, consisting of 4-5 nodes with terminal bud and crown 2-3 leaves
- Treatment with 1500 ppm IBA to basal end encourages rooting.
- Takes 30-35 days for rooting
- After 8-10 weeks used for transplanting.

10.7.2. Direct planting

- 15-20 cm long cuttings
- 2-3 cuttings planted per planting point.

10.8. Spacing

- 60 x 60 cm

10.9. Irrigation

- Watered every day for 3-4 days
- Subsequently on alternate days for 10-15 days.

10.10. Weed management

- Weeding at six weeks of transplanting followed by one hoeing.
- Diuron 3 kg/ha as pre-emergence.

10.11. Manures and fertilizers

- 12 t/ha of FYM
- 150:50:50 kg NPK/ ha.
- N is applied with five split doses after every harvest.

10.12. Growth regulators

- Foliar spray of triacontanol (0.05%), auxin (0.25%) + cytokinin or phenol compound (3 ppm) at 2 weekly intervals or 2,4-D (0.5%) at 4 weekly intervals to enhance plant growth and yield

10.13. Intercropping

- Grown as intercrop with coconut, arecanut, rubber, coffee, banana and other plantation crops.

10.14. Harvest

- First harvest: 5 months of transplanting.
- Foliage becomes pale green to light brown and emits patchouli odour
- Subsequent harvests are done every 3-4 months
- Length of cut tops: 40-60 cm.
- Crop maintained for 3 years.
- Harvested material spread out under shade in thin layers and turned periodically to ensure proper drying
- Drying for 3 days.

10.15. Yield

- 2 t/annum - dry leaves and 50-60 kg of oil per ha

11

Lavender (*Lavendula angustifolia*) Lamiaceae, 2n=42 or 48

11.1. Origin and distribution

- Indigenous to Mediterranean region, growing wild in higher hills of Southern France, Italy and Bulgaria.

11.2. Economic importance

- True lavender (*Lavendula angustifolia* Mill sub. sp *augustifolia*)
- Spike lavender (*Lavendula latifolia* Mill sub. sp *spica*)
- True lavender - used in perfumery industry.
- Lavandin - hybrid between true and spike lavender
- Lavender oil - delightfully clean, refreshing and sweet odour.
- Main constituent - ester, linalyl acetate 30 to 60%
- Another compound - ethyl-n-amyl ketone.
- Free linalool - major component
- Other compounds - α-pinene, camphene, β-pinene, limonene, cineole, cis-oscimene, transocimene, camphor terpinen-4-01, caryophyllene, lavandulyl acetate, lavandulol, α-terpineol and borneol.
- Lavender water - mixture of oil in water and alcohol
- Best quality oil (50% esters) - used in preparation of high grade perfumes
- Second quality (38 to 42% esters) - lavender water, toilette waters etc.
- Low grade (30 to 35% esters) oil - scenting of soaps and talcum powders.
- Oil used in medicine as flavoring agent and as a carminative.
- Lavender flowers used in making sachets and potpourri.
- Oil of lavendeis used as substitute for true lavender oil.

11.3. Varieties

- Lanvander has two forms: Spike vero and vero spike.
- Bulgaria - Kazanluk, Karlovo, Hemus,Aroma, Svezhest and Vanets

- Soviet varieties - Steinway, Goranya, Prima, Record, French Bareme and Lambris.
- High yielding clones - AM-1, AM-2, AM-3, A-5, A-7, A-8, A-9, B-2 and B-11

11.4. Climate

- Temperate, photophilous plant.
- It does well in cold winters and cool summer.
- Grown successfully on arable lands at very high altitudes.
- Resistant to drought and frost.

11.5. Soil

- Light, well aerated, dry and calcareous soils with pH of 7.0 to 8.4
- Poor carbonate and sandy soils and avoid water logged soils.

11.6. Season

- October-November

11.7. Propagation and planting

11.7.1. Seed propagation

- Sown in nursery beds in autumn (November-December)
- Seed rate: 0.2-2.5 g/sq m

11.7.2. Vegetatively by cuttings

- Annual and biennial branches of mother plantation are cut during October-November in plains and in February-March in hills.
- 8-10 cm long cutting are treated with 500 ppm IBA

11.7.3. Transplanting

- 1.20-1.40 m between the rows and 3.5-4 cm within row.
- 20,000 plants/ha give the highest yields

11.8. Irrigation

- In dry periods, irrigate frequently

11.9. Weed management

- Regular weeding and hoeing

11.10. Fertilizer requirement

- 100:40:40 kg NPK/ha.
- N is applied in four splits including basal application in equal dose of 20 kg N/ha.

11.12. Harvest

- Harvest at blooming stage after three year of planting.
- Whole plant leaving basal one third of plant for natural regeneration
- Fresh herbage - used for steam distillation.
- Essential oil content - 0.5% to 1.1% with 0.8% as average oil content.

11.13. Yield

- 10 t/ha of fresh herbage.
- Lavender oil: 100 kg/ha.

12

Eucalyptus (Lemon-Scented Gum) (*Eucalyptus citriodora*), Myrtaceae, 2n=22

12.1. Origin and distribution

- Native of Australia and Malaysia.
- Introduced by British people in Nilgiri hills of Tamil Nadu for fuel and timber.
- Commercially grown in Seychelles, Java, South Africa, Guatemala, Congo and Brazil.
- *Eucalyptus tereticornis* - Commercial species
- Major growing states - UP, Tamil Nadu, Kerala and Karnataka.

12.2. Economic importance

- Known as spotted gum tree.
- Source of citronellol, hydroxycitronellal and menthol.
- Contains 0.5-4% essential oil; rich in citronella.
- Essential oil - Therapeutic and ayurvedic benefits.
- Used in soaps, mouth wash liquids, massage oils, air fresheners and bug repellents
- Helps in curing common cold, flu and fever
- Used as an antiseptic.
- Act as a stimulant in bronchitis.
- Helps in treating sinus and allergies
- Wound treatment and relieving muscle pains

12.3. Varieties

- EC 153, EC 261 and EC 3315.
- IFGTB EC 1, IFGTB-EC2, IFGTB EC3, IFGTB EC4

12.4. Climate

- Wide variety of climatic conditions
- Thrives in tropical to temperate climatic areas
- Altitude - 600 m above msl.

- Rainfall - 4 to 40 cm
- Temperature - 15°C and 30°C.

12.5. Soil

- Gravelly, podzols or arid residual podzols of lateritic origin with organic soil.
- Acidic is a good for the species.
- Deep rich and well drained loamy soil with pH of 6 to 7
- Avoid water logging, alkaline and saline condition

12.6. Season

- February-September.

12.7. Propagation

- Commercially by seeds and vegetative method
- Shoots are treated with IAA, IBA and IPA at 50 ppm and 100 ppm.

12.8. Spacing

- 75-90 x 75 cm between pits.

12.9. Irrigation

- Drip irrigation
- 25-30 irrigations during entire period

12.10. Weed management

- Manual weeding and harrowing - 2 to 4 times in winter and summer

12.11. Fertilizer requirement

- 120 kg N and 30 kg P/ha.

12.12. Crop rotation and intercropping

- First two years: Most crops except paddy
- Thereafter, shade loving crops like ginger, turmeric.

12.13. Harvest and yield

- 2 to 3 years rotation
- First coppicing is about 30-45 cm above ground and subsequently, 75-90 cm from ground level.
- Three harvests from irrigated crop per annum from the coppiced trees.
    - First year: 7 t/ha.
    - Second year: 30 t/ha
    - Third year: 40 t/ha
- 30 tons from unirrigated crop/year/ha

Section III
Facts and Figures of Medicinal and Aromatic Crops

1

Common Name, Scientific Name, Family, Chromosome Number, Origin, Inflorescence and Fruit Type of Medicinal and Aromatic Crops

Common name	Scientific name	Family	Chromosome number	Origin	Inflorescence	Fruit type
Senna	*Cassia angustifolia*	Caesalpiniaceae	2n=28	Native of Yemen, South Arabia, Sudan and Sinai	Raceme	Legume pod
Periwinkle	*Catharanthus roseus*	Apocynaceae	2n=16	Madagascar	Solitary axillary or dichasial cyme.	Follicle
Coleus	*Coleus forskholii*	Lamiaceae	2n=30	South East Asia	Raceme	Achene like nutlets
Ashwagandha	*Withania somnifera*	Solanaceae	2n=48	Africa, India and Mediterranean region.	Axillary or umbellate cyme	Berry
Glory lily	*Gloriosa superba*	Colchicaceae	2n=22	Tropical Asia and Africa.	Head	Capsule
Sarpagandha	*Rauvolfia serpentina*	Apocynaceae	2n = 22	South East Asia	Terminal umbelliform cyme	Drupe
Dioscorea	*Dioscorea floribunda*	Dioscoreaceae	2n = 36	South East Asia, Central America.	Terminal raceme/ spike	Capsule
Aloe	*Aloe vera*	Liliaceae	2n = 14	Eastern and Southern Africa, Canary Islands and Spain.	Raceme	Capsule
Phyllanthus	*Phyllanthus amarus*	Euphorbiaceae	2n = 26	Tropical America	Axillary	Capsule
Creat or Green Chireta	*Andrographis paniculata*	Acanthaceae	2n = 50	India and Tropical South East Asia.	Axillary and terminal raceme or panicle	Capsule
Medicinal solanum	*Solanum khasianum*	Solanaceae	2n = 24	India	Axillary cluster	Berry
Isabgol	*Plantago ovata*	Plantaginaceae	2n = 8	Persia and neigh-bouring tracts of Western and Southern Asia	Awn or bristle	Capsule
Opium poppy	*Papaver somniferum*	Papaveraceae	2n = 22	Turkey Western Mediterranean region	Cymes, also solitary or raceme	Capsule

Safed musli	*Chlorophytum borivilianum*	Liliaceae	2n = 16	Africa	Raceme	Capsule
Stevia	*Stevia rebaudiana*	Asteraceae	2n = 22	Seirra Amambay in North eastern Paraguay and adjacent areas in Bolivia and Brazil.	Terminal and combined into cyme.	
Horse eye/ Velvet bean	*Mucuna pruriens*	Fabaceae	2n = 22	Asia	Axillary raceme	Pods
Vallarai	*Centella asiatica*	Apiaceae	2n = 18	India, China, Indonesia, Madagascar and Africa.	Umbel	Cremocarp
Brahmi	*Bacopa monnieri*	Scrophulariaceae	2n = 64	Southern and Eastern India, Australia, Europe, Africa, Asia, and North and South America	Solitary, axillary	Capsule
Black nightshade	*Solanum nigrum*	Solanaceae	2n = 24	Eurasia	Tight umbel like cluster	Berry
Sappan wood	*Caesalpinia sappan*	Caesalpiniaceae	2n = 24	Indo-Malayan region	Axillary or terminal panicle or raceme	Pod
Gymnema	*Gymnema sylvestre*	Apocynaceae	2n = 22	Tropical Asia and China	Umbellate cyme	Follicle
Sacred basil/ Holy basil	*Ocimum sanctum*	Lamiaceae	2n = 32	India	Verticillaster/ Raceme/terminal spike	Nutlets
Noni	*Morinda citrifolia*	Rubiaceae	2n = 44	South Eastern Asia (Indonesia) to Australia	Globose heads	Syncarp/drupe
Tippili	*Piper longum*	Piperaceae	2n = 52	Hills of South Western India, Indo-Malaya region.	Spike	Drupe/pod
Digitalis/ Wooly	*Digitalis lanata*	Scrophulariaceae	2n = 56	Western and South Western Europe.	Spike	Capsule

Pyrethrum	*Chrysanthemum cinerariaefolium*	Asteraceae	2n = 18	Dalmatia (Yugoslavia), Herzegovina and Montenegro.	Head/capitulum	Achene
Palmarosa	*Cymbopogon martini*	Poaceae	2n = 40	South East Asia especially India	Spike	Caryopsis
Lemon grass	*Cymbopogon flexuosus*		2n = 20	India	Panicle/spike	
Citronella	*Cymbopogon winterianus*			Sri Lanka	Terminal panicle	
Vetiver	*Vetiveria zizanioides*			India	Panicle/spike	
Geranium	*Pelargonium graveolens*	Geraniaceae	2n = 22	Cape Province in South Africa.	Pseudo-umbel	Capsule
Davana	*Artemisia pallens*	Asteraceae	2n = 18	Southern India	Racemose panicle/ capitulum	Berry
Mint	*Mentha* spp	Lamiaceae	2n = 36		Raceme or cyme	Nutlets
Japanese mint/corn mint/field mint	*M. arvensis*		2n = 36, 72	China and Brazil.		
Pepper mint	*M. piperita*		2n = 36			
Bergamot mint	*M. citrata*		2n = 96			
Spear mint	*M. spicata*		2n = 48			
Rosemary	*Rosmarinus officinalis*	Lamiaceae	2n = 20	Mediterranean regions of Europe, Asia Minor and North Africa.	Terminal and axillary short raceme	Schizocarp/ nutlets
Thyme	*Thymus vulgaris*	Lamiaceae	2n = 30	Southern Mediterranean	Spike	Schizocarp
Patchouli	*Pogostemon patchouli*	Lamiaceae	2n = 34	Philippines.	Spike	

Lavender	*Lavendula angustifolia*	Lamiaceae	2n = 42 or 48	Mediterranean region, growing wild in higher hills of Southern France, Italy and Bulgaria.	Spike	Schizocarp/ nutlets
Eucalyptus	*Eucalyptus citriodora*	Myrtaceae	2n = 22	Australia and Malaysia.	Axillary panicle	Capsule/gumnets

2

Alkaloids Present in Medicinal and Aromatic Crops and their Uses

Crop	Alkaloid	Oil (%)	Uses
Glory lily	Colchicine, colchicoside (Seeds), gloriosine	0.5-0.7	Gout, anti inflammatory, anticancer, snake bites and scorpion stings
Senna	Sennoside A, B, C, D (Leaves, pods)	Indian senna: 1.5-3% Alexandrian: 4-4.5%	Laxative, constipation
Periwinkle	Roots: Ajamalacine, serpentine Leaves: Vincristine, vinblastine Roots: High vincristine	Vincristine: Roots: 0.75-1.2% Leaves: 0.6-0.65%	Anticancer
Pyrethrum	Pyretherin	South India hills: 0.42-1.9% Average: 1.19%	Synthetic insecticide
Fox glove	Digitoxin, gitoxin, digoxin	1-1.4%	Heart diseases
Lemon grass	Citrol (80%)	1-1.2%	Ionones, cosmetics, synthetic vitamin A
Citronella	Citronellal (30-35%), citronellol and geraniol	1.2-1.5%	Mosquito repellents, deoscented soaps
Palmarosa	Geraniol (95%)	0.43-0.52%	Soap, perfume
Geranium	Geraniol and citronellol		Perfumes, creams
Mint	Menthol (80-90%)	0.1-0.2%	Flavouring agent, coolant, carminative, digestive, pain killer.
Patchouli	Patchouliol	2-6%	Perfumes, flavoring agent
Vettiver	Vetiverol	1-1.5%	Soap making, cosmetics, perfumery, agarbathis, soft drinks, pan masala
Ashwagandha	Withanine and somniferine (Leaves and roots)	0.13 -0.31	Restorative tonic, stress, nerval disorder
Sarpagandha	Rescinamine, deserpidine, reserpine, serpentine, ajmaline, ajmalicine and rauvolfinine.	1.7 - 3% (Dried roots)	Enhance lactation, general weakness, cough
Tulsi	Eugenol (about 71%), eugenol methyl ether (20%), nerol caryophyllene, selinene, α-pinene, β-pinene, camphorcineole, linalool and carvacrol (3%) – (Leaves/seed)		Cough, cold, bronchitis

Stevia	Stevioside and rebaudioside A (Leaf)		Diabetes
Aloe vera	Aloin	4.5 to 25%	Wound healing, moisturizing, emollient properties, dermatitis, cutaneous leishmaniasis
Medicinal solanum	Solasodine, a nitrogen analogue of diosgenine		Anti-inflammatory, anabolic and antifertility properties
Coleus	Forskolin (Dried roots)		Treating hypertension, glaucoma, asthma, congestive heart failures, cholesterol and cosmetics
Phyllanthus	Phyllanthin and hypophyllanthin		Treatment of liver, kidney and bladder, diabetes and intestinal parasites
Cinchona	Cinchonidine, quinidine and cinchonine	Leaves contain 1% total alkaloids	Anesthetic, skin against sunburn, bitter tonic, stomachic and appetizer.
Belladonna	Tropane alkaloids	0.13 - 0.70 %	Manufacture of tinctures and plasters, gout, rheumatism, treatment of renal and bilary colic, stomach disorders
Dioscorea	Diosgenin		Steroid hormones including sex hormones, cortisones, other corticosteroids and active ingredient in the oral contraceptive pills.
Isabgol	Colloidal mucilage mainly consisting of xylose, arabinose, galacturonic acid.		Anti diarrhoeal drug, chronic dysenteries of amoebic and bacillary origin, treating constipation
Opium poppy	Morphine, codeine, narcotine, papaverine and linoleic acid		Lowers blood cholesterol, relieve pain and induce sleep, use against cough
Rosemary	1, 8-cineole (26.54%), α-pinene (20.14%), camphor (12.88%), camphene (11.38%) and β-pinene (6.95%)		Culinary, medicine, perfumery, cosmetics, excellent fixative material.
Thyme	Thymol, linalool, linalyl acetate, burneol and α-terpineol		Laxative, stomachic and tonic properties, insect-repellent property, perfumery and liquor distillery
Davana (Artemisia)	Artimisinin, artimisilene, arteannuin and artemusinic acid		Anti malarial activity, perfumery, cosmetics and in dermatology

3

An Abstract on Production Technology of Medicinal and Aromatic Plants

Crop	Varieties	Season	Seed rate	Spacing and fertilizer	Yield (Oil and herbage yield) Irrigated	Rainfed
Senna	KKM, ALFT-1, Sona	Early summer (February-March) or June-July September-October in South India	15-20 kg of seeds/ha	45 x 30 cm 80:40:40 kg NPK/ha/year	Dried leaves - 2t/ha Dried pods - 400 kg/ha	Dried leaves- 1t/ha Dried pods- 150-200 kg/ha
Periwinkle	Nirmal, Dhawal	June-July September-October	Seeds for direct sowing: 2.5 kg/ha Seeds for transplanting: 0.5 kg, Softwood cuttings from lateral shoots of 10-15 cm length with 5-6 nodes treated with NAA at 25 or 50 ppm	45 x 30 cm 25:50:70 kg NPK/ha/year	Roots - 1.5 t/ha Stems - 1.5 t/ha Leaves - 4 t/ha	Roots- 750 kg/ha Stems - 750 kg/ha Leaves- 2 t/ha
Coleus	TNAU - medicinal coleus CO-1, K 8, Aisiri, Manganiperu Garmai, Maimul	June-July	Terminal cuttings (10 cm) or rooted cuttings of 10-12 cm long with 3-4 pairs of leaves	60 x 45 cm - Normal soil, Low fertile soils - 60 x 30 cm Drip irrigation system - broad row ridges of 90 × 60 cm and planting at 60 x 45cm; 40:60:50 kg NPK/ha	Fresh tubers: 15-20 t/ha; Dry tubers: 2-2.2 t/ha	
Ashwagandha	Poshita, Rakshita, Jawahar-20, Jawahar, Ashwagandha B4, WSR, Nagori	Late rainy season: August-September	Broadcasting: 10-12 kg/ha Transplanting: 5 kg/ha, 6 weeks old seedlings are transplanted	60 x 60 cm	Dried roots: 400-1200 kg/ha Seeds: 200-500 kg/ha.	
Glory lily	Singaleri	June-July	Underground, V-shaped biforked	45 x 30 cm; 120:	Seed - 200-250 kg/ha; Tuber - 300	

			tubers - 2.5 to 3.0 tons tubers/ha.	50:75 kg NPK/ha	kg/ha
Sarpagandha	RS1	June-July	Seed (6kg/ha). Root cuttings (100 kg of root cuttings/ha), root stumps of 5 cm, hardwood stem cutting	30 × 30 cm 30:60:30 kg NPK/ha	2-2.5 t/ha at 30 months. 2.2 t/ha air dried roots from 2 years old plantation. 3.3 t/ha from 3 years old plantation.
Dioscorea	FBC-l, Arka Upkar, Pusa-1	June-July	Tuber pieces (50-70 g weight), single node stem cuttings or seed	60 × 30 to 45 cm 300:150:150 kg NPK/ha	Fresh tubers; First year: 15 to 20 t/ha, Second year: 40 to 50 t/ha
Aloe	Zanzibar, Uganda, Natal, Hyderabad aloe	June-July September-October	Root suckers or rhizome cuttings (5-6 cm length with 2-3 nodes). 10000 suckers/ha and 15-18 cm long root suckers	60 x 30 cm or 60 x 45 cm, Mixture of NPK at 150 kg/ha	1 to 15 t /ha
Phyllanthus	Navyakrit	June-July	1 kg/ha seed, 8 lakh seedlings/ha	10 × 15 cm 50:25:25 kg NPK/ha	Fresh herbage - 17.5 t/ha Dry herbage -1750 kg/ha
Creat or green chireta	Local lines	October-November and January-May	Seed: 650-750 g/ha. Transplanting of six weeks old seedlings of 10-25 cm long. Direct sown crop by broadcasting - 1.5 kg/ha. Vegetative methods: Cuttings	45 × 30 to 60 × 45 cm; 30 x 15 cm, 15 x 15 cm. 75:75:50 kg NPK/ha	3.5-4 t/ha
Medicinal solanum	Arka Sanjeevini, Arka Mahima Galxo strain, BARC strain, Pusa-1, RRL 20-2, RRL-G, SL-6	June to September or October	Seeds: 1.25 kg/ha. Transplanting: 4-5 weeks after sowing, 10-12 cm high seedlings with 6 leaves	50 x 50 cm, 75 x 75 cm and 90 x 120 cm 50:40:80 kg NPK /ha	Fresh berries: 10 t/ha. Dried berries: 2-2.5 t/ha. Solasodine content: 2.0 to 2.5%
Isabgol	Gujarat Isabgol 1, Gujarat Isabgol 2, TS-1-10, EC 124345, Niharika,	November - December to March-April	Seed: 4 - 8 kg/ha	15cm apart in beds (30-45 cm) 50:25:30 kg NPK/ ha	700-1000 kg/ha.

	Haryana Isabgol 1-5, Jawahar Isabgol - 4				
Opium poppy	Shama, Shweta, BROP 1, Kirtiman (NOP-4), Chetak (U.O.285), Jawahar Aphim - 16 (JA-16), NBRI-3, Sujatha, Shubhra	October-November	Seed rate: 7-8 kg/ha for broad-casting, 4-5 kg/ha for line sowing.	30 x 30 cm 80:50:30 kg NPK/ha	Raw opium: 25-30 kg/ha. Seeds: 400-500 kg/ha.
Safed musli	Jawahar Safed Musli 405, Raj Vijay Safed Musli 414, MDB-13, MDB-14, RC-2, RC-16, RC-36, RC-20, RC-23, RC-37, CT-1	Kharif crop: June-August	Fleshy root rhizome fingers or seeds or tissue culture plantlets. 80,000 fingers 1-1.2 t/ha; 250 – 300 kg of fleshy roots/ha.	30 x 15 cm. FYM at 20-25 t/ha, 20-40 kg P/ha	Fresh roots: 2 t/ha. Dry safed musli: 350 to 400 kg/ha
Stevia	S.R.B 123, S.R.B 512, S.R.B 128	March-May	Stem cuttings of 2-4 inches (15 cm) long, two leaf buds above ground with 2 or 3 small leaves. Transplanting - 25 to 30 days of rooting	40 x 45 cm 27.5:112.5: 112.5 kg NPK/ha	Green herb: 15 t/ha Dried leaves: 3 to 3.5 t/ha
Vallarai (Indian pennywort)	Vallabhmedha, Arka Prabhavi (CA-13), Arka Divya (CA-1)	October-November	Stem cuttings - Rooted nodes with leaves. 1,00,000 plants/ha	30 x 30 cm 100:60:60 kg NPK/ha	Fresh herb: 5.5 t/ha for single harvest. Dry herb: 2 t/ha Asiaticoside: 20 kg/ha
Brahmi	Subodhak, Pragyashakthi, CIM - Jagriti	July-August	Soft herbaceous cuttings. Plants cut into small divisions of 5-10 cm long with few leaves and nodes and planted in sunken beds.	10 x 10 cm 100:60:60 kg NPK/ha	Fresh herbage: 30 t/ha. Dry herbage: 6 t/ha. Ratoon dry herbage: 4 t/ha

			62,500 cuttings/ha		
Black nightshade	-	April - May	Seeds raised in the nursery, Transplant 30-45 days old seedling	60×60 cm 75:40:40 kg NPK/ha.	Fresh herbage: 12 to 20 t/ha
Sappan wood	-	June - July	Seed: 600g/ha. 30-45 days old seedlings are transplanted in pits of 45 cm^3.	4×4 m 100:50:50 kg NPK/ha	Inner pulp: 80 kg/tree. Pods: 2-2.5 t/ha. Seeds: 200-250 kg/ha.
Gymnema	-	June - July	Seed rate: 2-3 kg/ha.Trans-planting: 40-50 days old seedlings.	95:45:35 kg NPK/ha	Dried leaves: 5-6 kg/plant from 3-4 years old plant, Dried leaves: 10-15 t/ha. Cultivated for 10-15 years
Tulsi	Green type (Sri tulsi), Purple type (Krishna tulsi).	April - June	Seed rate: 350-450 g/ha sown in nursery beds. Seed beds: 4.5 x 1.0 x 0.2 m size. Transplanting of 6 weeks old seedlings at 4-5 leaf stage. Terminal cuttings of 8-10 nodes and 10-15 cm length with 8-10 nodes.	40 × 40 cm, 40 × 50 cm or 50 × 30 cm 120:60:60 kg NPK/ha	Fresh herbage: 10 t/ha/year. Essential oil: 0.1 to 0.23 %; 10-20 kg/ha. Irrigated tulsi: Herbage yield (up to 20 t/ha) and oil yield (up to 40kg/ha).
Noni		Summer	Seed, 20-40 cm stem cutting	4 x 4 m 3 x 3 m (High density or dwarf varieties), 50:225:50 kg NPK/ha	Fruit: 80 t/ha or more
Tippili (Long pepper)	Vellanikara -1 (Visvum tippili), Yercaud (Pl)-9	June-July and September - October	Rooted cuttings/suckers (2/pit).	Hills: 2 x 2 m Plains: 60 x 30 cm 50:20:70 kg NPK/ha.	First year: 750 kg dried fruits/ha Second year: 1500 kg dried fruits/ ha. Roots: 500 kg/ha in the third year.
Digitalis/ fox clove	EC1156	May-June for direct sowing and February for nursery	Seed rate: 5-8 kg/ha for direct sowing, 1-2 kg/ha for transplanting.	45 x 30 cm 170:30:30 kg NPK/ha.	Dried leaves: 2-3 t/ha Active principle in dried leaves: 0.44 to 0.71%.

		sowing.			
Pyrethrum	KKL-1, Hansa, Jhelum, Sel-2	April-May October-November.	Seed rate: 2-3 kg/ha. Seedlings are transplanted after 6-8 weeks.	45 x 60 cm 40 - 60 kg N, 40-50 kg P and 50-80 kg K /ha.	Dried flowers: 500 kg/ha.
Palmarosa	RRL (B)-71, RRL-82 (Jam rosa) Trishna, Motia, Sofia, Tripta, PRC-1, IW 31245, IW 3629, IW 3244, OPD-1, OPD-2	June-July	2.5 kg seeds. Slips - 28000	60 x 60 cm 60:50:40 kg/ha	Herbage - 20-30t/ha Oil - 0.2-0.3% I year - 20 kg/ha II year - 60 kg/ha III year - 70 kg/ha
Lemon grass	Pragathi, Praman, Cauveri, Krishna, CKP-25, OD-19 (Sugandhi)	Transplanting - June to July and October to November	Seed rate: 3 - 4 kg/ha. Seedlings: 30 x 30 cm with a plant density of 1, 11,000/ha. 55,600 slips/ha at 60 x 45 cm. 37, 000 slips/ha at 90 x 60 cm.	60 x 30 cm 150:60:60 kg NPK/ha/year	Herbage - 20-30 t/ha
Citronella	Java-2, Jerlab-2, Java Citronella, Java II, Ceylon Citronella, BIO-13, Mandakini, Manjusha Jor Lab C-2, Jalpallavi, Manjari.	May-July	Propagated by clumps or slips: 28,000 slips/ha	Slips: 60 x 60 cm 60 × 45 cm 100:40:40 kg NPK/ha	Herbage: 20-30 t/ha. Oil: 0.2-0.3% I year: 50-100 kg/ha. II year: 100-150 kg/ha. III year: 250-300 kg/ha
Vettiver	Pusa Hybrid-7, Hybrid-8, CIMAPKS-2, Sugandha, KH-8, KH-40, ODV-3.	June-July March-April with irrigation facility	Tillers and slips/seeds. Tillers - longer time for growing; 1,50,000 to 2,25,000 slips with 2 to 3 slips/pit	45 × 30 cm 60:22.5:22.5 kg NPK/ha	Dry root: 3 – 4.5 t/ha. 1.4 – 1.8 t/ha under North India. Oil: 15 to 16 kg/ha

	Dharini, Gulabi, Nilamber, Kesari. ODV-3, KS-1 (Bharatpur type).				
Geranium	Algerian, Reunion, IIHR Sel. 8, KKL1, Hemanti, Bipuli, Kunti, Kelkar, Ooty	June-July	Propagated by cuttings: 30,000 cuttings/ha. Terminal cuttings of 20 cm long with 8 nodes	60 × 60 cm, 60 x 45 cm 150:150:100 kg NPK/ha.	Herbage: 20-25 t/ha. Oil: 15-20 kg/ha
Davana	Jeevanrasha, Suraksha, Asha	November-December	Seeds: 1.5 kg/ha. Transplanting of seedlings having 10 cm height on 6-8 weeks from DAS.	15 × 7.5 cm 120:40:40 kg NPK/ha	Fresh herbage: 12-13 t/ha Oil content: 0.29% Oil: 12-13 kg/ha
Mint	**Japanese:** Shivalik, MAS1, MS2, Kalka Hybrid 77, Kosi **Spearmint:** MSS1 & 5, Punjab spearmint, Arka and Neera **Bergamot mint:** Kiran **Peppermint:** Kukrail	June-July	Creeping stolons or suckers or terminal cuttings. Pepper mint and bergamot mint: Runners 400 kg stolons/ha	40 × 40 cm 75:75:50 kg NPK/ha	Fresh herb: 25-30 t/ha. Oil: 100-150 kg/ha
Rosemary	Ooty-1	September-November	Stem cuttings. Seeds: 0.2 to 2.5 g/sq. m	45 x 45 cm 50,000 seedlings/ ha . 100:40:40 kg NPK/ha	12 to 15 t/ha/year of fresh herbage yielding 85 to 100 kg of oil. 2.5 t/ha of dry leaves
Thyme	Creeping Thyme, Lemon Thyme, Golden Scented	May-June	Seeds and vegetatively by cuttings or by layering of side shoots	90 × 30-45 cm for seeds 60 × 30-45 cm	Dry herb: 1.1-2.2 t/ha

	Thyme, Garden Thyme, St. Louis, Common Thyme			for cuttings 75:40:40 kg NPK/ha	
Patchouli	Singapore, Johore, Indonesia, CIM – Shresta	August-October	Rooted stem cutting	60 x 60 cm 150:50:50 kg NPK/ ha.	2 t/annum of dry leaves 50-60 kg of oil/ha
Lavender	**Bulgaria:** Kazanluk, Karlovo, Hemus, Aroma, Svezhest, Vanets. **Soviet varieties:** Steinway, Goranya, Prima, Record, French Bareme, Lambris. **High yielding clones:** AM-1, AM-2, AM-3, A-5, A-7, A-8, A9, B-2, B-11	October-November	Seed rate: 0.2-2.5 g/sq. m Vegetatively by cuttings	100:40:40 kg NPK/ha	10 t/ha of fresh herbage Lavender oil: 100 kg/ha
Eucalyptus	EC 153, EC 261, EC 3315. IFGTB EC 1, IFGTB-EC2, IFGTB EC3, IFGTB EC4	February-September	Commercially by seeds and vegetative method	75-90 x 75 cm between pits.	First year: 7 t/ha. Second year: 30 t/ha. Third year: 40 t/ha. 30 t from unirrigated crop/year

Unit X
Post-Harvest Technology of Horticultural Crops
Section I: Post-Harvest Management of Horticultural Crops

1

Importance of Post-Harvest Technology of Horticultural Crops

1.1. Introduction

- Horticulture contributes 30% GDP from 11.73% of its arable land area.
- Fruits and vegetables
    - Second largest producer
    - Immense significance to man.
    - Given place of honour on being offered to God at every festival and also mentioned in epics like Mahabharata, Ramayana and writings of Sushrutha and Charaka.
    - Rich source of carbohydrates, minerals, vitamins and dietary fibres
    - Provide variety in taste, interest and aesthetic appeal.
- Benefits kidney functions, prevention of cancer and cardiac disorders through contribution of ascorbic acid, β-carotene and non-starch polysaccharides besides biochemical constituents like phenols, flavonoids and alkaloids.
- Very high post-harvest losses in fruits and vegetables (20-40%).
- Small fraction of fruits and vegetables are utilized for processing (less than 1%) and export (Fruits-0.5% and vegetables-1.7%) compared to other countries.
- About 10-15% fresh fruits and vegetables shrivel and decay, lowering market value and consumer acceptability.
- Improper handling and storage cause physical damage due to tissue breakdown.
- Mechanical losses include bruising, cracking, cuts, microbial spoilage by fungi and bacteria.
- Physiological losses include changes in respiration, transpiration, pigments, organic acids and flavour.

1.2. Nature and causes of post-harvest losses

- Losses after harvesting - post harvest losses.
- Starts first from field, after harvest, in grading and packing areas, in storage, during transportation and in wholesale and retail markets.
- Losses due to poor facilities, lack of know how, poor management, market dysfunction or simply carelessness of farmers.

1.2.1. Important sites of post-harvest losses

- Sites where post-harvest losses
  - Farmer's field (15-20%)
  - Packaging (15-20%)
  - Transportation (30-40%)
  - Marketing (30-40%)

1.2.2. Estimated loss of fruits

Crop	Estimated loss (%)
Papaya	40-100
Grapes	27
Banana	20-28
Citrus	20-95
Avocado	43
Apple	14

1.2.3. Estimated loss of vegetables

Onion	25-40
Garlic	8-22
Potato	30-40
Tomato	5-34
Cabbage and cauliflower	7-25
Chilli	4-35
Radish	3-5
Carrot	5-9

1.2.4. Causes of post-harvest losses

- **Metabolic**
  - All fresh horticultural crops are live organs.
  - Natural process of respiration involves breakdown of food reserves and aging of organs.

- **Mechanical**
  - Owing to tender texture and high moisture content, fresh fruits and vegetables susceptible to mechanical injury.
  - Poor handling, unsuitable containers, improper packaging and transportation can easily cause bruising, cutting, breaking, impact wounding and other forms of injury.
- **Developmental**
  - Sprouting, rooting, seed germination lead to deterioration in quality and nutritional value.
- **Parasitic diseases**
  - Micro-organisms attack fresh produce easily and spread quickly, because of plenty of nutrients and moisture to support microbial growth.
- **Physiological deterioration**
  - Cells still alive after harvest and continue their physiological activity.
  - Physiological disorders due to mineral deficiency, low or high temperature injury or high humidity, enzymatic action leading to over ripeness and senescence.
- **Causes of defects in various Horticultural produce are categorised as follows**

Defects	Damage
Insect pests	Holes and misshapen
Microorganisms	Black scurf, canker, scab, blight, blisters, sooty blotch, rotten/decay
Improper cultural practices	Green spot potato
Environmental factors	Russeting, water core, discoloration, staining, dried leaves, hail damage, dull/ pale look, shriveling / wrinkling, sun burn, sun scald, superficial sun scald, lanky, torn leaves, black/brown calyx, water core, fresh rooting, splitting, cracks, natural growth cracks, water berry, scales on surface
Birds / animals	Bird damage
Delayed harvesting	Mature/over mature, fibrous, over ripe
Handling	Black edges, handling damage, packing damage, pressure damage, shatter or loose berry, damage, soft, bruises and broken
Mechanical means	Healed dark brown marks, chipped, hole, punctured skin, cuts, mechanical damage
Improper washing, cleaning, trimming and pruning	Long stalk, dirty outer leaves, secondary roots, wrapper/ extra covered leaves, unclean
Physiological disorders	Bitter pit, puffiness, cracking
Improper cold storage	Dried berries, sprouting, hollow heart
Genetic abnormalities	Misshape and double

- **Defects that are common to pre-harvest and post-harvest stages**

Pre-harvest	Post-harvest	Both
Misshaped	Black edges	Damage
Bird damage	Black/brown calyx	Dull look/ Pale look
Bitter pit	Broken	Fungal infection
Black scurf	Bruising	Hole
Blight	Chipped	Rotten/ Decay
Blister	Cuts	Black/ brown spots on surface
Bottle neck	Dirty outer leaves	Dried berries
Canker	Dried leaves	Over ripe
Cracks	Dull look	Punctured/ damaged skin
Dark/healed brown marks	Fresh rotting	
Discoloration	Handling damage	
Double	Long stalk	
Dry circular crevices	Mechanical damage	
Fly speck	Packing damage	
Green spot	Pressure damage	
Hail damage	Shattered or loose berry	
Healed brown marks	Shriveling / wrinkling	
Hollow heart (potato)	Slanky	
Insect damage and presence of insects like scales, mealy bugs etc. on the surface of the produce	Soft	
Mature	Sprouting	
Natural growth cracks	Unclean	
Pale look	Wilted	
Puffy	Wrapper/Extra covered leave	
Riciness		
Russeting		
Scab		
Scar		
Secondary roots		
Seed stem		
Sooty blotch		
Splitting		
Staining		
Sun burn		
Sun scald		
Superficial scald		
Torn leaves		
Water core		
Yellow tip		

- **Lack of market demand**
  - Inadequate market facilities, market intelligence and market information service (MIS)
  - Occurs most frequently in areas where transportation and storage facilities are inadequate.
- **Consumption**
  - Due to inadequate preservation methods at home, methods of cooking and preparation such as peeling, consumption styles etc.
- **Others**
  - Lack of clear concept of packing house operations.
  - Lack of awareness among the growers, contractors and even the policy makers.
  - Lack of infrastructure.
  - Inadequate technical support.
  - Inadequate post-harvest quality control.
  - Absence of pre-cooling and cold storage.

1.2.5. Impact of post-harvest losses

- Affect both nutritious status of population and economy of country
- **Nutrition**
  - Vitamins and minerals wasted in transit from harvest to consumer.
- **Economy**
  - Careless harvesting and rough handling cause bruise and scar skin, thus reducing quality and market price
  - Damaged produce fails to attract international buyers results in huge economic losses.
  - Government provides basic infrastructure like storage, handling, grading, packing, transport and marketing facilities and technical expertise.

1.3. Technologies for minimizing losses

- Scientific harvesting and handling to reduce losses.
- Proper storage conditions with suitable temperature and humidity to lengthen storage life.
- Training of farmers, creation of infrastructure for cold chain with common facilities for sorting, grading, packing and post-harvest treatments in all major markets.

- Technologies for extension of shelf life of fruits and vegetables
  - o **Waxing:** Protective coating helps in reduction in loss of moisture and rate of respiration and ultimately results in prolonged storage life.
  - o **Evaporative cool storage:** Best short-term storage at farm level. Reduce shriveling and extend their storage life.
  - o **Pre-packaging:** Controls rate of transpiration and respiration and keeps commodity in fresh condition both at ambient and low temperature.
  - o **Cold storage:** Used to store for long period and employ principle of maintaining a low temperature, which reduces rate of respiration and thus, delays ripening.
  - o **Modified atmosphere packaging (MAP):** Modify atmosphere composition inside package by respiration. Fruits ripen within 3 - 4 days after arrival when placed at ambient temperature.
  - o **Controlled atmosphere (CA) storage:** Higher concentration of CO_2 and lower concentration of O_2 than normal atmosphere. Reduces rate of respiration and thus, delays aging.
  - o **Cold chain:** Helps in reducing wastages and retention of quality of commodities.
  - o **Irradiation:** Exposure to ionizing radiation (such as gamma rays) at optimum dosage delays ripening, minimizes insect infestation, retards microbial spoilages, controls sprouting, and rotting of onion, garlic and potato during storage. Also used as a disinfection treatment and controls fruit fly on citrus, mango seed weevil and papaya fruit fly.
  - o **Edible coatings:** Continuous matrices prepared from edible materials such as proteins, polysaccharides and lipids. Used as film wraps and when consumed with food, become an ingredient of food. Not only minimize post-harvest losses, but also need for energy intensive operations and controlled atmosphere storage. Can control migration of gases, moisture, oil, fat, and solutes, as well as retain volatile flavouring compounds. Edible coating improves structural integrity and mechanical handling and carry product so that they help to maintain quality and inhibit microbial growth causing deterioration of product.
  - o **Others:** Facilities/ services like grading, washing, cleaning, and scientific harvesting.
  - o Cold storage facilities extended to tropical fruits and vegetables.
  - o Handling protocols established for crops other than mango, citrus, grapes and capsicum to improve shelf life and export.

- Issue relating to increasing shelf life of horticultural products needs to be addressed.
- Appropriate packaging material for export of fresh fruits, vegetables and for modified atmosphere packaging should be developed.
- Value addition needs to be viewed in wider perspective to ensure better return to producer/ farmer, besides providing better quality product to consumer.
- Development of natural food columns, fibre, single cell protein and food grade enzymes from processing wastes will be useful.

2

Structure and Composition of Fruits and Vegetables

2.1. Structure of fruits and vegetables

2.1.1. Fruit

- Fruit is a part of a flowering plant that derives from specific tissues of the flower, mainly one or more ovaries (Fig. 1). Plant product with aromatic flavors, which are either naturally sweet or normally sweetened before eating.

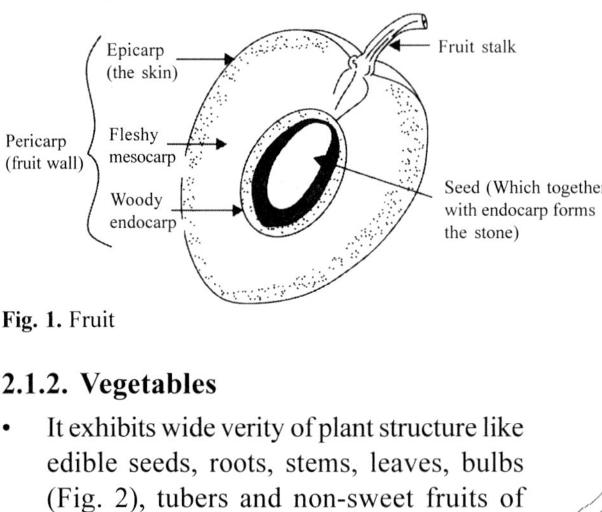

Fig. 1. Fruit

2.1.2. Vegetables

- It exhibits wide verity of plant structure like edible seeds, roots, stems, leaves, bulbs (Fig. 2), tubers and non-sweet fruits of herbaceous plants.
- Vegetables are grouped into 3 main categories
  - Seeds and pods-contain natural wax coating. Eg. Many legumes
  - Bulbs, roots and tubers with no coating, but tuber has suberisation
  - Flowers, buds, stems and leaves having low shelf life

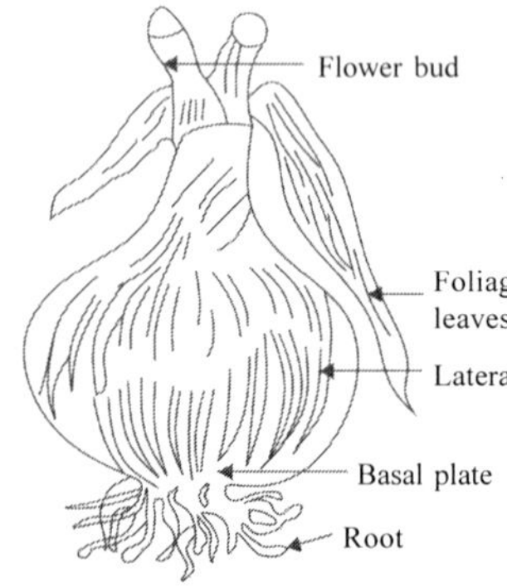

Fig. 2. Vegetable (Bulb)

- All fruits and many vegetables being storage organs or sink which are abundant in photosynthates.

2.1.3. Difference between fruit and vegetables

Fruits	Vegetables
Generally consumed as raw, not during meals (Dessert purpose)	Consumed as cooked during meals
More sweet	Less or nil sweet
Juicy, pulpy and luscious	Hard and mostly not juicy
Fruits are developed from flower / flower part/inflorescences	It can be leaf/stem/root/flower
Mostly woody perennial	Mostly non woody annuals or biennials
Mostly propagated asexually	Mostly by seeds
Fruits are acidic and are commonly called 'high acid' foods (pH < 4.5)	Vegetables are less acidic than fruits and hence, classified as 'low acid' foods (pH >4.5)
Acidity naturally controls growth of microorganisms	Microorganisms are able to grow in moist low acid products, which may lead to spoilage and the possibility of food poisoning.
The spoilage microorganisms are moulds and yeasts, which if consumed, rarely cause illness	To prevent and minimise microbes like bacteria, various methods are employed like processing
Processing may be achieved by using preservatives such as sugar, salt and vinegar and by drying, concentration or fermentation.	Processing by means of heating to destroy bacteria or by pickling, salting, or drying to inhibit bacterial growth (Possibility of transmitting food poisoning bacteria to consumers).

2.2. Composition of fruits and vegetables

2.2.1. Water

- Most of the fruits and vegetables contain 70-80% moisture, while some vegetables like leafy vegetables and melons contain almost 92-95% moisture.
- Tuber crops like cassava, yam and corms contain less moisture (around 50%) and are more starchy.
  - Eg. 95% - cucumber, lettuce, melons
  - >80% - many fruits and vegetables
  - 50% - starch tubers and seeds like yam, cassava and corn

2.2.2. Carbohydrates

- Major constituent after water, which account for 2-40% in tissues with lowest found in cucurbits and highest found in cassava.
- They occur mainly as starches and structural polysaccharides like pectins, celluloses, hemicelluloses.

- Eg. Most abundant group after water, accounts for 2-40 g 100^{-1} g
- Low in cucumber and high in cassava (20g 100^{-1} g)
- In fruits and vegetables, carbohydrates contribute mainly for its calorific value.
- Sugars constitutes major carbohydrates in fruits particularly after ripening

2.2.3. Protein

- Vegetables like brassica group contains 3-5% proteins and legumes (5g)
- Fruits and vegetables contain not more than 1-2%.
- These proteins are present mainly as enzymes.

2.2.4. Lipids

- Lipids are not more than 1% in majority of fruits and vegetables except some like avocado (20%) and olive (15%).
- In most of them, it is present as protective cuticle layer on surface.
- However, nuts contain considerable amount of fats.
- Generally, low fat levels seen in fruits and vegetables make it healthier foods to combat heart related diseases and disorders like hyper lipidaemia.

2.2.5. Minerals

- Fruits and vegetables are good sources of minerals.
- Calcium is present as calcium pectate in cell walls which appears to be linked to control of enzyme activities, respiration and ethylene production.
- Some fruits like banana are rich in potassium.

2.2.6. Vitamins

- Fat soluble vitamins A, D, E and K and water soluble vitamins C and B group are found in fruits and vegetables.
- **Vitamins and their sources**

Vitamin A	Leafy vegetables, radish tops, mango, papaya, carrots etc.
Thiamine (B1)	Fresh peas & beans, cabbage, bael, pomegranate, jamum, etc.
Riboflavin (B2)	Banana, litchi, papaya, radish top, pineapple, cowpea etc.
Niacin (B3)	Banana, strawberry, peach, cherry, green vegetables etc.
Vitamin C	Anola, guava, citrus fruits, cashew apple, leafy vegetables, green chilli, drumstick etc.
Vitamin D	Cabbage, carrot
Pyridoxine (B6)	Vegetables
Folic acid (B9)	Fresh GLV, lady's finger, cluster beans
Cyanocobalamin (B12)	Yeast, fermented foods.

2.2.7. Pigments

- Attractive colour of the many fruit is due to sugar derivates of anthocynidins.
- At the time of ripening, loss of chlorophyll and accompanied by synthesis of anthocyanins or carotenoids which present in vacuole and chloroplast respectively.
  - Anthocyanins – gives colour from red to blue
  - Carotenoids are synthesized in green tissue. Eg. beta-carotene and lycopene

2.2.8. Phenolics and antioxidants

- Comprising of anthocyanins, leucoanthocyanins, anthoxanthins, hydroxy benzoic acids, glycosides, sugar esters of quinic and shikimic acids, esters of hydroxy cinnamic acids and coumarin derivatives
- Phenols are impotent in determining the colour and flavour of the fruit. Phenols are by products of the metabolism of the amino acids and contribute the sensory qualities of the fruits (colour, astringency, bitterness and aroma) and play the vital role in the resistance to attack of pathogen and stress. It is known for its antioxidant activities.

2.2.9. Organic acids

- Imparts taste and flavour.
- Major acids are malic (apple), citric (citrus), tartaric (grape), quinic, succinic and shikimic acids.
- Organic acids play important role in photosynthesis and respiration, synthesis of phenolic compounds, lipids and volatile aromas.

2.2.10. Volatiles (Flavour) compounds

- Important in producing characteristic flavor and aroma.
- Compounds are esters, alcohols, acids, aldehydes and ketones.
- Ethanol is common to all fruits and vegetables, where as others are specific.
- Esters present in ripe fruits.
- Sulphur is present in *Brassica* sp and tomato.

2.2.11. Texture

- Governed by structural polysaccharides.

3

Maturity Indices and Harvesting Methods in Horticultural Crops

3.1. Maturity

- During process of maturation, fruit receives regular supply of food material from plant.
- When mature, abscission or corky layer forms at stern end stops inflow.
- Afterwards, fruit depends on own reserves, carbohydrates dehydrated and sugars accumulate until sugar acid ratio form.
- In addition, typical flavour and characteristic colour also develop.
- Stage of maturity at the time of picking influence storage life and quality of fruit
- Fruits harvested over mature or full ripe are easily susceptible to microbial and physiological spoilage and storage life reduction.
- Necessary to pick up fruits or vegetables at correct stage of maturity to facilitate proper ripening, distant transportation and maximum storage life.
- **Principles of harvest maturity**
  - Harvested commodity should have its peak acceptable quality when it reaches the consumer.
  - Produce should develop an acceptable flavour or appearance.
  - Produce should have optimum size and shape required by the market.
  - It should not be toxic
  - Harvest maturity should have adequate shelf life.

3.1.1. Horticultural maturity

- Developmental stage of fruit on tree, which will result in satisfactory product after harvest.
- Eg. Fresh vegetables for canning/ dehydration/ IQF-Individual Quick Frozen/ harvesting for local or distant market
- **Horticulture maturity is classified into 3 different groups**

- Physiological immature
- Firm and mature
- Harvest ripe

  i) **Physiological immature**: Vegetables such as cucumber, peas, beans, carrot, beetroot, baby corn and bhendi are harvested when they are tender, crisp and fibre free.

  ii) **Firm and mature**: Fruits and vegetables which attain characteristic size, shape and maturity are harvested.

  iii) **Harvest ripe**: In non-climacteric fruit, maturity is referred as full ripening. Eg. Citrus, pineapple, grape and cherry.

- **Judging the maturity in fruits crops**
  - Culinary maturity
  - Dessert maturity
  - Shipping maturity
  - Processing maturity

  i) **Culinary maturity**: Fruits like papaya, jack fruit, tomato, fig and bread fruits when used as vegetable, harvesting is done at immature/ suitable stage.

  ii) **Dessert maturity**: For local market and fresh consumption. Eg. Jack fruit, watermelon, mango and orange.

  iii) **Shipping maturity**: For long distance transportation, fruits will be harvested much earlier than for local consumption (before ripening) which prolongs shelf life. Eg. Mango and banana.

  iv) **Processing maturity**: For processing, harvest time depends on the distance of orchard from the processing units and the type of fruit/vegetable and product to be prepared.

3.1.2. Physiological maturity

- Stage in development of fruits and vegetables when maximum growth and maturation have occurred.
- Usually associated with full ripening in fruits.
- Physiological mature stage is followed by senescence
- Eg.: Fruits and vegetables produced for seed production

3.1.3. Commercial maturity

- State of plant organ required by market.
- Commonly bears little relation to physiological maturity.

3.1.4. Harvestable maturity

- Defined in terms of physiological maturity and horticultural maturity
- Stage which will allow fruits/vegetables at peak condition when it reaches to consumers and develop acceptable flavour or appearance and adequate shelf life.

3.2. Criteria of maturity for harvesting of fruits and vegetables

Fruit	Physical	Chemical
Mango	Olive green colour with clear lenticels, shoulder development size, specific gravity, days from fruit set.	Starch content, flesh colour
Banana	Skin colour, drying of leaves, brittleness of floral ends, angularity of fruit, days from emergence of inflorescence	Pulp/peel ratio, starch content
Citrus	Colour break of skin from green to orange, size	Sugar/acid ratio, TSS
Grapes	Peel colour, easy separation of berries, characteristic aroma	TSS: 12-18% for Thompson seedless,12-14% for Bangalore Blue, 14-16% for Anab-e-shahi
Apple	Colour, size	Firmness measured by pressure tester/ penetrometer
Papaya	Yellow patch or streaks	Jelliness of seed, seed colour

3.2.1. TSS content of fruits

Fruit	TSS (%)
Apple	11.50-14.50
Citrus	12-14
Grapes	12-20
Mango	12-18
Papaya	11-12
Pineapple	13.00
Kiwi	8.00
Pear	12.92-12.99

3.2.2. Maturity indices of vegetable crops

Vegetables	Maturity indices
Root, bulb and tuber crops	
Radish and carrot	Large enough and crispy
Potato, onion and garlic	Tops beginning to dry and topple clown
Yams, bean and ginger	Large enough
Fruit vegetables	
Cowpea, snap bean, sweet pea, winged bean	Well filled pods that snap readily

Lima bean and pigeon pea	Well filled pods that are beginning to lose their greenness.
Okra	Desirable size reached and the tips of which can be snapped readily
Snake gourd	Desirable size reached and thumbnail can still penetrate flesh readily
Eggplant, bitter gourd, slicing cucumber	Desirable size reached, but still tender
Tomato	Seeds slipping when fruit is cut, or green colour turning pink
Muskmelon	Easily separated from vine with a slight twist leaving clean cavity (full slip stage).
Watermelon	Dull hollow sound when thumped

3.2.3. Time taken from pollination to horticultural maturity

Vegetables	Time to harvest maturity (days)
Ridge gourd	5-6
Squash	7-8
Brinjal	25-40
Okra	4-6
Pepper (green stage)	45-55
Pepper (red stage)	60-70
Pumpkin (mature)	65-70
Tomato (mature green)	35-45
Tomato (red ripe stage)	45-60
Peas	30-35

- Skin colour, shape, size and flavour, abscission and firmness are used to assess maturity

3.2.4. Skin colour

- Loss of green colour in citrus, red colour in tomato.

3.2.5. Shape, size and flavour

- Sweet corn is harvested at immature stage, smaller cobs marketed as baby corn.
- Okra and cowpea are harvested at mature stage (Pre fibre stage).
- In chilli, bottle gourd, bitter gourd and cluster bean, maturity is related to size.
- Cabbage head and cauliflower curd are harvested before unpleasant flavour.

3.2.6. Abscission and firmness

- Musk melon harvested at formation of abscission layer (Fig. 1).
- Cabbage and lettuce are harvested at firmness stage.

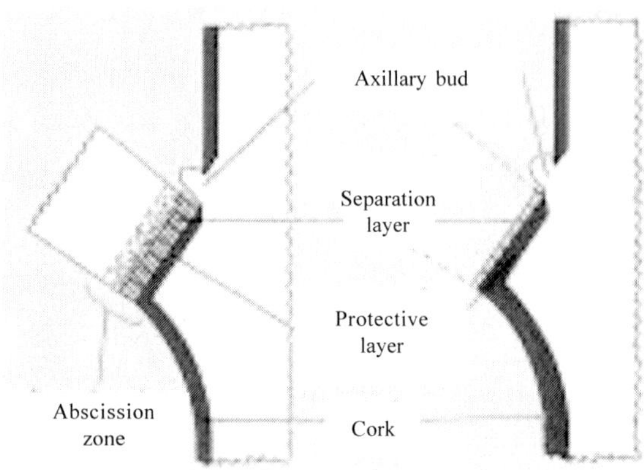

Fig. 1. Abscission layer

3.3. Factors affecting maturity

3.3.1. Temperature

- Higher temperature gives early maturity.
- Sun scorched fruits characterized by chlorophyll loss, yellowing, disappearance of starch and other alcohol insoluble material, increase in TSS content, decrease in acidity and softening.

3.3.2. Soil

- Soil on which fruit tree is grown affects time of maturity.

3.3.3. Planting material

- Propagated fruits affect fruit maturity.

3.3.4. Closer spacing

- Hastened maturity in hill bananas.

3.3.5. Pruning intensity

- Enhanced maturity of Flordasun and Sharbati peaches.

3.3.6. Girdling

- Process of constricting periphery of stem blocks downward translocation of CHO, hormones, etc.

3.4. Maturity indices

- Judged by visual means (colour, size, and shape), physical means (firmness, softness), chemical analysis (sugar content, acid content), computation (heat unit and bloom to harvest period), and physiological method (respiration).
- **Visual indices**
  - Most convenient index.
  - Drying of top leaves in banana
  - Yellowing of last leaf of peduncle in jackfruit.
  - Papaya - latex becomes almost watery.
  - Banana and sapota - floral ends become more brittle and shed with gentle touch or even on own.
  - Sapota - brown scurf on fruit skin starts propping.
  - In mango, lenticels become more prominent and waxy bloom gradually disappears.
  - Grapes - develop translucent bloom.
- **Seed development**
  - Also used as index of fruit maturity
  - Endocarp hardening for stone and fibre development for dessert in mango.
- **Start of bud damage**
  - Used as index of fruit maturity in mango.
- **Calendar date**
  - Time of flowering depends on temperature and variation in days from flowering to harvest
- **Heat units**
  - Heat requirement for fruit growth and development calculated in terms of degree days
  - Maturity at higher temperature is faster as heat requirement is met earlier.
  - Helps in planning, planting, harvesting and factory programmes for corn, peas and tomato for processing.

3.4.1. Maturity indices for fruits

- **Banana**
  - Harvested when ridges on surface of skin change from angularity to round i.e. after attainment of 3% full stages.

  - Dwarf banana: ready for harvest within 11-14 months after planting
  - Tall cultivars: 14-16 months to harvest.
  - Peel colour from dark green to light green
  - Remaining style ends dry, brittle and fruits less angular in shape.
- **Guava**
  - TSS acid ratio, specific gravity and colour determine maturity
  - Eg. Allahabad safeda - 35.81
  - Apple colour guava - 26.39
  - Chittidar guava - 28.13
  - Lucknow - 49-34.25
  - Specific gravity - Less than I
  - Colour - Light green to yellow
- **Ber**
  - Maturity judged by colour (Yellow), specific gravity (Less than 1) and TSS
- **Pomegranate**
  - Sugar 12-16% and acid 1.5-2.5%
  - Variety Ganesh - harvest when seed colour becomes pink
  - TSS 12.5% and sugar acid ratio 19.5%
- **Bael**
  - Fruits ripen after one year of flowering
- **Mango**
  - When one or two mangoes ripen on tree fall on ground on own accord
  - TSS 10-14%.
  - Harvesting criteria for mango (Fig. 2)

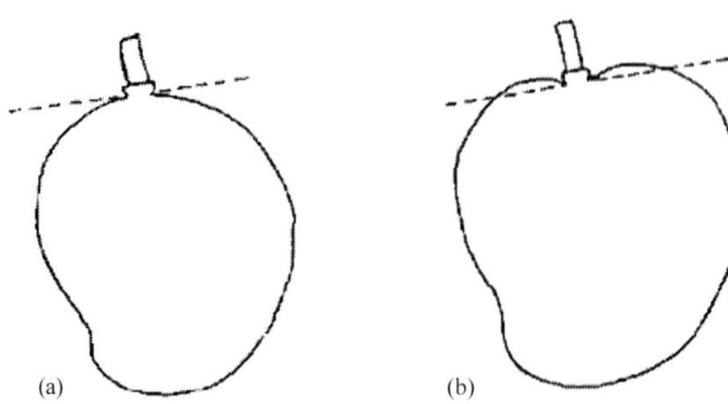

Fig. 2. Harvesting index in mango a) Immature, b) Mature-Fruit with marked shoulders

3.5. Harvesting

3.5.1. Hand harvesting

- People accurately determine product quality, allowing accurate selection of mature product.
- Particularly, important for crops having wide range of maturity and harvested several times during season.
- Properly trained workers pick and handle product with minimum of damage.
- Many fresh market products have short shelf life, if bruised or damaged during harvest and handling.
- Requires a minimum of capital investment.
- Main problem - labour management.

3.5.2. Mechanical harvesting

- Used for fresh market crops (Roots, tubers, or rhizomes) and for nut crops.
- Vegetables grown below ground (Radishes, potatoes, garlic, carrots, beets and others) - harvested only once and soil used to cushion product from machine caused mechanical injury.
- Tree nuts and peanuts are protected by shell and easily withstand mechanical handling.
- Tomatoes, wine grapes, beans, peas, prunes, peaches and some leafy green vegetables are machine harvested because damage does not significantly affect quality of processed product.
- Use of tree nut harvesters
  - Reduces management problems associated with workers.
  - Will not be feasible until crop or production techniques can be modified to allow one time harvest.
  - Causes damage (Bark damage from tree shaker)
  - Quite expensive

3.6. Post-harvest handling

- Fruits and vegetables continue to respire even after harvesting.
- In addition to degradation of respiratory substrates, number of changes in taste, colour, flavour, texture and appearance which make unacceptable for consumption by consumers.
- It combines biological and environmental factors during process of value addition of commodity.

3.6.1. Pre cooling

- Removal of field heat from commodities.
- Reduces bruise damage from vibration during transit.
- Methods: Cold air (Room cooling, forced air cooling), cold water (Hydro cooling), direct contact with ice (Contact icing), evaporation of water from produce (Evaporative cooling, vacuum cooling) and combination of vacuum and hydro cooling (Hydrovac cooling).
- **Commodity wise cooling methods**

Cooling methods	Commodities
Room cooling	All fruits and vegetables
Forced air cooling	Fruits and fruits type vegetables, tubers and cauliflowers
Hydro cooling	Stems, green leafy vegetables, fruits and fruit type vegetables
Package icing	Roots, stems, cauliflowers, green onion, brussel sprouts
Vacuum cooling	Stems, Leafy and flowers type vegetables

- **Difference between forced air cooling and vacuum cooling**

Forced air cooling	Vacuum cooling
Air passes over the surface of crop, cooling the outside while the inside is cooled by heat transfer from inside to the outside for the crop.	In cooling chamber, pressure (reduced) is exactly the same around the produce and in the centre of the produce. This means the cooling is very even and quick throughout the crop.

- **Crops usually pre-cooled by forced air**

Anona	Citrus	Litchi	Plantain
Avocado	Coconut	Mango	Pomegranate
Banana	Cucumber	Mangosteen	Prickly pear
Barbados cherry	Eggplant	Melons	Pumpkin
Berries	Fig	Mushroom	Rhubarb
Breadfruit	Ginger	Okra	Sapota
Brussels sprouts	Grape	Orange	Snap beans
Cactus leaves	Grape fruit	Papaya	Strawberry
Capsicum	Guava	Passion fruit	Summer squash
Carambola	Kiwi fruit	Persimmon	Tomato
Cassava	Kumquat	Peas	Tree tomato
Cherimoya	Lima bean	Pineapple	Yam

- **Crops suitable for ice cooling (Fig. 3)**

Belgian endive	Chinese cabbage
Broccoli	Carrot
Brussels sprouts	Green onions
Cantaloupe	Kohlrabi
Leek	Sweet corn
Parsley	Pea
Spinach	

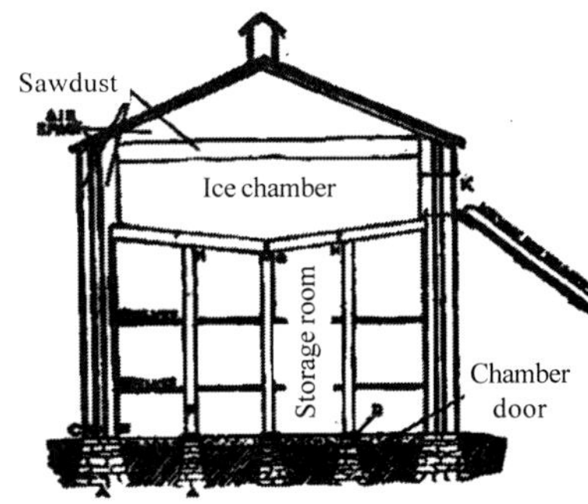

Fig. 3. Ice cooling

- **Crops that can be vacuum cooled (Fig. 4)**

Belgian endive	Celery
Brussels sprouts	Escarole
Carrot	Leek
Cauliflower	Lettuce
Chinese cabbage	Lima bean
Mushroom	Swiss chard
Snap bean	Watercress
Peas	Spinach
Sweet corn	

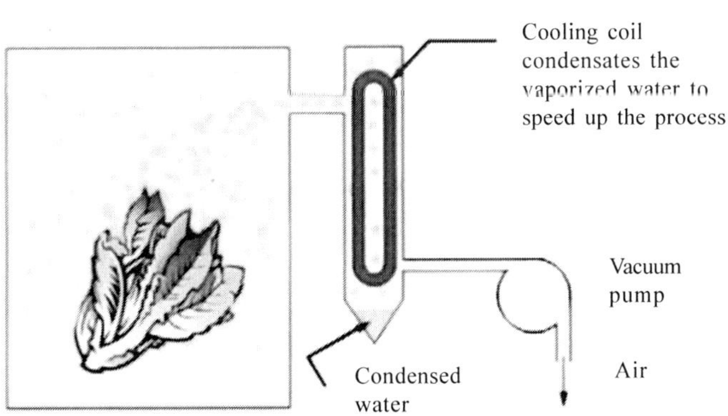

Fig. 4. Vaccuum cooling

- Nutrients/growth regulators/ fungicides are mixed with water used in hydro cooling to prolong shelf life by improving nutrient status of crop and preventing spread of post-harvest diseases.

- **Crops normally hydro cooled (Fig. 5)**

Artichoke	Carrot	Kiwi fruit	Potato (early)
Asparagus	Cassava	Kohlrabi	Pomegranate
Beet	Celery	Leek	Radish
Belgian endive	Chinese cabbage	Lima bean	Rhubarb
Broccoli	Cucumber	Orange	Snap bean
Brussels sprout	Eggplant	Parsley	Spinach
Cantaloupe	Green onions	Parsnip	Summer squash
Cauliflower	J. artichoke	Peas	Sweet corn

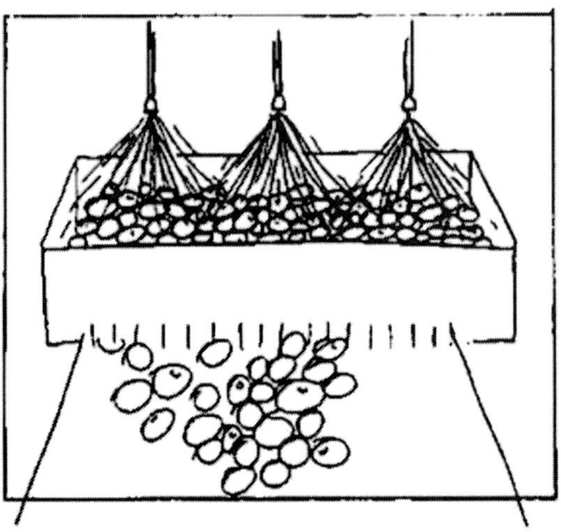

Fig. 5. Hydro cooling

3.6.2. Washing, cleaning and trimming

- Removal of soil dust, adhering debris, insects and spray residues.
- Chlorine in fresh water is used as disinfectant to wash commodity.
- Fungicides like diphenyl amine (0.1-0.25%) or ethoxyquin (0.2-0.5%) used to control disorders.
- Clean fruit type vegetables (Melon, brinjal, tomato, cucumber) by wiping with damp cloth.
- Many vegetables need trimming, cutting and removal of unsightly leaves or other vegetative parts.

3.6.3. Sorting, grading and sizing

- Done by hand to remove damaged produces by insects, diseases or mechanical injuries which are unsuitable to market or storage.

- Separated into two or more grades on basis of surface colour, shape or visible defects.
- Apples in India are graded in to extra fancy, fancy and standard.
- After sorting and grading, sizing is done either by hand or machine.
- Machine sizers work on two basic principles (Weight and diameter).
- Sizing based on shape and size is most effective for spherical (Oranges, tomato, certain apple cultivars) and elongated (Delicious apples and European pears of non-uniform shape) commodities, respectively

3.6.4. Curing

- Effective operation to reduce water loss during storage from hardy vegetables *viz.*, onion, garlic, sweet potato and other tropical root vegetables.
- Curing methods employed for root crops entirely different than bulbous crops (Onions and garlic).
- Helps in healing of harvest injuries, reduces loss of water and prevents infection by decay pathogens.
- Onions and garlic cured to dry necks and outer scales.

3.6.5. Waxing

- Coating applied to surface of fruit called waxes
- Waxes - esters of higher fatty acid with monohydric alcohols and hydrocarbons and free fatty acids.
- Paraffin wax, carnauba wax, bee wax, shellac, wood resins and polyethylene waxe are used commercially.
- Incorporated with fungicides, growth regulators/ retardants, preservative for reducing microbial spoilage, sprout inhibition etc.
- Does not improve quality of any inferior horticulture product.
- **Advantages**
  - Improved appearance of fruit
  - Reduced moisture losses and retards wilting and shrivelling during storage of fruits.
  - Less spoilage specially due to chilling injury and browning.
  - Creates diffusion barrier as a result of which it reduces the availability of O_2 to the tissues thereby reducing respiration rate
  - Protects fruits from microbiological infection
  - Considered a cost effective substitute when refrigerated storage is unaffordable.

- Used as carriers for sprout inhibitors, growth regulators and preservatives.

- **Disadvantage**
  - Development of off - flavour, if not applied properly.
  - Adverse flavour changes attributed to inhibition of O_2 and CO_2 exchange resulting in anaerobic respiration and elevated ethanol and acetaldehyde contents.

3.6.6. Packaging

- Proper or scientific packaging houses (Fig. 6) are needed to reduce wastage of commodities by protecting from mechanical damage, pilferage, dirt, moisture loss and other undesirable physiological changes and pathological deterioration during course of storage, transportation and subsequent marketing.
- Providing uniform quality to packed produce provided commodities are carefully supervised and sorted prior to packaging.
- Helps in maintaining produces against hazards of journey.

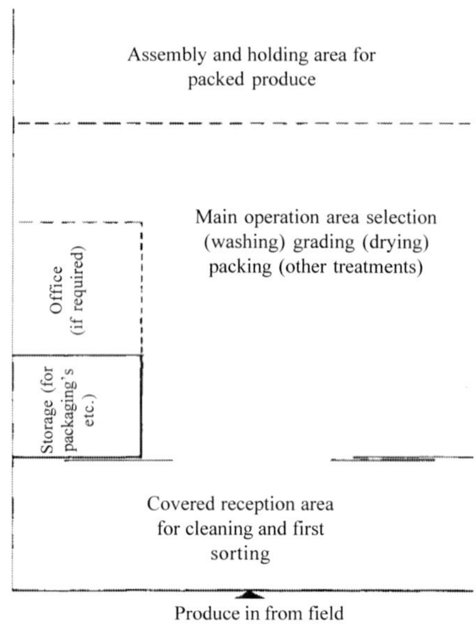

Fig. 6. Lay out of a small packaging house

- Gunny bags, grasses and stem leaves are replaced by wooden boxes, baskets woven from bamboo or twigs, sack/jute bags and corrugated fibre board (CFB) boxes (Fig. 7).

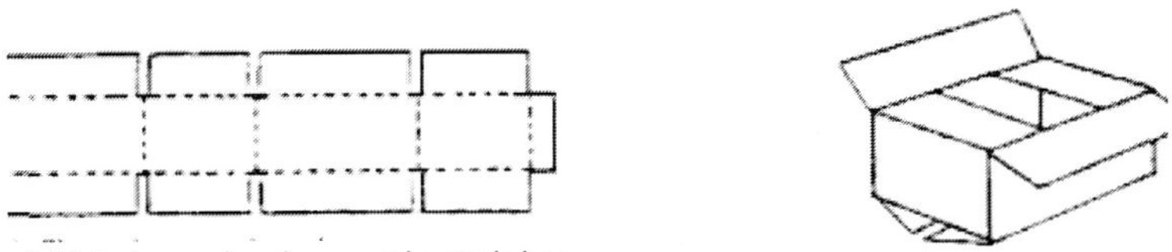

a) Slotted, one-piece box must be stapled or glued shut

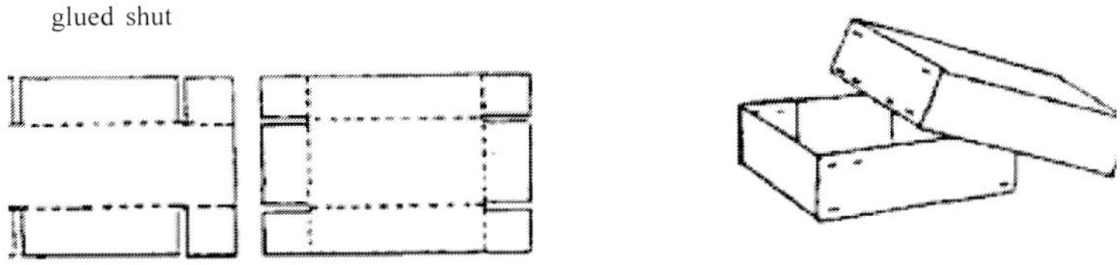

b) Telescopic box has top and bottom to be glued or stapled in assembly

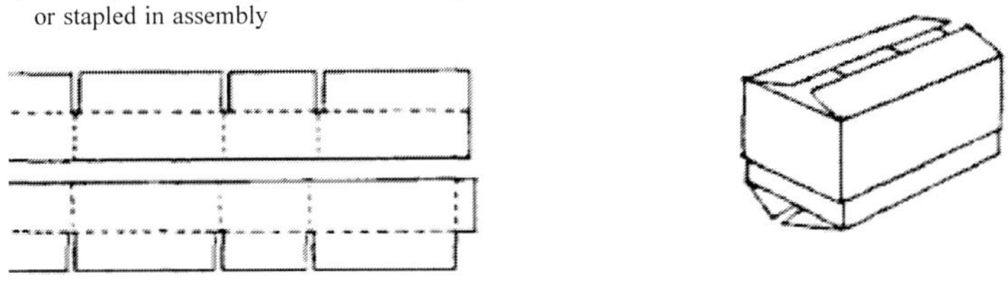

c) Deeper telescopic box takes bigger load of less fragile produce

d)

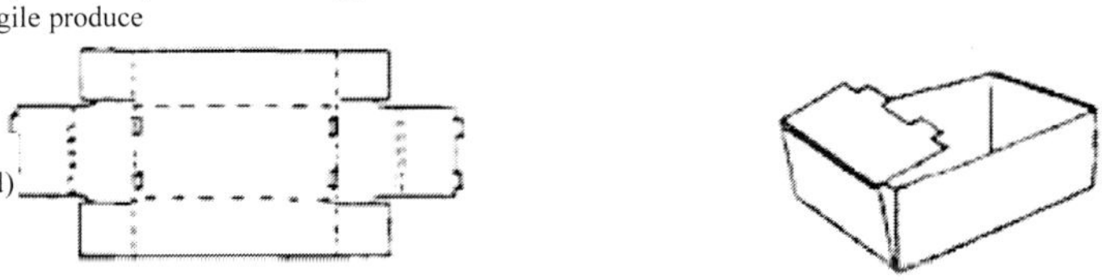

e)

d) e) These are one piece trays made rigid by engaging tabs in slots

Fig. 7. Packaging boxes (Cardboards cartons)

3.6.7. Storage

- Storage techniques are ground storage, ambient storage, refrigerated storage, air cooled storage, zero energy storage, modified atmospheric storage, hypobaric storage and controlled atmospheric storage.

4

Pre-Harvest Factors Affecting Post-Harvest Life of Fruits and Vegetables

4.1. Quality of post-harvest product

- Post-harvest quality - market, edible, transport, table, nutritional, internal and appearance qualities
- Consumers consider good quality in relation to colour, flavour and nutrition.
- Final manifestation of inter relation between commodity and its environment.
- Genetic characteristics and physiological status of commodity determine typical post-harvest behaviour and quality of the produce
- Temperature, RH, water potential, light, cultural practices and pest management techniques determine the quality.

4.2. Pre-harvest factors

4.2.1. Related to plants

- **Crops:** Quality varies from crop to crop E.g. jackfruit, bael, potato, onion, pumpkin, garlic etc. having good quality in relation to shelf life
- Apple, mango, cherry, strawberry, tomato, capsicum, okra, Brussels sprout, Chinese cabbage, carrot, radish attract more consumers due to attractive appearance.
- **Cultivars:** Quality of seed or planting material is an important factor that controls quality of fruits and vegetables.

4.2.2. Related to cultural practices

- **Planting period:** Many plants are very sensitive to environmental conditions and quality is not optimized when crop produced under adverse conditions.
- **Planting density:** High density planting increases competition between plants, reduces light availability and decrease quantity.
- Low density planting produces large size and better coloured fruit.
- Larger fruits are more sensitive to physiological disorders.

- **Irrigation:** Irregular watering usually reduces fruit size, increases splitting, physiological disorders, reduces water content in plant.
- Moisture stress increases the rate of transpiration over the rate of absorption and irregular irrigation/ moisture regime leads to fruits/vegetable cracking (potato and pomegranate cracking).
- **Fertilization:** Poor management of fertilizers increases physiological disorders due to deficiencies or toxicity.
- **Nitrogen**
    - High N fertilization reduces, while moderate to high K improves post-harvest life and quality of many horticultural produces.
    - Application of K in watermelon tends to decrease the respiration.
    - High levels on N tend to decrease flavor, TSS, firmness and color of the fruit and in stone fruits, it increases physiological disorders and decrease fruit colour.
    - Generally, crops that have high levels of nitrogen typically have poorer keeping qualities than those with lower levels. High nitrogen increases fruit respiration, faster tissue deterioration thereby reducing their storage life.
- **Phosphorous**
    - Application of phosphorous minimizes weight loss, sprouting and rotting in bulb crops compared with lesser application.
    - Phosphorous nutrition can alter the post-harvest physiology of some produce by affecting membrane lipid chemistry, membrane integrity and respiratory metabolism.
    - Respiration rate of low phosphorous fruits will be higher than that of high phosphorous fruits during storage.
- **Potassium**
    - K improves keeping quality, its deficiency can bring about abnormal ripening of fruits and vegetables.
    - Potassium helps in reducing some physiological storage disorders. E.g. superficial rind pitting in oranges.
- **Calcium**
    - Ca deficiency in fruits and vegetables reduce its storage life.
    - Calcium treatment delays ripening, senescence, reduces susceptibility to chilling injury, increase firmness and reduces decay subsequent to storage in avocados and improves the quality.

- Physiological disorders of storage organs related to low Ca content of the tissue are:
    i) Bitter pit in apples
    ii) Cork spot in pears
    iii) Blossom end rot in tomato
    iv) Tip burn in lettuce and hallow heart in potato
    v) Red blotch of lemons

- **Pruning:** Reduces load and increases growth and chemical use after harvest
- **Thinning:** Reduces competition between plants and promotes good balance between vegetative and fruit parts, and improves quality. Increases fruit size, but reduces total yield.
- **Protection:** Pathogens and insects have negative effect on quality.
- Poor management led to very poor quality and reduced yield.

4.2.3. Related to environments

- **Temperature**: Affects quality due to very low or very high temperature
- Wind and rain cause negative effects

4.2.4. Related to chemicals

- Hormones and growth regulators affect quality
- Pre harvest application chemicals like CCC on onion field prevent them sprouting during storage.

4.3. During harvest factors

4.3.1. Season

- Greatly influenced by season. E.g. winter harvest have more shelf life
- Off season fruits and vegetables give more remunerative price.
- Harvesting during or immediately after rains favours multiplication of microorganisms.
- Citrus become susceptible to damage, if harvested during rains as rind becomes turgid and prone to easy bruising, sun scald etc.
- If long days, onion (temperate) grown during short day (tropics) condition it leads to very poor storage quality.

4.3.2. Time of harvest

- Harvest when temperature is mild.
- Higher temperature leads to faster respiration.

- Morning harvest prefers local market because of freshness.
- Evening harvest is preferred for distant market due to higher accumulation of reserved carbohydrates and less amount of moisture that give better quality of produce to consumer.
- Leafy vegetables harvested in late morning or late afternoon break less easily and leaves are more resistant to tearing because of loss of water through transpiration

4.3.3. Method of harvesting

- Method of harvesting (hand vs mechanical) can also have significant impact on the composition and post-harvest quality of fruits and vegetables.
- Sharp tools/ secateur /harvester/hand gloves/digger/vibrater/ trimmer/ any such items should always be used to detach the fruits/vegetable/flowers from the mother plant (Fig.1).

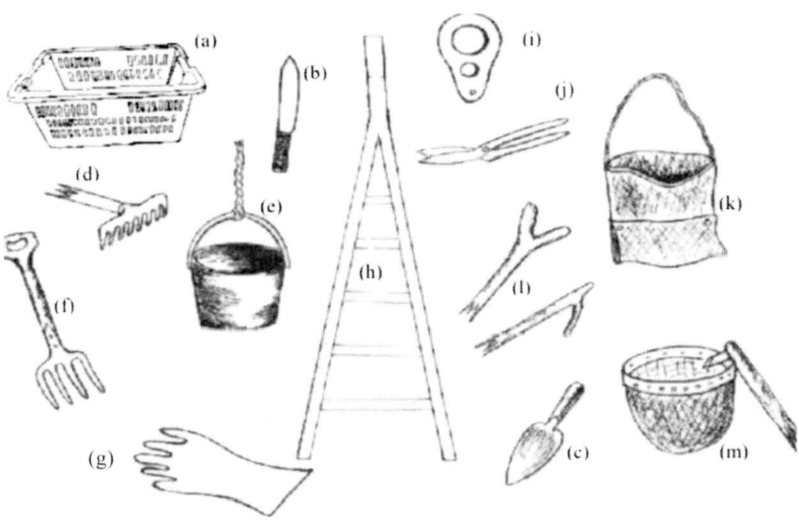

Fig. 1. Harvesting tools: (a) Plastic field box, (b) Knife, (c) Spade, (d) Rake, (e) Bucket, (f) Fork, (g) Glove, (h) Ladder, (i) Sizer (j) Scissors, (k) Bag, (l) Fork sticks, (m) Pole with collection bag.

- Bruises or injuries during harvesting results in black or brown patches making unattractive.
- Latex coming out of stem in mango creates a black spot on fruits.
- Injury to peel become entry point for microorganisms, causing rotting.

4.3.4. Stage of harvesting

- Harvest at right stage of maturity.

- Poor quality at harvest leads to rapid deterioration due to harvesting of immature produce.
- Immature or over mature vegetables have poor keeping quality, susceptible to decay, and deteriorates fast.

4.3.5. Consumer demand

- Harvest time and maturity are altered by consumers demand which affect quality of produce.

4.4. Post-harvest factors

4.4.1. Curing

- Curing is done immediately after harvesting to strengthen skin.
- Induced at higher temperature and humidity, involving suberization of outer tissues followed by development of wound periderm acts as effective barrier against infection and water loss.
- Potato, sweet potato, colocasia, onion and garlic cured prior to storage or marketing.
- Reduces moisture especially in onion and garlic and protects from microbial infection in storage

4.4.2. Degreening

- Process of decomposing green pigment (Chlorophyll) to fruit.
- Carried out in special treating rooms with controlled temperature and humidity in which low concentration of ethylene (20 ppm) is applied.
- Applicable to banana, citrus and tomato.

4.4.3. Pre cooling

- High temperature is detrimental to keeping quality when harvesting is done during hot days.
- Removing of field heat.
- Slows down rate of respiration, minimizes attack of microorganisms, and reduces water loss.
- Peas and okra deteriorate fast that need prompt pre cooling.

4.4.4. Washing and drying

- Washing after harvesting improves appearance, prevents wilting and removes primary inoculum load of microorganism.
- Fungicide/bactericide is used in washing water.
- Improves shelf life of banana by delaying ripening.

4.4.5. Sorting and grading

- Done on basis of size and colour while grading is performed as per defect or on basis of marketable and unmarketable produces.

4.4.6. Disinfection

- Papaya, mango, melon and other fruits are susceptible to fruit fly attack.
- Disinfect either by vapour heat treatment (VHT) at 43°C with saturated air with water vapour for 6-8 h or by ethylene dibromide fumigation.
- **List of post-harvest diseases of fruits**

Fruit	Disease	Casual organism
Banana	Crown rot	*Acremonium* sp, *Curvularia* sp, *Colletotrichum musae*, *Fusarium semitectum*, *Verticillium* sp
Ber	Fruit rot	*Alternaria* sp, *Phomopsis* sp, *Colletotrichum* sp
Citrus	Black rot	*Alternaria citri*
	Grey mold	*Botrytis cinerea*
	Green mold	*Penicillium digitatum*
	Stem-end rot	*Diaporthe citri, D. medusae, D. natalensis*
Guava	Anthracnose	*Colletotrichum gloeosporioides*
Kiwi fruit	Stem rot	*Botrytis cinerea*
	Ripe rots	*Botryosphaeria dithodea*
Litchi	Skin injuries	*Aspergillus* sp., *Penicillium* sp., *Rhizopus* sp
Mango	Anthracnose	*Colletotrichum gloeosporioides*
	Stem-end rot	*Botryodiplodia theobromae*
	Alternaria rot	*Alternaria alternata*
	Black mold	*Aspergillus niger*
Papaya	Chocolate spot	*C. gloeosporioides*
	Wet rot	*Phomopsis* sp
	Dry rot	*Mycosphaerella* sp.
	Alternaria spot	*Alternaria alternata*
	Fusarium rot	*Fusarium solani*
	Internal yellowing	*Enterobacter clocae*
	Anthracnose	*C. gloeosporioides, C. dematium, C. capsici, C. circinans, C. papaya*
Pear	Blossom end rot	*Alternaria* sp., *Botrytis* sp., *Penicillium* sp.
Pineapple	Black rot	*Chalara paradoxa*
	Fruitlet core rot	*Penicillium funiculosum, Fusarium moniliforme, Candida guilliermondi*
Pomegranate	Heart rot	*Aspergillus niger, Alternaria* sp. *Aspergillus* spp
Stone fruits	Brown rot	*Monilinia fructicola*
Onion	Black mold	*Aspergillus* spp
	Neck rot	*Botrytis allii*

- **List of insect pests affecting post-harvest quality**

Crop	Pest name	Scientific name
Fruit flies		
Cucurbits	Lesser pumpkin fly	*Dacus ciliatus*
Cucurbits and tomato	Melon fly	*D. cucurbitae*
Most fleshy fruits and vegetables	Oriental fruit fly	*D.dorsalis*
Mites		
Leafy vegetables	Red legged earth mite	*Halotydeus destructor*
Apple and other deciduous fruits	European red mite	*Panonchus ulmi*
Potato, tomato, brinjal	Potato tuber moth	*Phthorimaea operculella*
Mealy bugs		
Citrus, grape	Citrus mealy bug	*Planococcus citri*
Pineapple	Pineapple mealy bug	*Dysmicoccus bevipes*
Moths		
Apple, pear, peach, quince, prunus, walnut	Codling moth	*Cydia pomonella*
Legumes	Bean pod borer	*Maruca testulalis*
Scale insects		
Citrus	Red scale	*Aonidiella aurantii*
	Purple scale	*Lepidosaphes beckii*
Deciduous fruits	San Jose scale	*Quadraspidiotus preniciosus*
Weevils		
Sweet potato	Sweet potato weevil	*Cylas formicarius*
Mango	Mango seed weevil	*Sternochaetus mangiferae*

- **Chemicals used to control spoilage and quality in fruit and vegetables**

Crop	Chemicals
Apple	Sodium-phenyl phenate
Banana	Thio bendazole, Benomyl
Citrus	Sodium carbonate, Borax, SOPP, Biphenyl, 2,4-D, N Cl 3 fumigation
Mango	Hot water, Benomyl
Grapes	SO_2 fumigation
Papaya	Hot water
Pomegranate	Ethyl oleate
Potato	Hypochlorite
Carrot and cabbage	Thiobendazole, Benomyl
Onion	Benomyl
Sweet potato and tomato	2,6-dichloro-4-nitroaniline

4.4.7. Waxing

- Extra discontinuous layer of wax applied artificially with sufficient thickness and consistency to prevent anaerobic condition within fruits provides necessary protection against decay organism.
- Improves appearance and glossiness, making more acceptable.
- **List of commercial waxes**

Waxes	
Shellac	
Carnauba wax	
Bee wax	
Polyethylene	12% used in Israel for mango
Wood resins	
Paraffin wax	

- **Methods of wax application**
- **Spray waxing**
  - Fruits and vegetables which move on the roller conveyor are sprayed with water wax emulsion.
  - Waxed produce is dried in a current of air at 55°C.
  - Two types of spray waxing namely low pressure spraying and high pressure atomizing.
- **Dipping**
  - Fruits are dipped in water wax emulsion of required concentration for 30-60 seconds
  - Fruits or vegetables could be waxed by keeping them in wire boxes holding about 100 fruits (30 kg) and dipping in 30 litre capacity tank containing wax emulsion.
  - Fruits are then removed and allowed to dry under electric fan or in the open air or with warm air at 54 to 55°C. The produce should be turned periodically while drying.
- **Foam waxing**
  - Foaming leaves a very thin coating of wax on the fruit after the water has evaporated.
  - A foam generator is mounted over a suitable brush head, and water is applied to the fruit or vegetable in the foam.
- **Flooding**
  - Flooding is similar to dipping and is a safe and convenient method of application.

- **Trade name of some extensively used waxes**
  - Citrashine @ from DECCO, India UPL
  - Waxol-12 - Oil/ water emulsion wax containing 12% solids
  - Tal-Prolong
  - Semper fresh
  - Frutox - Emulsion of different waxes with 12% solids.
- **Fruits suitable for waxing**
  - Immature fruit vegetables - Cucumbers and summer squash
  - Mature fruit vegetables - Eggplant, peppers, tomato, potato, pumpkin, carrot, snake gourd, coccinia and capsicum
  - Fruits - Apple, avocado, banana, citrus (orange, mandarin, lemon, grapefruit), guava,mangoes, melons, papaya, peaches pine apple
- **Use of wax concentration on the fresh produce**

Wax concentration (%)	Crops
12	Carrot, brinjal, snake gourd, potato, cucumber, coccinia, capsicum, ribbed gourd, pine apple, guava and papaya
9	Tomato, lime, orange
8	Apple
6	Mango and musk melon

- **Quantity of wax emulsion at 12% concentration required for one metric ton of commodity**

Crop	Wax emulsion (12%) in L
Orange	3.6
Mosambi	5.4
Mango (300-350g)	3.6
Potato	7.9
Apple	5.4
Guava	5.7
Tomato	5.0
Banana	9.0
Lime	3.0

4.4.8. Packing

- Not only protects, but also makes a favourable impression on buyers and fetch higher income

4.4.9. Delivery

- Moving harvest produce from farm to customer in good condition is important.

- All efforts up to delivery are invalid, if fresh fruits and vegetables reach destination in poor condition.
- Care taken to protect produce and becomes necessary when mixing load of fruits and vegetables to prevent violating compatibility factors.

5

Ethylene Action on Ripening of Fruits and Ethylene Management

5.1. Fruit ripening

- Genetically programmed stage of development overlapping with senescence.
- Ripe when attains full flavour and aroma
- **Optimum ripening condition for fruits**

Temperature	18-25°C (< 18 delay ripening > 25 microbes)
RH	85-90%
Ethylene concentration	10-100 ppm
Duration	12-72 h
Air circulation	Sufficient to maintain the air temperature
Ventilation	Sufficient to prevent accumulation of CO_2

- **Conditions for controlled ripening of fruits at RH of 85-90%**

Produces	**C_2H_4 (ppm)**	**Temperature (°C)**	**Duration (h)**
Avocado	10-100	15-18	12-48
Banana	100-150	15-18	24
Honeydew melon	100-150	20-25	18-24
Kiwi fruit	10-100	0-20	12-24
Mango	100-150	20-22	12-24
Stone fruits	10-100	13-25	12-72
Tomato	100-150	20-25	24-48

5.1.1. Climacteric fruits

- Apple, apricot, avocado, banana, ber, cherimoya, fig, guava, kiwifruit, mango, melons, pear, peach, plum, persimmon, papaya, tomato, sapota and passion fruit

5.1.2. Non climacteric fruits

- Carambola, cherries, citrus, grape, litchi, loquat, olive, pineapple, pomegranate and strawberry

Difference between climacteric and non-climacteric fruits

Climacteric fruit	Non-climacteric fruit
Normally they ripen after harvest	Fruit that does not ripen after harvest. Ripen on the plant itself.
The quality of fruit changes drastically after harvest characterized by softening, change in colour and sweetness (except in avocado, which will ripen only after detached from the plant)	The quality do not change significantly after harvest except little softening. Do not change to improve their eating characteristics
Exhibits a peak in respiration	Does not exhibit a peak in respiration
More ethylene is produced during ripening	Little/No ethylene production
Significant increase in CO_2 production	No significant increase in CO_2 production
Decrease in internal oxygen concentration	More in internal oxygen content
Low concentration of ethylene 0.1-1.0 μ L/L/day is sufficient to hasten ripening	Not much response is seen to exogenous application of ethylene.

5.2. Changes during fruit ripening (Fig. 1)

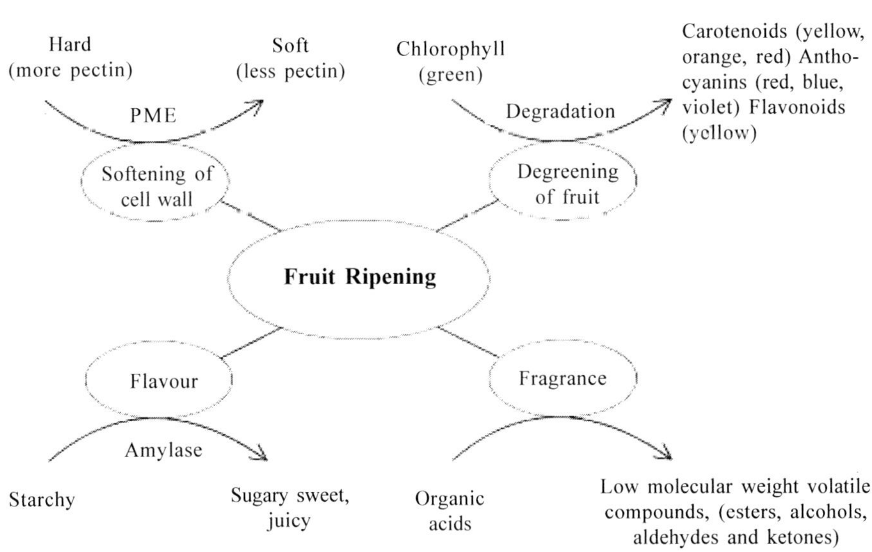

Fig. 1. Flow chart showing physical and physiological changes during fruit ripening

5.2.1. Cell wall changes

- Cell wall consists of pectic substances and cellulose along with small amounts of hemi cellulose and noncellulosic polysaccharides.

- Changes particularly in middle lamella rich in pectic polysaccharides which are degraded and solubilised during ripening.
- During softening, there is a loss of neutral sugars (Galactose and arabinose, major components of neutral protein) and acidic pectin (Rhamno galacturonan) of cell wall.
- Major enzymes in softening are pectine esterase, polygalacturonase, cellulase and β-galactosidase.

5.2.2. Starch

- Sugar levels tend to increase due to either increased sugar importation from plant or to mobilization of starch reserves within fruit
- With advancement maturity, accumulated starch is hydrolysed into sugars (Glucose, fructose or sugars) known as characteristic event for fruit ripening.
- Further, breakdown of sucrose in to glucose and fructose mediated by invertase.
- In potato and peas, higher sucrose remains high at fresh immature stage, converts into starch with approach of maturity.

5.2.3. Organic acids

- With onset of ripening, a downward trend in levels of organic acids is noticed.
- Decline in organic acids during fruit ripening and increase in membrane permeability which allow acids to store in respiring cells, formation of salts of malic acid.
- Reduction in amounts of acid translocated from leaves.
- Reduced ability of fruits to synthesize organic acids with fruit maturity, translocation into sugars and dilution effect due to increase in volume of fruit.

5.2.4. Colour

- Degradation of chlorophyll accompanied by synthesis of anthocyanins or carotenoids.
- Give rise to wide range of colours (From red to blue).
- Chloroplasts in green immature fruit lose chlorophyll on ripening and change into chromoplasts which contain carotenoid pigments.
- Carotene and lycopene are synthesized during ripening.

5.2.5. Flavouring compounds

- Characteristic flavour derived from production of specific flavouring volatile substances.
- Mainly esters, alcohols, aldehydes, acids and ketones.

5.2.6. Ascorbic acid

- L-ascorbic acid (Vitamin C), naturally occurring ascorbic acid in fruits.
- Reduced amount noticed in pome, stone and berry fruits at time of harvest.
- Increase in ascorbic acid with increase in fruit growth
- Declined with advancement of maturity and onset of fruit ripening in pear, sweet potato, potato, asparagus and okra during post-harvest handling.

5.2.7. Phenolics

- Declines from high levels during early growth to low levels when fruit is physiologically mature and thereafter, susceptible to induction of ripening.

5.2.8. Amino acids and proteins

- Decrease in free amino acid often reflects an increase in protein synthesis.
- During senescence, level of free amino acids increases reflecting breakdown of enzymes and decreased metabolic activity.

5.3. Ethylene production and ripening

- Physiological events responsible to ripening process
  - Ethylene production
  - Rise in respiration

5.3.1. Ethylene production

- In climacteric fruits (Mango, banana), ethylene production is increased and causes
  - Rise in respiration and temperature
  - Rise in activity of hydrolytic enzymes (Fig. 2).
- Ethylene is produced from essential amino acid methionine.

The hormone $\frac{\text{ethylene}}{H_2C = CH_2}$ initiates the ripening response :

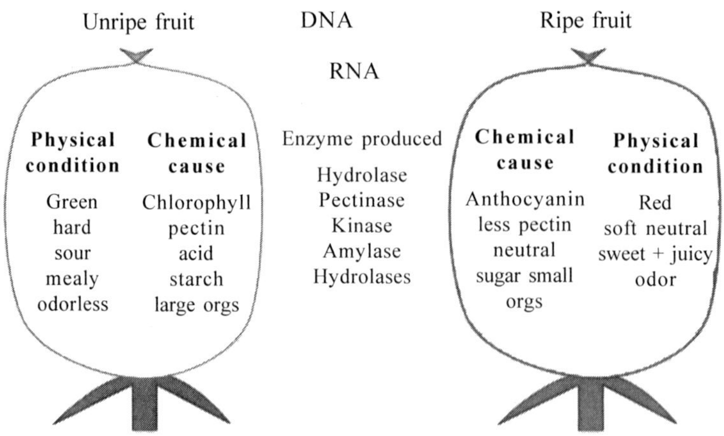

Fig. 2. Role of ethylene on enzyme production during ripening

- On the basis of ethylene production rate, horticultural commodities are classified into following way

Class	Range at 200 C (μ C_2H_4 release kg^{-1} hr^{-1})	Crops
Very low	< 0.1	Artichoke, asparagus, cauliflower, cherry, citrus fruits, grape, cut flowers, leafy vegetables, pomegranate, potato, root vegetables, strawberry
Low	0.1-1.0	Brinjal, chilli, cucumber, green capsicum, okra, pineapple, pumpkin, water melon
Moderate	1.0-10	Banana, guava, fig, litchi, melon, mango, tomato
High	10-100	Apple, apricot, avocado, kiwi fruit (Ripe), papaya, peach, plum, pear
Very high	> 100	Sapota, passion fruit

5.3.2. Rise in respiration

- Respiration required for releasing energy and substrate for synthesis of several organic compounds required in ripening process.
- During ripening in climacteric fruits, rise in respiration is observed.
- Climacteric peak obtained very fast when temperature is relatively high.
- Respiration - most deteriorating process of harvested fruits and vegetables which leads to oxidative breakdown of complex materials (Carbohydrates or acids) of cell into simpler molecules (CO_2 and water) with concurrent production of energy required by cell for completion of chemical reactions (Fig. 3).

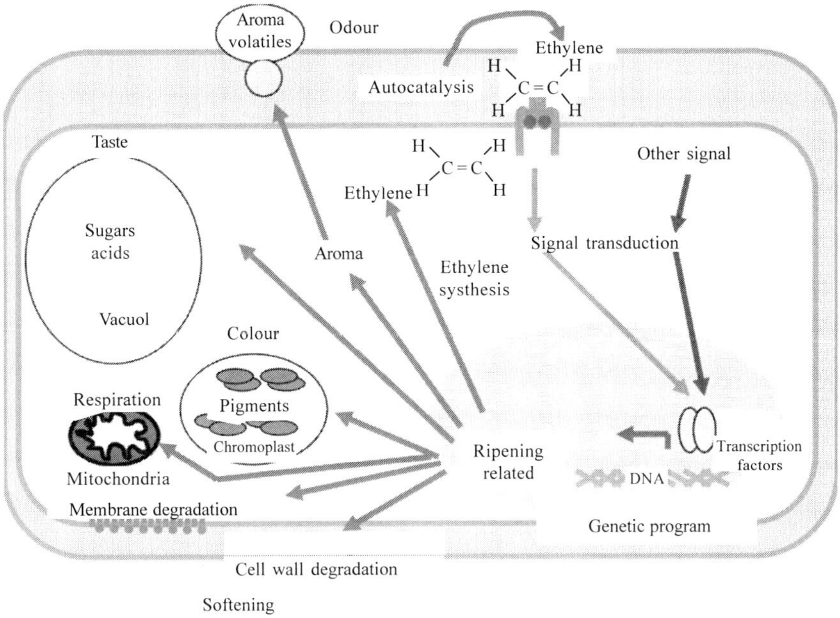

Fig. 3. Ethylene production and respiration

5.3.3. Sources of ethylene

- **Ethylene gas**: Pure C_2H_4 gas enclosed in the can/cylinder is sprayed / injected into chamber. Ethylene of 3 g is sufficient to ripe 2-6 tons of produce (Fig. 4).

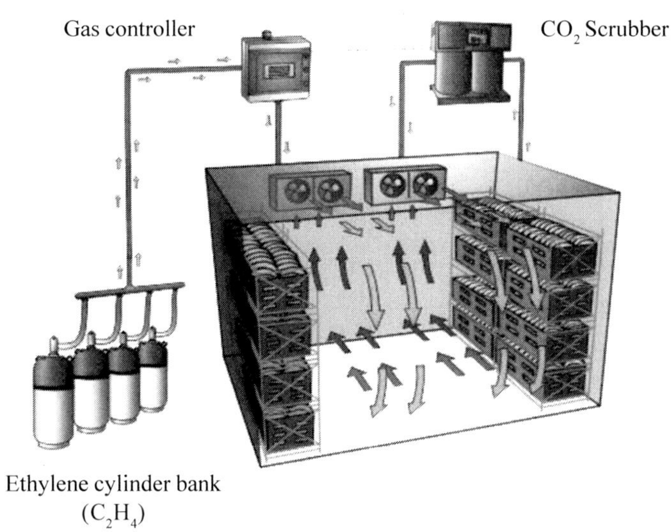

Fig. 4. Ethylene gas machine

- **Ethephon**: Used as spray/ dip, acidic in water releases C_2H_4
- **Ethylene mixture**: C_2H_4 + inert gas like CO_2. Inert gas because not enough O_2 remains in the chambers to provide an explosive mixture. Eg, Ripe gas contains 6% C_2H_4
- **Ethylene generators**: Widely used method where in liquid spirit produces C_2H_4 when heated in the presence of catalyst platinised asbestos.
- **Use of ripe fruits**: Cheap and simple, where in ripe fruit with high C_2H_4 producers such as apple, banana, mango, sapota and tomato is used at home to ripe / degreen

5.3.4. Use of chemicals for increasing shelf life of fruits and vegetable

- **Ethylene absorbent**
    - Ethylene is responsible for decreasing shelf life
    - Putting $KMNO_4$ at 100 ppm soaked filter paper minimized ripening and increased shelf life.
- **Antifungal agents**
    - Sodium ortho phenyl phenate
    - Diphenyl wraps protection against moulds, stem end rot
    - Dibromotetrachloroethane and esters gives better flavor
- **Use of inhibitors**

Treatment	Crop	Chemical	Concentration
Post-harvest	Mango	CCC	1000-2000 ppm
After fruit formation	Apple	2-Dimethyl hydrazide	10,000 ppm

- **Use of auxins**
    - Also helpful to advance in ripening and increase shelf life.

Chemical	Concentration (ppm)	Crop	Stage
2, 4-D	5	Grape	Pre harvest
2, 4, 5-T	25	Fig	Pre harvest
2, 4, 5-T	100	Mango	After harvesting

- Preserved by lactic acid increases shelf life
- Post-harvest dipping of papaya either 100 ppm GA_3 or $CaCl_2$ at 2% extended shelf life up to 9 days without any decline in quality

6

Pre-Cooling, Treatments Prior to Transport and Waxing

- Different steps involved in post-harvest processing of horticultural produces are depicted in Fig. 1.

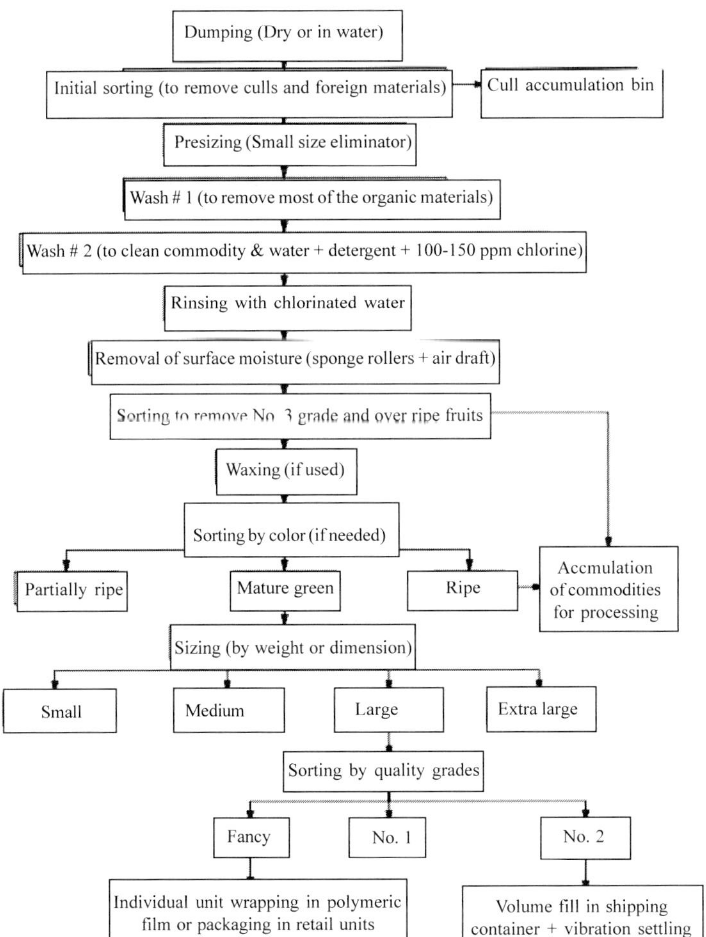

Fig. 1. Flow chart on post-harvest processing of horticultural produces.

6.1. Pre sorting

- Done to eliminate injured, decayed and unwanted produce before cooling and handling.

6.2. Pre cooling

- Removal of field heat after harvest at 32°C for 1 h.

6.2.1. Advantages

- Reduces rate of respiration and ripening, moisture, bruise damage during transits, production of ethylene and growth of spoilage organisms
- Eases load on cooling system (Refrigeration) of transport or storage chamber
- Extends product shelf life

6.2.2. Factors

- Air temperature during harvesting
- Time between harvest and pre cooling
- Nature of crop (High perishable crop require immediate pre cooling)
- Difference in temperature between crop and cooling medium
- Nature/velocity of cooling medium

6.2.3. Mechanism

- Two main heat transfer mechanisms used for cooling of produce are conduction and convection.
- With conduction, heat is transferred within produce to coldest surface.
- Direct movement of heat from one object to another by direct methods (From fresh produce to water or warmer to cooler).
- With convection, heat is transferred away from surface of produce *via* cooling medium such as moving water or air.
- Potato, apple, cauliflower, orange and vegetables require more time to pre cool than produce having smaller mass and large surface area like lettuce, green onion, carrot tops, pea, corn and brussel sprouts.

6.2.4. Methods

- Cold air
  - Room cooling - All fruits and vegetables
  - Forced air cooling (Pressure cooling) - Fruits and fruit type vegetables, tubers and cauliflowers
- Cold water / hydro cooling - Stems, green leafy vegetables, fruits and fruit type vegetables

- Top icing (Direct contact with ice) - Roots, stems, cauliflower, green onion, brussel sprout
- Evaporation of water from produce
    - Evaporative cooling
    - Vacuum cooling - Stems, leafy and flowers type vegetables
    - Hydrovac cooling - Combination of hydro and vacuum cooling

6.2.5. Precautions

- Done as soon as possible after harvest.
- Harvest in early morning hours to minimize field heat and refrigeration load on pre cooling equipment.
- Harvested produce are protected from sun with covering until placed in pre cooling facility.

6.3. Washing

- Done using washing tank (Fig. 3) to remove adhering dirt, stains, insects, molds and sometimes spray residues.
- Not only help in cleaning and making vegetables/ fruits fresh, but also improves appearance
- Helps in extending shelf life of produce.
- Done manually under tap water or in wash tank using soft muslin cloth.
- 100 - 150 ppm hypochlorite/chlorine or soap or calcium hydroxide.
- Thumb rule is to use 1 to 2 ml of chlorine bleach per liter of water
- 100-150 ppm of Cl. pH of the water must be around 6.5 to 7.5. Most efficient detergent - Sodium meta bisulphate.

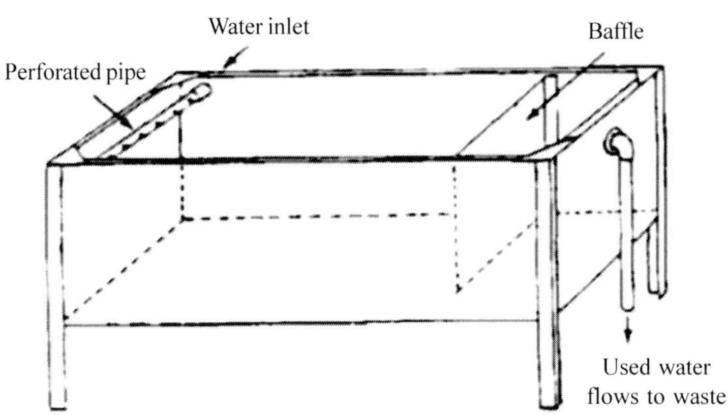

Fig. 2. Washing tank

6.4. Dry cleaning

- Done by dry brushing instead of washing.
- Removal of white cottony mealy bugs attached in between surface holes of custard apple fruits.
- Fruits and vegetables which are not suitable for washing are onion, garlic, okra, grapes, strawberry, mushrooms

6.5. Dressing

- Removal, trimming and cutting of all undesirable leaves, stem, stalks, roots and other non-edible or unmarketable parts
- Makes vegetables attractive and marketable.

6.6. Trimming

- Done to remove unwanted, discoloured, rotting and insect damaged parts (Eg. Cabbage, cauliflower, spinach, lettuce, rose, chrysanthemum, gladiolus, tuberose etc.).
- In grapes, trimming is done to remove undersize, immature, dried, split and damaged berries.
- Enhances visual quality, minimizes water loss and other deteriorative processes.
- Reduces likelihood of diseases or spread, facilitates packaging and handling and reduces damage.

6.7. Water spray

- Helps in compensating water loss and maintaining quality for longer period.
- Produce covered with gunny sack soaked in cold water, if it stores for longer period before sale.
- Example: Green leafy vegetables

6.8. Sorting

- Done by hand before washing to remove unsuitable fruits and vegetables to market or storage due to damage by mechanical injuries, insects, diseases, immature, over mature, misshapen etc. (Fig. 3).
- Reduces losses by preventing secondary contamination.
- Done either at farm level or in pack house.
- In sorting, only sensory quality parameters are taken.
- Sizing is also done by machine. Machine sizers work on two basic principles; weight and diameter. Sizing on the basis of fruit shape and size are most

effective for spherical (Oranges, tomato, certain apple cultivars) and elongated (Delicious apples and European pears are of non-uniform shape) commodities, respectively.

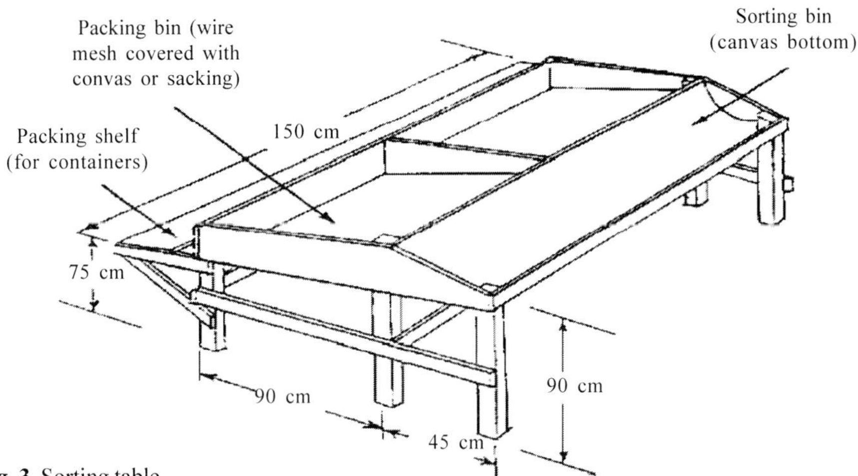

Fig. 3. Sorting table

- **Mechanisms/Types of sizing**
  - **Diverging belts/rope grader** - The different speed of belts makes produce rotate besides moving forward to a point where produce diameter equals belt/rope separation. Eg. Cucumber, gherkin, pineapple and large root vegetables.
  - **Sizing rollers** - With increased spaces between rollers. Eg. Citrus
  - **Hand held template**-Sizing can be performed manually using rings of known diameter.
  - **Sizing by weight** - Sorting by weight is carried out in many crops with weight sensitive trays. These automatically move fruit into another belt aggregating all units of the same mass. Individual trays deposit fruit on the corresponding conveyor belt. Eg. Citrus, apple and pear and irregular fruits
  - **Mesh screens** - Eg. Potato, onion, anola

6.9. Grading

- Separated manually or mechanically into two or more grades on basis of colour, shape, size, weight, soundness, firmness, cleanliness, maturity.

6.9.1. Systems of grading

- **Static systems**
  - Common in tender and/or high value crops.

- **Dynamic system**
  - Moves along belt in front of sorters which removes units with defects
  - Eg. Onion grading
- **New innovation in grading systems**
  - Computerized weight grader - Operates on the basis of tipping buckets that drops to release the pre weighed item at a particular position. Eg. Apple, citrus
  - Video image capture and analysis - Used for size, colour and external defect grading. Eg. Coffee, apple
  - NIR Spectrometer - To assess the TSS non - destructively in apple and stone fruits
  - X-ray imaging and computer aided tomography
  - MRI - Magnetic Resonances Imaging
  - Spectroscopy
  - Acoustic methods
  - Volatile emission analysis

6.10. Curing

- When roots and tubers stored for long periods, curing is necessary to extend shelf life.
- To reduce water loss during storage from hardy root and tuber vegetables such as potato, sweet potato, yam and other tropical vegetables where cuticle are poorly developed.
- Relatively susceptible to mechanical wounding during harvesting and handling.

6.11. Degreening

- Treatment of citrus fruits with ethylene under controlled conditions to hasten loss of chlorophyll (Fig. 4)
- Consists of chlorophyll degradation to allow expression of natural pigments masked by green colour.

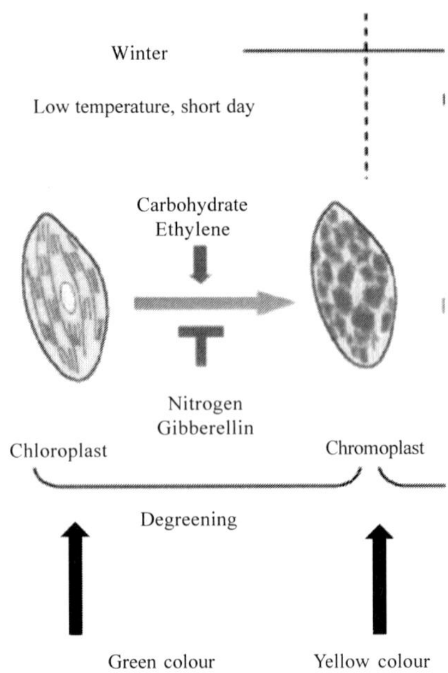

Fig. 4. Degreening

6.12. Waxing

- Waxing is important, if tiny injuries and scratches on surface of fruit or vegetable are present.
- Waxes - Esters of higher fatty acid with monohydric alcohols and hydrocarbons and some free fatty acids.
- Protective edible coat to protect from transpiration losses and reduce respiration rate.

6.12.1. Skin coating (Protective coating)

- Artificial application of very thin film of wax or oil or other materials to surface of fruits or vegetables as addition or replacement for natural wax coating.

6.12.2. Advantages of wax application

- Improves appearance
- Reduces moisture losses and retards wilting and shrivelling during storage
- Reduces weight loss
- Prevents chilling injury and browning
- Protects produce from bruising

6.12.3. Fruits suitable for waxing

- Immature fruit vegetables - Cucumbers and summer squash
- Mature fruit vegetables - Eggplant, pepper, tomato, potato, pumpkin, carrot, snake gourd, coccinia and capsicum
- Fruits - Apple, avocado, banana, citrus (Orange, mandarin, lemon, grapefruit), guava, mango, melons, papaya, peach, pineapple etc.

6.12.4. Commonly used waxes

- Artificial waxes like solvent waxes, water waxes and paste or oil waxes.
- Shellac
- Carnauba wax
- Bee wax
- Polyethylene
- Wood resins
- Paraffin wax

6.12.5. Methods of wax application

- Spray waxing
- Dipping
- Foam waxing
- Flooding

7

Storage Methods

7.1. Storage

- Helpful in checking market glut, providing wide selection of fruits, vegetables and flowers to consumer through most part of year i.e. especially during off season.
- Helps in orderly marketing and increases profit to producers/farmers.
- Done to maintain freshness, quality, reduce spoilage and extend their usefulness.
- Huge post-harvest losses due to lack of proper storage facilities.
- Basic principle of storage - to reduce rate of physiological processes like respiration, transpiration, ripening and other biochemical changes.
- Proper storage aims to control disease infection and preserve best quality for consumers.
- **Chemical used to extend the shelf life and quality of produce**

Crop	Chemical	Concentration
Apple	Calcium	1-4%
Mango	Calcium nitrate	1.0%
	Calcium chloride	0.6%
Ber	Calcium	0.17%
Banana	GA_3	50 ppm
	Kinetin	20 ppm
Guava	Calcium nitrate	1.0%

7.1.1. Factors affecting storage

- Pre harvest factors
- Maturity at harvest
- Harvesting and handling practices
- Pre storage treatments
- Temperature and humidity in storage room
- Overall hygiene

7.1.2. Principles of storage

- Control of respiration
- Control of transpiration
- Prolonging dormancy period/control of sprouting and rooting
- Control of spoilage

7.2. Improved storage methods

7.2.1. Low temperature storage (Refrigeration/cold storage)

- Best known, effective and most widely used method for extending storage life and long term storage of fruits, vegetables and flowers.
- Temperature management is most important aspect to maintain quality, reduce losses and extend storage life of perishable commodities.
- Cold storage - System with thermal insulation and refrigeration in which perishable commodities stored for set period of time under controlled conditions of temperature and humidity (Fig. 1).

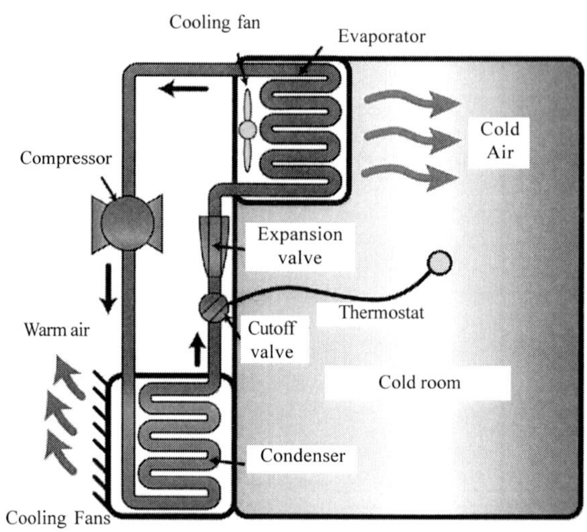

Fig. 1. Cold storage unit

- **Factors involved for effective cold storage of produce**
  - **Product quality**: Free from physical damage, optimum maturity and free from infections.
  - **Temperature**: Low temperature storage recommended for perishables as it retards respiration and metabolic activity, aging due to ripening, softening and textural and colour changes, moisture loss, spoilage due to diseases and undesirable growth (Sprouting / cooling).

- **Relative humidity**: If too low, wilting or shrivelling occur; if too high, favour development of decay. RH of 85-90% is recommended for perishables.
- Air circulation and package spacing
- Respiration rate, heat evolution and refrigeration
- Weight loss in storage
- Sanitation and air purification
- Temperature management

7.2.2. Controlled atmospheric storage (CAS)

- Storage of fruits and vegetables in CA storage - One of most advanced methods of storage (Fig. 2).

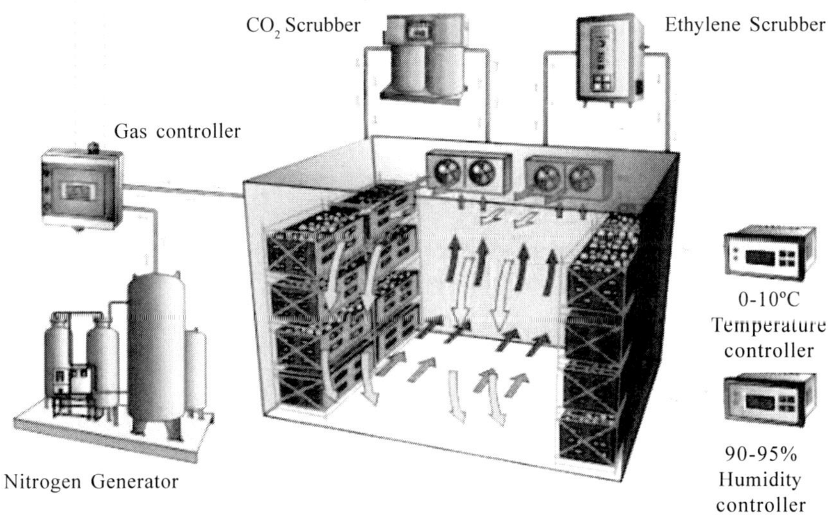

Fig. 2. Controlled atmosphere storage

- First suggested by W.R. Philips of Canada.
- **Physiological basis**
  - Oxygen is reduced and CO_2 is increased to slow down ripening and respiration rate.
- **Essential features**
  - Mechanical refrigeration is used to maintain temperature of -1 to 3°C.
  - CA storage room is constructed gas tight.

  - Reduction of O_2 - Nitrogen gas is introduced into storage by cylinder to reduce oxygen level after room is filled and sealed. CO_2 is added into storage from CO_2 gas cylinder.
  - Excess CO_2 is removed by dry hydrated lime, ethanolamine, aluminium calcium silicate, activated carbon, magnesium oxide and other CO_2 scrubbers.
  - Atmospheric composition is crop specific. General rule - most common combinations are 2-5% O_2 and 3-10% CO_2

    Storage room atmosphere samples are monitored daily for CO_2 and O_2.

- **Benefits**
  - Retardation of senescence and associated biochemical and physiological changes
  - Reduction of produce sensitivity to ethylene action at O_2 below 8% and/or CO_2 above 1%.
  - Useful tool for insect control in some commodities.
- **Limitations**
  - Causes certain physiological disorders such as black heart in potato, brown stain of lettuce.
  - Irregular ripening of produce such as banana, pear, tomato etc.
  - Development of off flavours and off odours at very low O_2 concentrations.
  - Timely non-availability of gas
  - Costly and technical knowhow is required
  - Blackening - Due to tissue asphyxia (Suffocation) of artichoke head caused by storing in an inadequate atmosphere

7.2.3. Modified atmospheric storage (MAS)

- Implies lower degree of control of gas concentration in atmosphere surrounding commodity (Fig. 3).

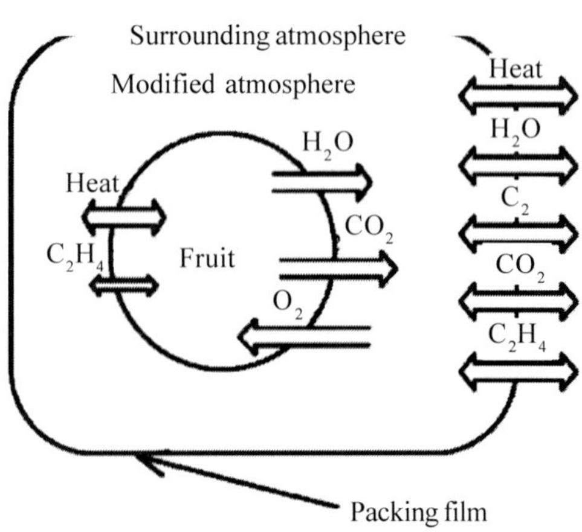

Fig. 3. Modified atmosphere storage

- MA and CA differ only in degree of control, CA is more exact.
- **Biochemical and physiological basis**
  - Rate of respiration and metabolism doubles for every 10°C rise in temperature.
  - Respiration reduced by decreasing temperature, O_2 level and/or increasing CO_2 level in storage atmosphere. Both O_2 and CO_2 levels exert independent effects on respiration.
  - Net effect may be additive or synergistic.
  - When O_2 concentration is reduced below 10%, respiration rate is decreased.
  - When O_2 concentration falls below 2%, anaerobic respiration may set in, thereby leading to accumulation of ethanol and acetaldehyde.
- **Environmental factors affecting MA storage**
  - Temperature and relative humidity
  - Light
  - Sanitation factors

7.2.4. Differences between CA and MA storage

CA storage	MA storage
High degree of control over gas concentration	Low degree
Longer storage life	Less
More expensive technology	Less
Atmosphere is modified by adding gas	Created by either actively (Addition or removal of gas) or passively
Specific temperature should be maintained	May or may not be maintained

8

Packaging Technologies Packaging Materials and Transport

8.1. Packaging

- One of the important steps in long and complicated journey from grower to consumer.
- Bags, crates, hampers, baskets, cartons, bulk bins, and palletized containers are convenient containers for handling, transporting, and marketing fresh produces.

8.2. Packaging points

- Recyclability / Biodegradability
- Variety
- Sales appeal
- Shelf life
- Containment
- Protection
- Identification
    - To provide information on produce name, brand, size, grade, variety, net weight, count, grower, shipper, and country of origin.
    - Universal Product Codes (UPC or bar codes) is included as part of labelling.
    - UPC is used in food industry which consists of a ten digit machine readable code.

8.3. Types of packaging materials

- **Wood**
    - A pallet size sheet (Slip sheet) of corrugated fibre board or plastic (or a combination of these materials) with a narrow lip along one or more sides.

- **Pallet bins**
  - Substantial wooden pallet bins of milled lumber or plywood (Fig. 1) used to move produce from field or orchard to packing house.

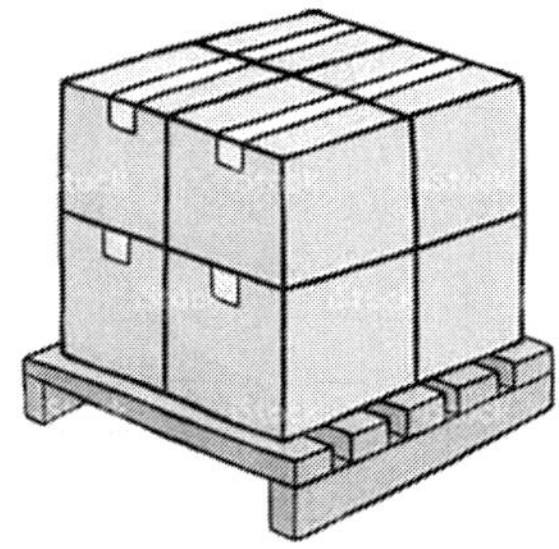

Fig. 1. Pallet bins

- **Wire bound crates**
  - Used extensively for snap beans, sweet corn and several other commodities that requires hydro cooling.
  - Sturdy, rigid and have very high stacking strength essentially unaffected by water.
- **Wooden crates and lugs**
  - Extensively used for apple, stone fruit, and potato (Fig. 2)

Fig. 2. Wooden crates

- **Wooden baskets and hampers**
  - Wire reinforced wood veneer baskets and hampers used for wide variety of crops from strawberries to sweet potatoes.
  - Durable and nested for efficient transport when empty.
- **Corrugated fibre board**
  - Corrugated fibre board (Often mistakenly called cardboard or pasteboard)
  - Basically two methods used to print corrugated fibre board containers
  - **Post printed**
    - When liner is printed after corrugated fibre board has been formed
    - Most widely used printing method for corrugated fibre board containers because economical and used for small press runs.
  - **Preprinted**
    - High quality, full colour graphics obtained by preprinting liner board before attached to corrugated paper board.
- **Pulp containers**
  - Made from recycled paper pulp and starch binder mainly used for small consumer packages of fresh produce.
  - Absorb surface moisture from product, which is benefit for small fruit and berries.

    Also biodegradable, made from recycled materials, and recyclable.
- **Paper and mesh bags**
  - Consumer packs of potato and onion for good ventilation

- Also sweet corn, cabbage, turnip, and citrus are packed in mesh bags.
- Supermarket managers like small mesh bags because attractive displays that stimulate purchases.

- **Plastic bags**
  - Plastic bags (Polyethylene film) are good for fruit and vegetable consumer packaging.
  - Are clear, allowing for easy inspection of contents, and readily accept high quality graphics.
  - Plastic films available in wide range of thicknesses and grades
- **Shrink wrap**
  - Newest trends in produce packaging are shrink wrapping of individual produce items.
  - Potato, sweet potato, apple, onion, sweet corn, cucumber and a variety of tropical fruits.
  - Reduce shrinkage and mechanical damage, protect produce from disease and provide good surface for stick on labels.
- **Rigid plastic packages**
  - Inexpensive, versatile, provide excellent protection to produce and pleasing consumer package.
  - Most often used with consumer packs of high value produces like small fruits, berries, mushrooms or items easily damaged by crushing.
  - Used extensively with precut produce and prepared salads.

8.3.1. Packaging materials in use

- CFB boxes corrugated fibre board - Most widely used material for fruit and vegetable packages
  - Light in weight
  - Reasonably strong
  - Flexibility of shape and size
  - Easy to store and use
  - Good pointing capability
  - Economical
- **Wooden boxes**
  - Materials used for manufacture of wooden boxes include natural wood and industrially manufactured wood based sheet materials.

- **Sacks**
  - Traditionally made of jute fibre or similar natural materials.
  - Most jute sacks provided in a plain weave.
- **Palletisation**
  - Widely used for transport of fruit and vegetable packages, in all developed countries.
  - Labour cost in handling is greatly reduced.
  - Transport cost may be reduced.
  - Goods are protected and damage reduced.
  - Mechanized handling can be very rapid.
  - Through high stacking, storage space more efficiently used.
  - Pallets encourage introduction of standard package sizes.
  - Most common pallet size is 1200 mm x 1100 mm.

8.5. Ventilation of packages

- Reduction of moisture loss is main requirement of limited permeability packaging materials.
- Semi permeable plastic films are widely used.
- Airflow through ventilation holes allows hot fruit or vegetable to slowly cool and avoid buildup of heat produced by commodity in respiration.
- Holes are important in cooling fruit when packages are placed in a cold storage.
- Ventilation holes improve dispersal of ethylene produced.

8.6. Cushioning materials

- To fix commodities inside packages and prevent them from mixing about in relation to each other and package itself, when there is vibration or impact.
- Also provide packages with additional stacking strength.
- Made of wrapping papers, fibre board (Single or double wall), moulded paper pulp trays, moulded foam, polystyrene trays, moulded plastic trays, foam plastic sheet, plastic bubble pads, fine shredded wood, plastic film liners or bags.

8.7. Controlled and Modified Atmospheric Packaging (CAP and MAP)

- Normal composition of air is 77% nitrogen, 21% oxygen, 0.03% carbon dioxide (Fig. 3).
- Method for extending shelf life of perishable and semi perishable food products by altering relative proportions of atmospheric gases that surround

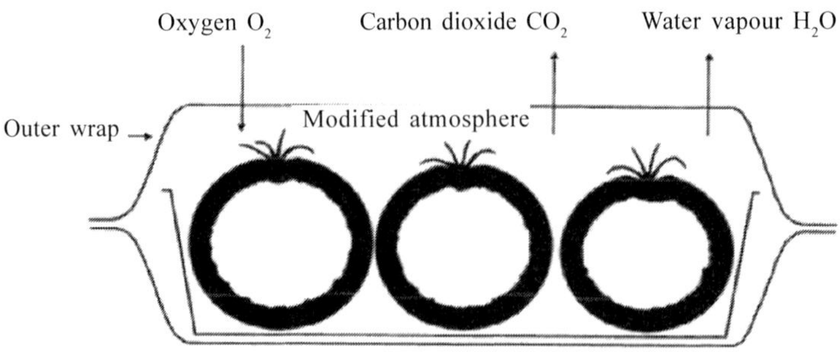

Fig. 3. Controlled and Modified Atmospheric Packaging

produce.

8.7.1. Controlled atmosphere (CA)

- Refers to storage atmosphere different from normal atmosphere in its composition, wherein component gases are precisely adjusted to specific concentrations and maintained throughout storage and distribution of perishable foods.
- Controlled atmosphere relies on continuous measurement of composition of storage atmosphere and injection of appropriate gases or gas mixtures into it, if and when needed.
- Tomato - Excessively low O_2 or high CO_2 prevents proper ripening even after removal of fruit to air, and enhances danger of chilling injury.

8.7.2. Modified atmospheric packaging (MAP)

- Modified atmosphere conditions created inside packages by commodity itself and / or by active modification.
- Commodity generated or passive MA evolved as consequence of commodity's respiration.
- Active modification - creating slight vacuum inside package and replacing it with desired mixture of gases, so as to establish desired EMA quickly composed to passively generated EMA.
- Another active modification technique - use of carbon dioxide or ethyl absorbers (scavengers) to prevent build-up of particular gas within package.
- Compounds like hydrated lime, activated charcoal, magnesium oxide are known to absorb carbon dioxide.

- Potassium permanganate, squakna and phenyl methyl silicone are used to absorb ethylene within packages.
- Scavengers held in small sachets within packages or impregnated in wrappers or into porous materials like vermiculite.
- With activity respiring commodities like fruits and vegetables, package atmosphere should contain oxygen and carbon dioxide at levels optimum to particular commodity.
- In general, MA containing 2-5% oxygen and 3.7% carbon dioxide to extend shelf life of wide variety of fruits and vegetables.
- Shelf life doubled,
- Few types of films - polyvinyl chloride, (PVC), polystyrene, (PS), polyethylene (PE) and polypropylene (PP).

8.9. Vacuum packaging

- Offers an extensive barrier against corrosion, oxidation, moisture, drying out, dirt, attraction of dust by electric charge, ultra violet rays and mechanical damages, fungus growth or perishability etc.
- Has commendable relevance for tropical countries with high atmosphere humidity.
- In vacuum packaging, product to be packed is put in vacuum bag (made of special, hermetic fills) then evacuated in vacuum chamber and then, sealed hermetically to provide total barrier against air and moisture.
- If product cannot bear atmosphere pressure due to vacuum inside package, then packages are flushed with inert gases like nitrogen and CO_2 after evacuation.

8.10. Edible packaging

- Thin continuous layer of edible material formed on, placed on, or between foods or food components.
- Integral part of food, which is eaten as part of whole food product.
- Selection of material based on properties to act as barrier to moisture and gases, mechanical strength, physical properties, and resistance to microbial growth.
- Types of materials used - lipids, proteins and polysaccharides or combination of any two or all of these.
- Many lipid compounds are animal and vegetable fats, acetoglycerides are used in formulation of edible packaging for fresh produces because of their excellent moisture barrier properties.
- Reduce, weight losses due to dehydration during storage by 40-70 per cent.

9

Principles and Methods of Preservation

9.1. Principles of food preservation by heat

- Pasteurization temperature at below 100°C
- Heating at about 100°C
- Sterilization temperature at above 100°C

9.1.1. Pasteurization

- Heat treatment kills part, but not all microorganisms present and temperature below 100°C.
- Heating by means of steam, hot water, dry heat or electric current and cooled after heat treatment.
- Surviving microorganisms are inhibited by low temperature or some other preservative method, if spoilage prevented.
- Preservative methods used to supplement pasteurization include
    - Refrigeration, e.g. milk
    - Keeping out microorganisms usually by packaging the product in a sealed container
    - Maintenance of anaerobic conditions as in evacuated, sealed containers
    - Addition of high concentration of sugars in sweetened condensed milk
    - Presence or addition of chemical preservatives, e.g. organic acids in pickles.
- **Methods of pasteurization**
    - HTST method - High temperature and short time (Above 70°C)
    - LTHT method - Low temperature and higher time or holding method (60-70°C)

9.1.2. Heating at about 100°C

- Boiling liquid food by immersion of container of food in boiling water or by exposure to flowing steam.

- Some very acid foods, e.g., sauerkraut preheated to temperature below 100°C, packaged hot and not further heat processed.
- Blanching fresh vegetables before freezing or drying involves heating briefly at about 100°C.

9.1.3. Sterilization at above 100°C

- All microorganisms completely destroyed due to high temperature.
- Time and temperature for sterilization vary with type of food.
- Temperature above 100°C obtained using steam pressure sterilizers (Pressure cookers and autoclaves).
- Fruits and tomato at 100°C for 30 min.
- Vegetables like green peas, okra, beans, etc., being non acidic and containing more starch than sugar, require higher temperature to kill spore forming organisms.
- Continuous heating for 30-90 min. at 116°C is essential for sterilization.
- Before using, empty canes and bottles are sterilized for about 30 min. by placing in boiling water.

9.1.4. Difference between pasteurization and sterilization

Pasteurization	Sterilization
Partial destruction of microorganism	Complete destruction of microorganism
Temperature below 100°C	Temperature at 100°C and above
Used for fruits	Used for vegetables

9.1.5. Aseptic canning

- Technique in which food sterilized outside and then, aseptically placed in previously sterilized canes and sealed in aseptic environment.

9.1.6. Hot pack or hot fill

- Filling of previously pasteurized or sterilized foods, while still hot, into clean, but not sterile containers and aseptic conditions.

9.2. Preservation by low temperature

- Microbial growth and enzyme reactions are retarded in foods.
- Lower the temperature, greater the retardation.
- Low temperature produced by
  - Cellar storage (About 15°C) - Root crops, potato, cabbage, apple, onion can be stored for limited period during winter months.

- Refrigerated or chilling (0 to 5°C) - Chilling ice or mechanical refrigeration is used for temporary preservation until some other preservative process is applied.
- Most perishable foods including eggs, dairy products, meats, sea foods, vegetables and fruits, held in chilling storage for limited time with little change from original condition.

9.3. Preservation by chemicals

- Preservative - Substance which capable of inhibiting, retarding or arresting growth of microorganisms.
- Inhibitory action is due to interfering with cell division, membrane permeability and activity enzymes.
- Chemically preserved squashes and crushes kept for fairly long time even after opening seal of bottle.
- Two important chemical preservatives are permitted to beverages according to FPO (1955).
  - Sulphur dioxide
  - Benzoic acid

9.3.1. Sulphur dioxide

- Widely used throughout world in preservation of juice, pulp, nectar, squash, crush, cordial and other products.
- Has good preserving action against bacteria and moulds and inhibits enzymes, etc.
- Acts as antioxidant and bleaching agent.
- Helps in retention of ascorbic acid, carotene and other oxidizable compounds.
- Also retards development of non-enzymatic browning or discolouration of product.
- Generally used in sulphite, bisulphate and metabisulphite forms.
- Potassium metabisulphite ($K_2O_2SO_2$ or $K_2S_2O_5$) - Commonly used as a stable source of SO_2.
- Fairly stable in neutral or alkaline medium, but decomposed by weak acids like carbonic, citric, tartaric and malic acids.
- When added to fruit juice or squash, it reacts with acid in juice forming potassium salt and SO_2, which liberated and forms sulphurous acid with water of juice.

9.3.2. Benzoic acid

- Only partially soluble in water
- Sodium benzoate is used.
- More effective against yeasts than against moulds.
- Does not stop lactic acid and acetic acid fermentation.
- Mostly used in coloured products of tomato, jamun, pomegranate, plum, watermelon, strawberry, coloured grapes etc.

9.4. Preservation by fermentation

- Decomposition of carbohydrates by microorganisms or enzymes.
- One of the oldest methods of preservation.
- Foods are preserved by alcohol or organic acid formed by microbial action.
- Keeping quality of alcoholic beverages, vinegars, and fermented pickles depend upon presence of alcohol, acetic acid and lactic acid, respectively.
- Wines, beers, vinegar, fermented drinks, fermented pickles etc.,
- Fourteen per cent alcohol acts as preservative in wines, because yeasts cannot grow at that concentration.
- About 2% acetic acid prevents spoilage.

10

Minimal Processing of Fruits and Vegetables: Pre Treatments Blanching, Canning and Irradiation

10.1. Preservation through canning

- Process of sealing food stuffs hermetically in containers and sterilizing them by heat for long storage
- In 1804, Appert in France invented this process.
- Also known as appertizing.
- Saddington in England - First to describe a method of canning of foods in 1807.
- In 1810, Peter Durand obtained first British Patent on canning of foods in tin containers.
- In 1817, William Underwood introduced canning of fruits on commercial scale in U.S.A.
- Fruits and vegetables are canned in season when raw material is available in surplus.
- Canned products are sold in off season for better returns to grower.

10.1.1. Principles and process of canning

- Principle - Destruction of spoilage organisms within the sealed container by means of heat.
- Different steps involved in fruit and vegetable canning process is furnished in Fig. 1.

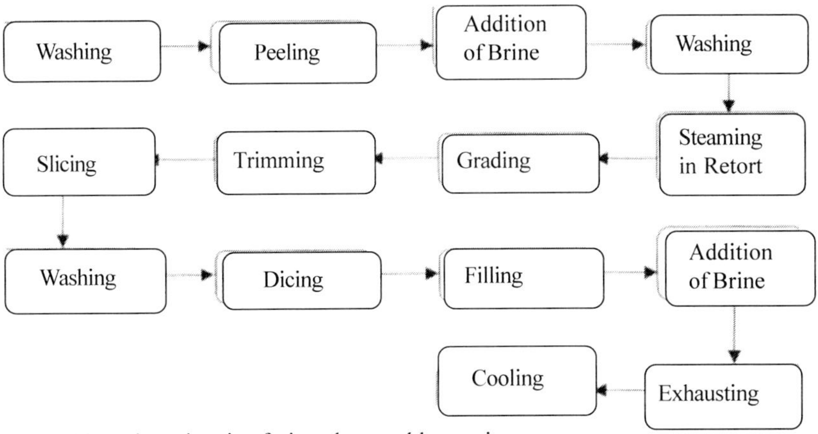

Fig. 1. Flow chart showing fruit and vegetable canning process

- **Selection of fruits and vegetables**
  - Fruits and vegetables should be fresh ripe, but firm and uniformly mature.
  - Over ripe and unripe fruits should be rejected.
  - All vegetables except tomato should be tender.
  - Tomato should be firm, fully ripe and of deep red colour.
  - Free from dirt, blemishes, insect damage or mechanical injury.
- **Grading**
  - Graded according to size and colour to obtain uniform quality.
  - Done by hand or by machines such as screw grader and roller grader
  - Fruits like berries, plums and cherries are graded whole; while peaches, pears, apricots, mangoes, pineapple, etc., are generally graded after cutting into pieces or slices.
- **Washing**
  - Important to remove pesticide residue and dust from fruits and vegetables.
  - Removal of microorganisms by washing with water is essential.
  - Root crops are washed by soaking in water containing 25 to 50 ppm chlorine.
  - Other methods of washing are spray washing, steam washing, etc.
- **Peeling**
  - Removal of outer layer.
  - Hand peeling - Done mostly in case of fruits of irregular shape, e.g., mango and papaya, where mechanical peeling is not possible.

  - Steam peeling - Free stone and clingstone peaches are steam peeled
  - Potato and tomato are peeled by steam or boiling water.
  - Mechanical peeling - Done in case of apples, peaches, pineapples and cherries and also for root vegetables like carrots, turnips and potatoes.
  - Lye peeling - Peaches, apricots, sweet oranges, mandarin oranges, carrots and sweet potatoes are peeled by dipping them in 1 to 2% boiling caustic soda solution (Lye) for 30 seconds to 2 minutes
  - Hot lye loosens skin from flesh by dissolving pectin.
  - Peel is then, removed easily by hand.
  - Any trace of alkali is removed by washing fruit or vegetable thoroughly in running cold water or dipping it for a few seconds in 0.5% citric acid solution.
  - Flame peeling - Used only for garlic and onion which have papery outer covering.
  - Vegetables like peas are shelled, carrots are scarped, and beans are snipped or trimmed.
- **Cutting**
  - Pieces of size required for canning are cut.
  - Seed, stone and core are removed.
  - Some fruits like plum from which seeds cannot be taken out easily are canned whole.
- **Blanching**
  - Also known as scalding, parboiling or precooking.
  - Usually done in case of vegetables by exposing them to boiling water or steam for 2 to 5 minutes, followed by cooling.
  - Generally, fruits are not blanched.
  - Heat treatment accomplishes the following
    - Inactivates most of the plant enzymes which cause toughness, discolouration (Polyphenol oxidase), and mustiness, off flavour (Peroxidase), softening and loss of nutritive value.
    - Reduces area of leafy vegetables (Spinach) by shrinkage or wilting, making their packing easier.
    - Removes tissue gases which reduce sulphides
    - Reduces number of microorganisms by 99%.
    - Enhances green colour of vegetables (Peas, broccoli and spinach).
    - Removes saponin in peas.

        - Removes undesirable acids and astringent taste of the peel, and thus improves flavour.
        - Removes the skin of vegetable such as beetroot and tomatoes which helps in their peeling.
    - **Disadvantages**
        - Water soluble materials like sugar and anthocyanin pigments are leached by boiling water
        - Fruits lose their colour, flavour and sugar
- **Cooling**
    - After blanching, vegetables are dipped in cold water for better handling and keeping them in good condition.
- **Filling**
    - Before filling, canes are washed with hot water and sterilized, but in developing countries, subjected to a jet of steam to remove dust and foreign material.
    - Automatic, large can filling machines used in advanced countries, but choice grades of fruits are normally filled by hand to prevent bruising.
    - In India, hand filling is common practice.
    - After filling, covering with syrup or brine is done and is called syruping or brining.
    - **Syruping**
        - A solution of sugar in water Syrup.
        - Normally sucrose syrup used in canning.
        - Syrup is added to improve flavour and to serve as heat transfer medium for facilitating processing.
        - Done only for fruits.
    - **Brining**
        - Solution of salt in water - Brine.
        - Only vegetables are brined.
        - Common salt of good quality free from iron is used.
- **Exhausting**
    - Process of removal of air from canes
    - After filling and lidding or clinching, exhausting is essential.
    - **Major advantages**
        - Corrosion of the tinplate and pin holing during storage is avoided.

    - Minimizes discolouration by preventing oxidation.
    - Helps in better retention of vitamins particularly vitamin C.
    - Prevents bulging of canes when stored in a hot climate or at high altitude.
    - Reduces chemical reaction between the container and the contents.
    - Prevents development of excessive pressure and strain during sterilization.
  - Containers exhausted either by heating or mechanically.
  - Heat treatment method is generally used.
- **Sealing**
  - Immediately after exhausting, canes are sealed airtight by means of cane sealer.
  - In case of glass jars, rubber ring placed between mouth of jar and lid, so that sealed airtight.
  - During sealing, temperature should not fall below 74°C.
- **Processing**
  - Heating of foods for preserving
  - In canning technology, processing means heating or cooling of canned foods to inactivate bacteria.
  - Many bacterial spores killed by either high or every low temperature.
  - All fruits and acid vegetables processed satisfactorily at temperature of 100°C, i.e., in boiling water.
  - Presence of acid retards growth of bacteria and their spores.
  - Vegetables (Except more acid ones like tomato) are nonacid in nature, have hard texture, and proximity to soil which infect with spore bearing organisms processed at higher temperatures of 115 to 121°C.
- **Cooling**
  - After processing, canes are cooled rapidly at about 39°C to stop cooking process and to prevent stack burning.
  - Cooling is done by following methods.
    - Dipping or immersing the hot canes in tanks containing cold water.
    - Letting cold water into the pressure cooker especially in case of vegetables.

        - Spraying canes with jets of cold water
        - Exposing canes to air.
        - Generally, dipping canes in cold water is used.
    - If canned products are not cooled immediately after processing, peaches and pears becomes dark in colour, tomato turn brownish and bitter in taste, peas become pulpy with cooked taste and many vegetables develop flat sour.
- **Storage**
    - After labelling, canes should be packed in strong wooden cases or corrugated cardboard cartons and stored in cool and dry place.
    - Outer surface of canes should be dry as even small traces of moisture sometimes induce rusting.
    - Storage at high temperature is avoided, as shortens shelf life and leads to formation of hydrogen swell.
    - Canned peach, grapefruit, pineapple, beans, spinach, pea, celery stored for about two years, while pear, apricot, carrot, beetroot, tomato can be stored for long period.

10.2. Preservation by bottling

- Bottles - very good containers for home preservation of fruits (Fig. 2).
- Fruits look attractive through glass and do not develop metallic flavour.
- Bottling does not need sealing machine.
- Products remain in very hygienic condition and do not come into contact with rubber or metal.

Fig. 2. Fruit preservation in bottles

10.2.1. Filling and syruping

- Bottles are thoroughly washed and sterilized.
- Fruit slices are filled leaving about 3cm space at top of jar or bottle.
- Sugar syrup recommended for different fruits is filled leaving ahead space of 1-1.5 cm.

10.2.2. Exhausting and sterilization

- Separate exhausting of bottles is not required and done simultaneously with sterilization by putting pad of cloth under bottles.
- Bottles should not be immersed in hot water, otherwise break because of sudden rise in temperature.

10.3. Preservation by freezing

- At temperature below freezing point of water, growth of microorganisms and enzyme activity are reduced to minimum.
- Most perishable foods preserved for several months.
- Fruits, vegetables, juices and fleshy foods are preserved
- Cold storage - Storage at temperature above freezing, and 16°C (60°F) down to -2.2°C (28°F).
- Good frozen storage means -18°C or below.
- Refrigerated or cold storage generally preserve perishable foods for days or weeks depending upon food.
- Frozen storage will preserve perishable foods for months or even years.
- Distinctions between refrigeration and freezing temperatures are related to microorganisms activity.
- Food spoilage microorganisms grow rapidly at above 10°C

10.3.1. Methods of freezing

- **Sharp freezing (Slow freezing)**
  - First used in 1861, involves freezing by circulation of air, either naturally or with aid of fans.
  - Temperature from -15 to -29°C and freezing from 3 to 72 hours.
- **Quick freezing**
  - Food attains temperature of maximum ice crystal formation (0 to -4°C) in 30 min. or less.
  - By direct immersion
  - By indirect contact with refrigerant
  - By air blast
- **Cryogenic freezing**
  - Few require ultrafast freezing.
  - Freezing at very low temperature (Below -60°C).
  - Refrigerant used at present in cryogenic freezing is liquid nitrogen and liquid CO_2.
- **Dehydro freezing**
  - Process where freezing is preceded by partial dehydration.
  - Dehydration does not cause deterioration and dehydro frozen food is relatively more stable.

- **Freeze drying**
    - Food is first frozen at -18°C on trays in lower chamber of freeze drier and frozen material is dried under high vacuum (0.1 mm Hg) in upper chamber.
    - Direct sublimation takes place without passing through intermediate liquid stage.

10.4. Preservation by dehydration / drying

- Practice of drying of food stuffs especially fruits and vegetables for preserving is very old.
- Term 'drying' and 'dehydration' means removal of water.
- Drying under influence of non-conventional energy sources like sun and wind.
- Quick removal of moisture prevents growth of microorganisms.
- Dehydration - Process of removal of moisture by application of artificial heat under controlled conditions of temperature, humidity and air flow.

10.4.1. Types of drier

- **Air convection driers**
    - Kiln drier
    - Cabinet, tray or pan drier
    - Tunnel drier
    - Continuous conveyor belt drier
    - Belt trough drier
    - Air lift drier
    - Fluidized bed drier
    - Spray drier
- **Drum or roller drier**
    - Atmospheric drier
    - Vacuum drier
- **Vacuum drier**
    - Vacuum shelf
    - Vacuum belt
    - Freeze drier

10.5. Preservation by irradiation

- Sterilization of food by ionizing radiation
- Unacceptable flavour of some irradiated foods
- When gamma rays or electron beams pass through food, collisions between ionizing radiation and food particles at atomic and molecular levels, resulting in production of ion pairs and free radicals.
- Reactions among themselves and with other molecules result in physical and chemical phenomena which inactivate microorganism in food.
- Irradiation is considered to be 'cold sterilization' i.e., food is free of microorganism without high temperature treatment.
- Radiation dose of upto 1 Mrad is not hazardous.
- Ionizing radiation used for sterilization of foods in hermetically sealed packs, reduces spoilage of organisms in perishable foods, delays ripening of fruits, inhibits sprouting of root vegetables and controls infestation of insects in stored cereals.

Section II
Value Addition in Horticultural Crops

1

Value Addition in Fruits

1.1. Mango

1.1.1. Mango squash

Step 1 : Selection of fruits

Step 2 : Washing

Step 3 : Pulping

Step 4 : Preparation of syrup (sugar + water + acid, heating just to dissolve)

Step 5 : Straining

Step 6 : Mixing with juice

Step 7 : Addition of preservative (0.6 g KMS (Potassium metabisulfite)/litre squash)

Step 8 : Bottling

Step 9 : Capping

Step 10 : Storage

1.1.2. Mango RTS (ready-to-serve)

Step 1 : Fruit (pulp/juice)

Step 2 : Mixing with strained juice

Step 3 : Preparation of syrup solution (sugar + water + acid, heated just to dissolve)

Step 4 : Homogenization

Step 5 : Bottling

Step 6 : Crown corking

Step 7 : Pasteurization (at about 90°C) for 25 minutes

Step 8 : Cooling

Step 9 : Storage

1.1.3. Mango jam

Ingredients

- Mango pulp - 1 kg
- Sugar - 750 g
- Citric acid - 20 g

Method

- Mix the mango pulp and sugar thoroughly in a clean stainless steel vessel.
- Add citric acid and boil mango pulp and sugar till jam consistency (68°B) is reached.
- Fill hot in sterilized bottles, cool and store.

1.1.4. Mango pickle

- Mango (peeled and sliced) - 1 kg, salt - 200 g, red chilli powder 10 g, asafoetida -5 g, fenugreek, black pepper, cardamom (large), cumin and cinnamon (powdered) each 10 g, clove (headless) 6 numbers.

Preparation

Step 1 : Mangoes (Mature, green)

Step 2 : Washing

Step 3 : Peeling

Step 4 : Slicing

Step 5 : Putting slices in jar, sparkling salt

Step 6 : Putting in sun for one week (shaking jar at least twice a day to mix the salt).

Step 7 : Mixing spices

Step 8 : Storage at ambient temperature (in cool and dry place).

1.1.5. Immature mango (Mavadu)

Ingredients

Raw mangoes - Small sour variety	100 piece
Salt	2 kg
Fenugreek	350 g
Red chillies	700 g
Turmeric powder	125 g
Mustard seeds	250g
Castor oil	100 ml

Method

- Wash and wipe mangoes.
- Apply castor oil to the mangoes.
- Mix with salt and keep aside.
- Roast and powder fenugreek seeds and mustard.
- Remove seeds from chillies, sun dry and powder.
- Peel garlic and grind along with all the other spices using salt solution drained from mangoes.
- Mix with sesame oil and stuff mangoes.
- Store in jars and pour remaining oil.
- Cover well; keep aside for 2 weeks before use.

1.1.6. Sweet mango chutney

- **Recipe:** Mango slices or shreds - 1.0 kg, sugar or gur - 1.0 kg, salt - 45 g, onions (chopped) - 50 g, garlic (chopped) - 15 g, ginger (chopped) - 15 g, red chilli powder - 10 g, black pepper, cardamom, cinnamon, cumin - 10 g each, cloves - 5 nos. and vinegar - 170 ml.

Step 1 : Mature mangoes

Step 2 : Washing

Step 3 : Peeling

Step 4 : Grating or slicing

Step 5 : Cooking with a little water to make highly soft

Step 6 : Mixing with sugar and salt and leaving for an hour

Step 7 : Keeping all ingredients (except vinegar) in cloth bag, tied loosely, putting in mixture and cooling on low flame.

Step 8 : During cooking, spice bag pressed occasionally.

Step 9 : Cooking to consistency of jam (upto 105ºC) with stirring occasionally.

Step 10 : Removal of spice bag after squeezing.

Step 11 : Addition of vinegar

Step 12 : Cooking for 2-5 minutes

Step 13 : Filling hot into bottles

Step 14 : Sealing (air tight)

Step 15 : Storage at ambient temperature

1.1.7. Protein enriched spicy mango bar

Method of preparation

- Green gram dhal was roasted slightly to remove the raw flavour.
- Then, ground into fine flour.
- Green gram dhal flour was steamed for 10 minutes, dried and passed through fine mesh to avoid lumps.
- Soy flour was steamed for 10 minutes to reduce raw beany flavour and to inactivate trypsin inhibitor, dried and passed through fine mesh to avoid lumps.
- Soy protein isolate was steamed for 10 minutes, dried and passed through fine mesh to avoid lumps.

Preparation of spicy mango bar

- Mango pulp (80 g) was heated for 3 minutes to inactivate enzymes
- Corn flour (4 g) was made into paste with little amount of water (to avoid formation of lumps)
- The pulp was mixed with corn flour paste, sugar (50 g), chilli powder (0.75 g), asafoetida (0.1 g) and green gram (20 g)/soy flour (20 g)/soy protein isolate (10 g)/skim milk powder (10 g)
- This mix was heated on medium flame by stirring continuously using a wooden laddle upto a final TSS of 45° brix.
- Concentrated pulp was cooled to room temperature and 0.1 per cent KMS was added and mixed thoroughly.
- Finished pulp was spread evenly in an aluminium tray to a thickness of 0.5 cm and dried at 60°C for 6 hr in a cabinet drier.

1.1.8. Mango slices (Amchur)

Step 1 : Mangoes (mature, green)

Step 2 : Washing

Step 3 : Peeling

Step 4 : Cutting into slices of suitable size

Step 5 : Dipping in 2% salt solution for an hour

Step 6 : Draining out and dipping in 2000 ppm SO_2 solution for 2 hrs

Step 7 : Spreading slices in thin layer on wooden trays

Step 8 : Sun drying

Step 9 : Filling into airtight containers

Step 10 : Storage in dry place

1.1.9. Mango candy

Ingredients

Mango	1 kg
Sugar	1.120 kg
Water	500 ml
Citric acid	6.4g
KMS	1.2 g

Method

- Selection of ripe mango fruit
- Cutting into pieces
- Soaking in 2% $CaCl_2$ solution
- Preparation of sugar syrup (Addition of sugar, water, citric acid)
- Cooking the fruit in sugar syrup till soft consistency (TSS 75° brix and temperature 106°C)
- Soaking the fruit for 7 days
- Boiling of sugar syrup to 60° brix
- Packing the mango preserve in glass jar
- Draining the sugar syrup
- Drying in shade to get mango candy

1.1.10. Flow Chart for Processing of Fruit Powder

Step	Mango	Sapota
Step 1	Selection of mango	Selection of sapota
Step 2	Washing	Washing
Step 3	Peeling	Peeling
Step 4	Destoning	Removal of seeds
Step 5	Slicing	Slicing
Step 6	Pulping	Pulping
Step 7	Pre treatment (0.05% KMS)	Pre treatment (0.05% KMS)
Step 8	Addition of foaming agents (Glycerol-mono-stearate/Methyl cellulose/Egg albumen)	Addition of foaming agents (Glycerol-mono-stearate/ Methyl cellulose/Egg albumen)
Step 9	Whipping (20 minutes)	Whipping (25 minutes)

Step 10 : Formation of foam

Step 11 : Spreading the foamed pulp on a tray

Step 12 : Drying at two stages (75°C for one hour and 65°C for further drying)

Step 13 : Cooling
Step 14 : Grading (mix)
Step 15 : Sieving (BS 60)
Step 16 : Packaging
Step 17 : Storing at ambient temperature

1.2. Banana

1.2.1. Clarification of banana juice

Step 1 : Selection (ripc, firm sound)
Step 2 : Washing (surface)
Step 3 : Peeling and slicing
Step 4 : Pulping (mix 3 minutes)
Step 5 : Pasteurizing (80°C for 10 minutes)
Step 6 : Cooling (40°C)
Step 7 : Adding enzyme (pectinase 0.5%)
Step 8 : Leaving over night (room temperature)
Step 9 : Separating juice from pulp
Step 10 : Filtering (vacuum pump)
Step 11 : Clarified banana juice

1.2.2. Preparation of banana flour

Step 1 : Fully ripe banana fruits
Step 2 : Peeling
Step 3 : Pulping
Step 4 : Homogenizing
Step 5 : Spray drying (30-32°C) / drum drying (<94°C)
Step 6 : Moisture content (2 - 4%)
Step 7 : Banana powder
Step 8 : Packing in polyethylene bags
Step 9 : Storage

1.2.3. Banana chips/ crisps

Step 1 : Matured Nendran fruits
Step 2 : Harvesting
Step 3 : Dehanding

Step 4 : Peeling

Step 5 : Treating with 1% KMS

Step 6 : Cut into slices (1.2-0.8 mm thickness)

Step 7 : Deep fat frying

Step 8 : Yellow coloured chips

Step 9 : Addition of salt

Step 10 : Packing in polyethylene bags

Step 11 : Storage

1.2.4. Banana fig

Step 1 : Fully ripe banana

Step 2 : Peeled

Step 3 : Treated with 1% KMS

Step 4 : Dried (sun/ oven at 50°C) for 24 hr

Step 5 : Cooling

Step 6 : Packing in polyethylene bags

Step 7 : Storage

1. Used as nutritious adjuvant in several food preparations like bread, cakes, biscuits, health drink and baby food formulations.
2. Blended with other cereals for making chappathi, rotties.

1.2.5. Banana puree

Step 1 : Ripe banana fruits

Step 2 : Peeling

Step 3 : Pulping

Step 4 : Homogenizing

Step 5 : Deaerating

Step 6 : Flash pasteurizing

Step 7 : Packing into sterile aseptic bags

- Used in manufacture of dairy products such as yoghurts, milk shakes, ice creams, baking breads and cakes, banana flavoured drinks.

On a small scale

Step 1 : Ripe banana fruits

Step 2 : Peeling

Step 3 : Pulping
Step 4 : Pasteurizing at 85°C
Step 5 : Treated with 2000 ppm KMS
Step 6 : Packing in sterile aseptic bags

1.2.6. Banana RTS juice

Step 1 : Banana puree
Step 2 : Treating with pectolytic enzyme
Step 3 : Filtration/centrifugation
Step 4 : Clear juice
Step 5 : Decanting
Step 6 : Diluting 2 ½ times with water
Step 7 : Adjusting TSS 16°B
Step 8 : Acidity 0.3%
Step 9 : Pasteurizing
Step 10 : Filling in sterile bottles
Step 11 : Sealing
Step 12 : Processing at 100°C for 20 minutes
Step 13 : Cooling
Step 14 : Storing

1.2.7. Banana fruit bar

Step 1 : Ripe banana fruit
Step 2 : Homogenizing
Step 3 : Pulp
Step 4 : Addition of sugar, citric acid, pectin
Step 5 : Dehydrating the mass in ghee coated at 70°C in oven
Step 6 : Sets into sheet
Step 7 : Cut into suitable size
Step 8 : Packed in polyethylene pouches

1.2.8. Banana biscuits

Step 1 : Banana flour + maida (60% + 30%)
Step 2 : Flour mixture
Step 3 : Addition of sugar, dalda, baking powder, milk powder, essence

Step 4 : Mixing
Step 5 : Made into dough shaping
Step 6 : Shaping
Step 7 : Baking

1.2.9. Banana jelly

Step 1 : Boiling clear strained fruit extract
Step 2 : Addition of required amount of sugar, citric acid and pectin
Step 3 : Perfect jelly (transparent, attractive and sparkling in colour)

1. 2.10. Banana jam

Step 1 : Fruit pulp
Step 2 : Addition of sugar, pectin and citric acid
Step 3 : Cooking to a good set

1.2.11. Banana wine

Step 1 : Banana juice (enzyme treated clear juice)
Step 2 : Addition of *Saccharomyces cerevisiae* var. *ellipsoideus*
Step 3 : Fermenting for 14 days at 24-26°C
Step 4 : Filtering
Step 5 : Clarify
Step 6 : Bottling
Step 7 : Pasteurizing
Step 8 : Storing in bottles for aging
Step 9 : Alcohol content (9-12%)

1.2.12. Alcohol from banana peel

Step 1 : Ripe banana peel
Step 2 : Macerating
Step 3 : Acidification
Step 4 : Saccharification
Step 5 : Fermentation
Step 6 : Filtration
Step 7 : Distillation
Step 8 : Alcohol

1.2.13. Health drinks and baby food

- A highly nutritious and tasty health drink formulation and a baby food formula has been developed at NCRB using banana flour/ powder after supplementing with natural source of proteins, minerals, vitamins and fat.
- Has got a shelf life of about 6 months and is suitable for children and adults.

1.3. Papaya

1.3.1. Papaya tutti fruity

Ingredients

Mature unripe papaya	1 kg
Sugar	¾ kg
Citric acid	10 g
Essence colour	

Method

- Harvest mature, sound raw papaya, make incisions and drain the papain extract.
- Wash the fruit well, peel, remove seeds and cut into small pieces.
- Soak in 2 per cent brine containing one per cent calcium chloride for 30 minutes.
- Drain, wash papaya pieces and add equal quantitics of sugar and boil for 5 minutes.
- Let soak at room temperature for 4 hours, mix the required essence, colouring and citric acid.
- Add more sugar to above mixture and boil upto a sugar concentration of 70° brix.
- Drain the candied papaya pieces, spread on a tray and shade dry.
- Pack in polyethylene bags, seal and store.

1.3.2. Papaya ready to serve beverage (RTS)

Ingredients

Papaya	1 kg
Sugar	500 g
Citric acid	10 g
Water	2 lit
Preservative (KMS)	¼ tsp

Method

- Wash the papaya fruits, peel, remove seeds, cut into small pieces and make to pulp.
- To the water, add sugar and citric acid, mix well and boil for a few minutes and cool.
- Strain the sugar syrup and mix with papaya pulp and heat juice upto 80°C.
- Add essence and preservative to the juice and fill the juice into sterilized bottles and seal.
- Further pasteurize the bottle at 80°C, cool, wipe dry and store.

1.3.3. Papaya - Pineapple RTS

Ingredients

Fruit pulp	1kg
Sugar	475 g
Citric acid	10 g
Water	2.8 litre
KMS /SB	200 mg

Method

- Wash and peel ripe papaya.
- Scoop out the flesh and pulp it.
- Mix sugar, water and citric acid, bring it to boil and filter.
- Mix with the pulp.
- Heat it to 80ºC.
- Bottle and crown cork.
- Pasteurize in a water bath for 30 minutes.
- Cool, label and store till use.

Note

- 200 ppm of sodium benzoate or KMS can be added for commercial and long storage.
- Sterilized bottle should be used for bottling the juice.
- While pasteurizing sufficient cloth bedding should be given in between the bottles and the vessel so that they are not broken or cracked during processing.
- To bring variation, RTS can be prepared with papaya blended with other fruits.
- Fruits suitable for mixing with papaya are mango and pineapple.

1.3.4. Papaya pickle

- Peeled papaya pieces 1 kg, salt 100 g, red chilli powder 10 g, cardamom (large), cumin and black pepper (powdered) each 10 g, vinegar 750 ml.

Processing of papaya pickle

Step 1 : Papaya (mature, green)
Step 2 : Washing
Step 3 : Peeling
Step 4 : Removal of seeds
Step 5 : Cutting into pieces
Step 6 : Boiling (to soften)
Step 7 : Removal of pieces from boiling water
Step 8 : Mixing with salt
Step 9 : Drying in shade for few minutes
Step 10 : Mixing spices
Step 11 : Filling in jars
Step 12 : Addition of vinegar (sufficient to cover the pieces)
Step 13 : Keeping for a week
Step 14 : Storage

1.3.5. Papaya jam

Ingredients

Papaya pulp	100 g
Sugar	75g
Citric acid	A pinch

Method

- Select a ripe fruit, peeled and pulping is done using mixie.
- Addition of sugar, citric acid to the pulp and cooked to jam consistency.
- Poured in sterilized bottles.

1.3.6. Papaya squash

Ingredients

Papaya pulp	1 kg
Sugar	1.80 kg
Water	1 litre
Citric acid	25 g

Step 1 : Selection of fruits
Step 2 : Washing
Step 3 : Pulping
Step 4 : Preparation of syrup (sugar + water + citric acid heating just to dissolve)
Step 5 : Straining
Step 6 : Mixing with juice
Step 7 : Addition of preservative (2.5 g KMS/litre squash)
Step 8 : Boiling
Step 9 : Capping
Step 10 : Storage

1.4. Gauva

1.4.1. Steps in extraction of guava pulp

Step 1 : Ripe fruits
Step 2 : Washing
Step 3 : Cutting into pieces
Step 4 : Mixing with water (1:1)
Step 5 : Passing through pulper
Step 6 : Pulp

1.4.2. Processing of RTS beverages

Step 1 : Guava (juice)
Step 2 : Mixing with strained syrup solution (sugar + water + citric acid heated just to dissolve) according to recipe
Step 3 : Homogenization
Step 4 : Bottling
Step 5 : Crown corking
Step 6 : Pasteurizing (at about 90°C for 25 minutes)
Step 7 : Cooling
Step 8 : Storage

1.4.3. Guava jelly

Ingredients

Guava	1 kg
Sugar	600 g
Citric acid	8 g

Method

1. Select well matured sound guava fruits, wash well and cut into small slices.
2. To this, add equal quantities of water and boil under low flame for 30 minutes.
3. Strain the pectin extracts, add sugar and citric acid.
4. Boil the pectin extract and sugar mixture up to 65°B and fill hot into sterilized bottles, cool and store.

1.4.4. Guava – lime - ginger RTS

Steps in the preparation of guava - lime - ginger RTS

	Guava	Lime	Ginger	Sugar + citric acid + water
Step 1	Selecting	Selecting	Selecting	Boiling
Step 2	Washing	Washing	Washing	Cooling
Step 3	Peeling	Cutting	Peeling	Filtering
Step 4	Cutting	Juice extraction	Cutting	Sugar syrup
Step 5	Steam blanching (3 min)	Filtering	Pulping	
Step 6	Pulping	Lime juice	Juice extraction	
Step 7	Seed removal		Filtering	
Step 8	Guava pulp		Ginger juice	
Step 9	Pasteurization			
Step 10	Mixing			
Step 11	Filtering			
Step 12	Filling in the bottles (head space 2”)			
Step 13	Corking			
Step 14	Pasteurization for 30 minutes			
Step 15	Packing			
Step 16	Storing at room and refrigeration temperature			

1.4.5. Guava - banana RTS

Steps in the preparation of guava - banana RTS

	Guava	Banana	Sugar + citric acid + water
Step 1	Selecting	Selecting	Boiling
Step 2	Washing	Peeling	Cooling
Step 3	Peeling	Cutting	Filtering

Step 4	: Cutting	Pulping	Sugar syrup
Step 5	: Steam blanching (3 min)	Banana pulp	
Step 6	: Pulping		
Step 7	: Seed removal		
Step 8	: Guava pulp		
Step 9	: Pasteurization		
Step 10	: Mixing		
Step 11	: Filtering		
Step 12	: Filling in the bottles (head space 2")		
Step 13	: Corking		
Step 14	: Pasteurization for 30 minutes		
Step 15	: Packing		
Step 16	: Storing at room and refrigeration temperature		

1.4.6. Mixed fruit squash

Steps in the preparation of mixed fruit squash (Guava, banana, mango)

	Guava	Banana	Mango	Sugar + citric acid + water
Step 1	: Selecting	Selecting	Selecting	Boiling
Step 2	: Washing	Peeling	Washing	Cooling
Step 3	: Peeling	Cutting	Peeling	Filtering
Step 4	: Cutting	Pulping	Cutting	Sugar syrup
Step 5	: Steam blanching (3 min)	Banana pulp	Pulping	
Step 6	: Pulping		Mango pulp	
Step 7	: Seed removal			
Step 8	: Guava pulp			
Step 9	: Mixing			
Step 10	: Filtering			
Step 11	: Filling in the bottles (head space 2")			
Step 12	: Capping			
Step 13	: Labelling			
Step 14	: Storing at room and refrigeration temperature			

1.4.7. Dehydration of guava pieces

Step 1 : Selection of guava

Step 2 : Washing

Step 3 : Cutting into eight portions vertically

Step 4 : Scooping out the seeds

Step 5 : Drying in cabinet drier (60°C)

Step 6 : Cooling to room temperature

Step 7 : Tempering overnight

Step 8 : Packing and storage

1.5. Pineapple

1.5.1. Pineapple ready to serve beverage (RTS)*

Ingredients

Pineapple	1 kg
Sugar	500 g
Citric acid	10 g
Water	2.5 lit

Method

- Selecting pineapple fruits, peel, cut into small pieces and make into pulp.
- To the water, add sugar and citric acid, mix well and boil for a few minutes and cool
- Strain the sugar syrup and mix with the pineapple pulp and heat the juice up to 80°C.
- Add the essence and preservative to the juice, and fill the juice into sterilized bottles and seal.
- Pasteurize the bottle at 70°C, cool, wipe dry and store.

1.5.2. Pineapple squash

Pineapple pulp	1 kg
Sugar	1.75 kg
Water	1 litre
Citric acid	25 g
KMS	2.25 g

Step 1 : Pineapple

Step 2 : Peeling

Step 3 : Cutting into pieces

Step 4 : Pulping

Step 5 : Preparation of syrup (sugar + water + citric acid, heating just to dissolve)

Step 6 : Straining

Step 7 : Mixing the pulp + sugar syrup

Step 8 : Addition of preservative (2.5 g KMS/litre squash)

Step 9 : Bottling

Step 10 : Capping

Step 11 : Storage

1.5.3. Pineapple jam

Pineapple pulp	1 kg
Sugar	750 kg
Citric acid	10 kg

Step 1 : Pineapple pulp

Step 2 : Cooking with sugar

Step 3 : Mixing well

Step 4 : Cooking till jam consistency (Temperature 105°C / TSS 68° B)

Step 5 : Filling in sterilized jar

Step 6 : Cooling

Step 7 : Storage

1.5.4. Spiced pineapple RTS beverages

Step 1 : Selecting pineapple fruits (mature, sound, fresh, ripe and firm)

Step 2 : Washing (tap water)

Step 3 : Removing scaly portions and eyes

Step 4 : Cutting in to small pieces

Step 5 : Extracting juice (pulping)

Step 6 : Filtering juice (muslin cloth)

Step 7 : Preparing sugar syrup (sugar + citric acid + water)

Step 8 : Filtering sugar syrup (muslin cloth)

Step 9 : Cooling (syrup)

Step 10 : Mixing with fruit juice

Step 11 : Adding salt and spice extract to the fruit juice

Step 12 : Mixing thoroughly

Step 13 : Filling in bottles (head space 2 cm)

Step 14 : Crown corking

Step 15 : Pasteurization at 80°C for 30 minutes

Step 16 : Cooling

Step 17 : Labelling

Step 18 : Storage (Room temperature)

1.6. Avocado

1.6.1. Steps in the preparation of avocado bread

Step 1 : Sieve the refined flour

Step 2 : Rub fat and fruit pulp into flour

Step 3 : Add sugar, salt, water and activated yeast

Step 4 : Knead into dough

Step 5 : Allows fermenting (2 to 3 hours)

Step 6 : Knead softly

Step 7 : Shape into bread

Step 8 : Place in bread moulds (keep for 15 minutes)

Step 9 : Glazing

Step 10 : Bake at 435°F (223°C) (25 to 30 min)

Step 11 : Cool and pack

Step 12 : Store at room temperature

1.6.2. Steps in the preparation of avocado chilli biscuits

Step 1 : Sieve flour with baking powder and powdered sugar

Step 2 : Homogenize fat and pulp

Step 3 : Mix with flour

Step 4 : Add chopped chillies and curry leaves

Step 5 : Make into smooth dough

Step 6 : Cut into desired shapes

Step 7 : Arrange in greased plates

Step 8 : Bake at 105°C (15 minutes)

Step 9 : Cool and pack

Step 10 : Store at room temperature

1.6.3. Steps in the preparation of avocado incorporated vermicelli

Step 1 : Sieve flour twice

Step 2 : Steam blanch (15 minutes)

Step 3 : Add pulp, salt and water

Step 4 : Mix well

Step 5 : Extrude it

Step 6 : Steam (15 minutes)

Step 7 : Dry in cabinet drier (60°C for 4 hrs)

Step 8 : Cool and pack

Step 9 : Store at room temperature

1.7. Ber

1.7.1. Osmo air dehydration of ber fruits

Step 1 : Ber fruits

Step 2 : Washing

Step 3 : Inspection

Step 4 : Weighing

Step 5 : Pre-treatments

Step 6 : Blanching and pricking 3 minutes in boiling water, 2 mm pricks - 16 per fruits

Pricking and sulfuring 2 mm pricks - 14 per fruit; 3 g sulphur per kg of fruit for 30 minutes

Lye peeling 5% caustic soda for 2 minutes; rinse with 0.1% citric acid washing

Step 7 : Osmosis

Step 8 : 70° B and 50°C for 70 hrs

Step 9 : Draining, washing, surface drying

Step 10 : Weighing

Step 11 : Cabinet drying

Step 12 : 50°C for 15 hrs - cabinet drier

Step 13 : Packaging

Step 14 : Storing

1.7.2. Dehydration of ber fruits

Step 1 : Ber fruits

Step 2 : Washing

Step 3 : Inspection
Step 4 : Weighing
Step 5 : Pre-treatments
Step 6 : Blanching and pricking 3 minutes in boiling water, 2 mm pricks - 16 per fruits
Lye peeling 5% caustic soda for 2 minutes; rinse with 0.1% citric acid
Step 7 : Osmosis
Step 8 : 70° B, 50°C for 70 hrs
Step 9 : Draining, washing, surface drying
Step 10 : Weighing
Step 11 : Cabinet drying
Step 12 : 50°C for 15 hrs - cabinet drier
Step 13 : Packaging
Step 14 : Storage

1.8. Jamun fruit

1.8.1. Extraction of jamun juice

Step 1 : Jamun fruit
Step 2 : Washing
Step 3 : Grading
Step 4 : Destoning
Step 5 : Pressing
Step 6 : Juice
Step 7 : Heating at 85°C
Step 8 : Cooling at 30°C
Step 9 : Addition of sodium benzoate
Step 10 : Squash

1.8.2. Extraction of wine from jamun fruit

Step 1 : Jamun fruit
Step 2 : Washing
Step 3 : Grading
Step 4 : Destoning
Step 5 : Pulping

Step 6 : Adding sugar and adjust to 22° B

Step 7 : Pasteurizing for 15 minutes (60°C)

Step 8 : Adding starter inoculums (10% v/v)

Step 9 : Fermenting for 72 hrs (28°C)

Step 10 : Filtering wine (4 fold thick filter paper)

Step 11 : Filling in bottle and sealing

Step 12 : Pasteurizing for 30 minutes (60°C)

Step 13 : Cooling

Step 14 : Ageing for 3 months (5-7°C)

1.9. Tamarind

1.9.1. Tamarind jam

Step 1 : Selection of ripe fruit

Step 2 : Shelling

Step 3 : Boiling the pulp for 10 minutes

Step 4 : Draining the pulp

Step 5 : Separating from the seeds

Step 6 : Adding 2 cups of brown sugar for 1 cup of pulp

Step 7 : Cooking the mixture with constant stirring

Step 8 : Cooking till thick consistency

Step 9 : Packing in sterilized jar

Step 10 : Sealing

1.9.2. Tamarind syrup

Step 1 : Boiling immature tamarind fruit pulp

Step 2 : Straining through cheese cloth

Step 3 : Addition of ½ teaspoon of baking soda for 1 cup of juice

Step 4 : Boiling the mixture to ½ the original quantity

Step 5 : Removing the scum

Step 6 : Straining the juice

Step 7 : Addition ½ cup of sugar

Step 8 : Boiling for 20 minutes

Step 9 : Cooling

Step 10 : Packing in sterilized jar

Step 11 : Sealing

1.9.3. Tamarind candy

Step 1 : Selection of tamarind fruit

Step 2 : Peeling

Step 3 : Pouring the boiling sugar syrup (62°C)

Step 4 : Boiling syrup prepared by mixing 3 parts of sugar with one part of water

Step 5 : Soaking for 3 days

Step 6 : Draining

Step 7 : Soaking in freshly prepared syrup

Step 8 : Repeating the process until the fruit is sweet enough

Step 9 : Arranging the sweetened fruit in bamboo racks

Step 10 : Drying in sun

Step 11 : Final drying in oven at low temperature

Step 12 : Cooling

Step 13 : Wrapping the cellophane

1.9.4. Tamarind pickle

Step 1 : Selection of tamarind fruit

Step 2 : Soaking in water for 12 hr

Step 3 : Separation of pulp

Step 4 : Addition of sugar to pulp (1:1)

Step 5 : Boiling and stirring the mixture

Step 6 : Addition of spice mixture
(Spice mixture – coriander, cumin seed, black cumin, clove, cardamom, cinnamon, chilli, salt, mustard oil, black pepper)

Step 7 : Ground to paste

Step 8 : Adding to boiling tamarind

Step 9 : Packed in sterilized glass jar

Step 10 : Sealing

Step 11 : Storing

1.10. Other fruits

1.10.1. Grape squash

Ingredient

Grapes	1 kg
Sugar	1 ¾ kg
Citric acid	40 g
Water	1¼ lit
Tonovin essence	1 tbs
Sodium benzoate	½ tbs

Method

- Wash grapes in water.
- Cook the grapes in just enough water.
- Cool and extract juice.
- Add citric acid with sugar.
- Add water and boil till the sugar dissolves.
- Cool the syrup and filter.
- Add fruit juice and mix
- Add essence and preservative, mix well and bottle it.

1.10.2. Processing of grape wine

Step 1 : Ripe grapes

Step 2 : Removal of stems

Step 3 : Crushing (basket press)

Step 4 : Filling jar up to three fourths

Step 5 : Addition of sugar (20-24% TSS)

Step 6 : Adjustment of pH (0.6 to 0.8% acid)

Step 7 : Addition of preservative (KMS 1.5 g/10 kg grapes)

Step 8 : Keeping for an hour

Step 9 : Addition of wine yeast (S.E 20 ml/5 kg grapes)

Step 10 : Fermentation (for 2 days)

Step 11 : Filtration

Step 12 : Fermentation (for 10 days)

Step 13 : Racking (syphoning off clear liquid)

Step 14 : Fining and filtration (bentonite)

Step 15 : Ageing (maturation) (6-8 months)

Step 16 : Pasteurization at 82°C (for 2 minutes)

Step 17 : Bottling

Step 18 : Crown corking

Step 19 : Pasteurization at 82°C (for 20 minutes)

Step 20 : Cooling

Step 21 : Storage

1.10.3. Lime pickle

- Lime - 1 kg, salt - 200 g, red chilli powder - 15 g, cinnamon, cumin, cardamom and black pepper (powdered) each 10 g, clove - 5 nos.

Process

Step 1 : Limes

Step 2 : Washing

Step 3 : Cutting into 4 pieces

Step 4 : Squeezing out juice from ¼ amount of fruit

Step 5 : Mixing spices and salt with juice

Step 6 : Mixing with lime pieces

Step 7 : Filling in jars

Step 8 : Covering with lid

Step 9 : Keeping in sun for 4-6 days (shaking jar at least twice a day)

Step 10 : Storage at ambient temperature

1.10.4. Apple sauce

- Apple - 1.0 kg, sugar - 250 g, salt - 10 g, onion (chopped) - 200 g, ginger (chopped) - 100 g, garlic (chopped) - 50 g, red chilli powder - 10 g, cloves - 5 nos. cinnamon, cardamom - 15 g (each), vinegar - 50 ml, sodium benzoate - 0.7 g/kg of finished product.

Step 1 : Apples

Step 2 : Washing

Step 3 : Peeling

Step 4 : Removal of core and seeds

Step 5 : Making into fine pulp

Step 6 : Straining of pulp

Step 7 : Cooking pulp with one third quantity of sugar

Step 8 : Putting spice bag in pulp and pressing occasionally

Step 9 : Cooking to one third of its original volume of pulp
Step 10 : Removal of spice bag (after squeezing in pulp)
Step 11 : Adding remaining sugar and salt
Step 12 : Cooking to one third of its original volume
Step 13 : Addition of vinegar and preservative
Step 14 : Filling hot into bottles
Step 15 : Crown corking
Step 16 : Pasteurization at 85-90°C (for 30 minutes)
Step 17 : Cooling
Step 18 : Storing at ambient temperature

1.10.5. Preparation of jelly from wood apple

Step 1 : Wood apple ripe fruits
Step 2 : Washing and cleaning
Step 3 : Breaking of fruits
Step 4 : Extraction of water (1:4)
Step 5 : Boiling for half an hour
Step 6 : Filtration of extract
Step 7 : Addition of sugar + preservatives (65° B)
Step 8 : Boiling up to 105°C
Step 9 : Jelly preparation
Step 10 : Cooling
Step 11 : Sealing
Step 12 : Storage

1.10.6. Preparation of jack fruit bar

Step 1 : Sound ripe jack fruit
Step 2 : Cutting into two halves
Step 3 : Removing the centre core
Step 4 : Separating carpel from fibrous portion
Step 5 : Removing seed with seed coat from carpel
Step 6 : Cutting the carpel into pieces
Step 7 : Pulping in mixie for 10 minutes
Step 8 : Passing through hand pulper
Step 9 : Mixing with sugar, corn flour and citric acid

Step 10 : Boiling the content

Step 11 : Concentrating the content to 50° B

Step 12 : Cooling (room temperature)

Step 13 : Adding KMS and mixing (product weight basis)

Step 14 : Spreading in an aluminium tray (thickness 1.0 cm)

Step 15 : Drying at 50° C for 12 hrs

Step 16 : Cooling and cutting into rectangular shape bar

Step 17 : Sealing

Step 18 : Labelling

Step 19 : Storing

2

Value Addition in Vegetables

2.1. Tomato

2.1.1. Tomato paste

Steps in tomato paste preparation

Step 1 : Tomato juice (strained)

Cooking to desired consistency (Kettle/vacuum pan)

Step 2 : Judging the end point for puree or paste (Tomato solids by hand refractometer or reduction in volume by measuring stick)

Step 3 : Filling hot into bottles or canes (82-88°C)

Step 4 : Pasteurization in boiling water for 20 minutes

Step 5 : Cooling

Step 6 : Storage at ambient temperature

Step 7 : Packing and storage

2.1.2. Tomato sauce / ketchup

- It is made from strained tomato juice or pulp and spices, salt, sugar and vinegar, with or without onion and garlic and contains not less than 12 per cent tomato solids and 25 per cent total solids.

Steps in tomato sauce/ ketchup preparation

Step 1 : Tomatoes (Fully ripe, red)

Step 2 : Washing and sorting

Step 3 : Cutting and chopping

Step 4 : Heating at 70-90°C for 3-5 minutes (to soften)

Step 5 : Straining tomato pulp/juice

Step 6 : Pulping

Step 7 : Cooking pulp with one third quantity of sugar

Step 8 : Putting spice bag

Step 9 : Cooking on one third of original volume of pulp/ juice

Step 10 : Removal of spice bag (after squeezing in pulp)

Step 11 : Cooking and adding of remaining sugar and salt

Step 12 : Judging of end point (tomato solids by hand refractometer / volume)

Step 13 : Addition of vinegar/ acetic acid and preservative

Step 14 : Filling and cooling

Step 15 : Pasteurization

Step 16 : Crown corking

Step 17 : Storage at ambient temperature (in cool and dry place)

2.1.3. Tomato chutney

Processing of tomato chutney

Step 1 : Tomatoes (fully ripe, red)

Step 2 : Washing

Step 3 : Sorting

Step 4 : Blanching for 2 minutes

Step 5 : Putting quickly in cold water (to crack skin)

Step 6 : Peeling

Step 7 : Crushing

Step 8 : Addition of ingredients except salt and vinegar and cooking gently to desired consistency

Step 9 : Addition of salt and vinegar and cooking for 5 minutes

Step 10 : Addition of preservative

Step 11 : Filling in hot bottles

Step 12 : Sealing

Step 13 : Storage at ambient temperature (in cool and dry place)

2.1.4. Tomato soup mix

Preparation of tomato powder

- Fully ripe and firm tomatoes were washed well in running tap water.
- Then, it is cut into small pieces and dried in the cabinet drier at 80ºC for 10 hrs.
- Dehydrated pieces were then, ground into powder in a mixie.

Preparation of onion powder

- Disease is free big onions are selected and skin is peeled off.
- Then, it is washed in the running tap water and cut into small pieces.
- It is dried in the cabinet drier at 60°C for 7 hrs.
- Dehydrated flakes are ground into powder and packed in polythene bags.
- Ten gram of the prepared tomato soup mix is added to 150 ml of boiled water and stirred thoroughly.

Preparation of tomato soup mix

Ingredients

Tomato powder	5.0 g
Onion powder	0.5 g
Corn flour	2.0 g
Cumin powder	0.5 g
Pepper powder	0.3 g
Salt	1.5 g

Method

- All the ingredients are mixed thoroughly and packed in polythene bags.

2.2. Bitter gourd

2.2.1. Guava, pineapple and bitter gourd juice

- Servings: 4-6

Ingredients

- 1 cup guava chunks
- 1 cup apple chunks (optional)
- 1 cup pineapple chunks
- 1 cup ice cubes
- 1 cup bitter gourd chunks (any variety of bitter gourd)
- Water
- 2 tsp of honey (or enough to taste)

Directions

- Rinse and wash the fruit and bitter gourd thoroughly.
- Remove pineapple skin and cut the fruit into small chunks.
- Cut bitter gourd lengthwise and remove the seed and pulp.
- Cut into chunks.
- Cut other fruit into chunks.
- Put the fruit and bitter gourd in a blender and then, add ice cubes and honey.

- Add enough water to fill half of the blender volume.
- Blend twice on high speed.
- Then, blend continuously on low speed for 2 minutes or until smooth. Serve chilled.
- Drinking bitter gourd fresh as a juice retains many of the nutrient and bioactive compounds beneficial for health that may be lost in different cooking methods.
- This beverage is not recommended for people with hypoglycemia.

2.2.2. Bitter gourd curry

- Servings: 2-4

Ingredients

- 2-3 medium-sized bitter gourds, deseeded and cut into thin slices (dark green variety).
- 1 onion, finely chopped
- 2 tomatoes, chopped
- 1 tsp ginger-garlic paste
- 1 tsp cumin seed
- ¼ tsp red chilli powder
- ¼ tsp turmeric powder
- ½ tsp coriander powder
- ¼ tsp dried mango powder
- Salt to taste
- 2 tbsp cooking oil
- 1 pinch sugar
- Optional: 2-3 green chillies, finely chopped ¼ tsp garam masala (mixed Indian spices).

Preparation

- Heat oil in a pan.
- Add cumin seed and once it begins to sizzle, add onions and green chillies.
- Fry onions until tender and brown.
- Add tomatoes, ginger - garlic paste and dried mango powder.
- Add red chilli powder, turmeric powder, coriander powder and salt.
- Sauté to blend in the spices and continue cooking until the oil separates.
- Add salt and sliced bitter gourd, cover and cook for 7-8 minutes or until the bitter gourd is cooked.
- Sprinkle garam masala on top and serve hot with basmati rice.

2.2.3. Bitter gourd salad

- Servings: 2
- Preparation time: 30 minutes

Ingredients

100 g	Bitter gourd
50 g	Onion
50 g	Tomato
2 tbsp	Lime juice
¼ tbsp	Chilli powder
½ tbsp	Pepper powder and salt to taste
2 tbsp	Chopped coriander leaves

Directions

- Wash and finely slice the bitter gourd and remove the seeds.
- Apply the salt on both sides of the bitter gourd.
- Set aside for 30 minutes and rinse with water twice.
- Slice the onion and tomato; add to the bitter gourd with chilli powder and salt.
- Mix well.
- Add lime juice, coriander and pepper powder to the salad when ready.

2.2.4. Stuffed karela

- Servings: 3
- Preparation time: 30 minutes

Ingredients

200 g	Bitter gourd (4 small)
50 g	Onion, finely chopped
50 g	Tomato, finely chopped
½ tsp	Ginger, grated
½ tsp	Garlic, crushed
1 tsp	Red chilli powder
1 tsp	Coriander powder
¼ tsp	Turmeric powder
75 g	Peanuts, crushed
¼ tsp	Cumin seeds
2 tsp	Vegetable oil
Salt to taste	

Preparation

1. Scrape bitter gourds lightly and split lengthwise along each gourd.
2. Remove seeds and pulp.

3. Coat bitter gourds thoroughly with salt and keep aside for 1 hour.
4. Then, wash bitter gourds well under running water.
5. Pat dry on kitchen towel.
6. Heat oil, add cumin seeds.
7. Once it crackles, then add onions and cook until translucent in color.
8. Add ginger, garlic paste and mix well.
9. Then, add tomatoes, all dry powders and salt.
10. Cover and cook until the stuffing thickens, then add ground peanut.
11. Fill each bitter gourd with the stuffing.
12. Using a small amount of oil lightly, fry bitter gourd until crispy and golden brown.
13. The dish is ready to be served.

2.2.5. Kerela roll

- Servings: 3
- Preparation time: 30 minutes

Ingredients

For roti

100 g	Wheat flour
½ tsp	Salt
Water to knead the dough	

For filling

250 g	Bitter gourd
50g	Onion
50 g	Tomato
½ tsp	Chilli powder
½ tsp	Garam masala
1 tbsp	Oil
Waters as needed	
Salt to taste	

Preparation

For roti

1. Mix flour and salt together, and gradually add water to form soft yet firm dough.
2. Cover and let the dough rest.
3. Make dough balls of even size (lemon sized).

4. Dust the working surface with flour and roll out the dough into flat circles.
5. Heat a frying pan, place the flattened dough and let it cook lightly on both sides.
6. Prepare all the rotis and place them in a casserole to keep soft before serving.

For filling

1. Cut bitter gourds lengthwise and remove seeds and pulp.
2. Cut into small pieces.
3. In a frying pan, heat oil and fry onion until it turns light brown in colour. Add ginger and garlic paste. Finally, add tomato and cook well.
4. When the filling is thoroughly cooked, add salt, chilli powder and garam masala. Mix well and add cooked bitter gourd.
5. Place 3-4 tbsp bitter gourd filling in the middle of the roti and roll the roti tightly to form a roll. Serve hot.

2.2.6. Bitter gourd finger chips

- Servings: 3
- Preparation time: 25 minutes

Ingredients

300 g	Bitter gourd
1	Fresh green chilli, cut into half
3 tsp	Vinegar
¼ tsp	Chilli powder
1 tsp	Vegetable oil
Salt to taste	

Preparation

1. Cut bitter gourd lengthwise, remove seeds and cut into thin pieces.
2. Add vinegar, salt, chilli powder and green chilli.
3. Mix well and set aside for 15 minutes. Drain excess liquid.
4. Heat oil in a pan and fry the bitter gourd pieces until they are cooked on the inside and crispy on the outside.
5. Let cool before serving.
6. Variation: Lemon juice could be used instead of vinegar.

2.2.7. Mixed vegetable curry

- Servings: 3
- Preparation time: 30 minutes

Ingredients

100 g	Bitter gourd
50 g	Carrot
50 g	Green peas
50 g	Cauliflower
100 g	Onion
2	Tomato
2 tsp	Vegetable oil
1 tsp	Red chilli powder
1 tsp	Coriander powder
1 tsp	Ginger paste
2 tsp	Garlic paste
½ tsp	Turmeric powder
5 g	Green chillies
½ tsp	Cumin seeds
A few coriander leaves	
Salt to taste	

Preparation

1. Cut bitter gourds lengthwise and remove seeds
2. Dice carrots, beans, bitter gourds, and cauliflower.
3. Chop onion, green chillies, green coriander and tomato puree
4. Heat oil in a frying pan and add cumin seeds.
5. Add green chillies and onion once the cumin seeds crackle and fry the onion to a light brown colour.
6. Add ginger paste, garlic paste along with the powdered spices. Mix thoroughly.
7. Cook on high heat until the curry bubbles, then turn to low heat and simmer for 1 minute.
8. Add tomato and other vegetables. Continue simmering until the oil leaves the masala.
9. Garnish with green coriander and serve with paratha or chappathi.

2.2.8. Aonla bitter gourd juice

- Servings: 3
- Preparation time: 10 minutes

Ingredients

150 g	Aonla (5 fruits)
75 g	Bitter gourd
½ tsp	Sugar
½ tsp	Salt
200 ml	Drinking water

Preparation

- Cut the aonla and bitter gourd into pieces, place in the blender and blend until smooth.
- Sieve the juice and discard the remains.
- Add sugar and salt to the juice, mix well and serve.
- Variation: Honey can be used instead of sugar to sweeten the taste.

2.2.9. Bitter gourd juice

- Servings: 3
- Preparation time: 20 minutes (Soaking time: 15 minutes).

Ingredients

100 g	Bitter gourd
1	Lemon
½ tsp	Salt
½ tsp	Turmeric powder
250 ml	Drinking water

Preparation

- Wash the bitter gourd thoroughly and cut it into small pieces. Remove the seeds and inner pith.
- Soak the cut pieces in water with salt and turmeric powder for 15 minutes. Discard the salt water.
- Cut the lemon in half and juice or squeeze out the juice. Set aside.
- Blend the bitter gourd pieces and water in the mixer until smooth.
- Sieve the mixture, add lemon juice and serve cold.

2.2.10. Bitter gourd fried rice

- Servings: 4
- Preparation time: 20 minutes

Ingredients

- 250g bitter gourd, chopped
- 1 tbsp vegetable oil
- 50 g tomato
- ½ tsp mustard seeds
- Curry leaves - a few
- 1 tsp chilli powder
- ½ tsp coriander powder

- ½ tsp garam masala powder
- 200 g steamed rice with cumin seeds
- 2 tbsp coriander leaves, chopped

Preparation

- Heat the oil in a kadhai (Wok) and add mustard seeds.
- Add curry leaves when the seeds begin to crackle and cook for a few seconds.
- Add the chopped bitter gourd and tomato, fry for a few more seconds.
- Add the chilli powder, coriander powder and garam masala powder.
- Mix well and cook on medium heat until the bitter gourd becomes soft.
- Add cooked rice, mix thoroughly and cook for 1 minute.
- Now, it is ready to be served hot.

2.2.11. Bitter gourd noodles

- Servings: 3
- Preparation time: 15 minutes

Ingredients

- 150 g bitter gourd
- 1 packet of noodles
- 25 g cabbage
- 50 g carrot
- 50 g green pepper, sliced
- 20 g tomato
- 2 tsp tomato sauce (optional)
- ½ tsp soya sauce (optional)
- 1 tbsp vegetable oil
- Salt to taste

Preparation

- Cut bitter gourd into thin slices and remove the seeds.
- Heat oil in a frying pan and fry the bitter gourd slices. Set aside.
- In a small saucepan, cook the noodles as directed on the noodle pack.
- Heat oil in a frying pan on medium heat; stir-fry all the vegetables for 3 minutes.
- Add tomato sauce, soya sauce, salt, noodles and fried bitter gourd. Mix thoroughly and serve hot.

2.2.12. Pineapple bitter gourd soup

- Servings: 6
- Preparation time: 25 minutes

Ingredients

- 100 g pineapple
- 100 g bitter gourd
- 1 tbsp fresh ginger
- 1 green chilli
- 500 ml water
- Juice from 2 limes
- 1 tbsp vegetable oil
- 1½ tsp cumin seeds
- Salt to taste
- 1 tbsp coriander leaves

Preparation

- In a medium sauce pan, add water, pineapple, bitter gourd and salt. Place over low heat.
- Grind the ginger and green chilli in a blender or chop finely.
- When the water boils, add the ground mixture into the soup.
- Add coriander leaves and remove saucepan from heat. Add lime juice.
- In a small pan, heat oil over medium heat and add cumin seeds.
- Fry until golden colour and add to the soup as a garnish.

2.3. Green leafy vegetables

2.3.1. Preparation of dehydrated greens powder

Step 1 : Screening (mature, fresh greens)
Step 2 : Trimming and inspecting
Step 3 : Washing
Step 4 : Separating
Step 5 : Greens and stems
Step 6 : Weighing
Step 7 : Drying (80°C for 15 min)
Step 8 : Powdering
Step 9 : Greens powder
Step 10 : Packing
Step 11 : Storing

2.3.2. Value added products from cabbage

- Select matured clean cabbage
- Shred it to 5 mm size and mix with 2.5% salt
- Pack it layer by layer in a container, 2/3 height
- Seal hermetically and allow for natural fermentation at room temperature.
- Blanch it at 70ºC for 10 min, after 28 days
- Store under shade in clean place and use it within four months.

2.3.3. Preparation of leaf powder

Step 1 : Leaf (Curry leaf /coriander leaf)
Step 2 : Cleaning
Step 3 : Washing
Step 4 : Microwave drying (70ºC for 12 min / 15 min)
Step 5 : Powdering
Step 6 : Packing
Step 7 : Storing (at ambient temperature)

2.3.4. Preparation of curry leaf powder/coriander leaf powder based mixes

Step 1 : Preparation of biscuit mix + maida + powder sugar + baking powder
Preparation of rava laddu mix + roasted semolina powder + powder sugar + cardamom
Step 2 : Final mixing
Step 3 : Packing
Step 4 : Sealing and labeling
Step 5 : Storing (at ambient temperature)

2.3.5. Preparation of greens based ready-to-use vegetable soup mix

Step 1 : Selecting vegetable greens / roots and tubers / other vegetables (sound, mature, fresh and firm)
Step 2 : Washing (tap water)
Step 3 : Pre-processing (peeling / trimming / stringing / scraping)
Step 4 : Cutting
Step 5 : Pre-treating (potato, onion, garlic and ginger)
Steaming (greens, carrot, potato, beans and peas)
Step 6 : Drying in the cabinet drier at 50-60ºC (till it reaches 4.0% moisture)
Step 7 : Cooling (room temperature)

Step 8 : Grinding into powder (mixie)

Step 9 : Sieving (BS 40 sieve)

Step 10 : Preparing greens based ready-to-use vegetable soup mix

Step 11 : Araikeerai base
Ponnanganni base

Step 12 : Adding thickening and flavouring agents (skimmed milk powder, corn flour, onion, tomato, ginger and garlic powder mix)

Step 13 : Mixing with taste improvers (citric acid, salt and sugar)

Step 14 : Mixing and sieving (soup mix)

Step 15 : Packing

Step 16 : Labelling

Step 17 : Storing (room temperature)

2.4. Roots and tubers: Beetroot

2.4.1. Preparation of beetroot ready-to-serve beverage

Step 1 :	Selection of beetroot	Selection of ripe grapes
Step 2 :	Washing and cutting	Washing
Step 3 :	Pulping in mixie for 5 min	Pulping in hand pulper
Step 4 :	Filter through muslin cloth	Filter through muslin cloth

Step 5 : Blending (Beetroot juice: grapes juice - 100; 90:10 and 80:20)

Step 6 : Adding sugar (14° B and citric acid 0.25%)

Step 7 : Heating the mixer to 80°C

Step 8 : Filling in sterilized bottles (leave head space ½)

Step 9 : Cork sealing

Step 10 : Pasteurizing in boiling water (30 min)

Step 11 : Cooling

Step 12 : Labeling

Step 13 : Storing (at both room and refrigeration temperature)

2.4.2. Preparation of beetroot sweet pickle

Step 1 : Selection of beetroot

Step 2 : Washing and peeling

Step 3 : Scrapping

Step 4 : Frying the spices in oil and powdering

Step 5 : Seasoning

Step 6 : Adding the powdered spices
Step 7 : Adding scrapped beetroot and sugar
Step 8 : Cooking for 15 minutes.
Step 9 : Adding vinegar and KMS
Step 10 : Cooking for 5 minutes.
Step 11 : Cooling
Step 12 : Packing
Step 13 : Storing (at room temperature)

2.5. Cassava/ Tapioca

2.5.1. Processing of fermented cassava rava

Step 1 : Cassava
Step 2 : Cleaning
Step 3 : Chipping (10 cm pieces)
Step 4 : Blanching (3 min. 60°C)
Step 5 : Cooling
Step 6 : Steeping (24 hrs)
Step 7 : Adding inoculum
Step 8 : Decanting
Step 9 : Drying (80°C for 8 hrs)
Step 10 : Milling
Step 11 : Sieving
Step 12 : Packing
Step 13 : Sealing
Step 14 : Storing

2.5.2. Processing of cassava fermented sour flour

Step 1 : Cassava
Step 2 : Cleaning
Step 3 : Chipping
Step 4 : Adding of inoculum
Step 5 : Fermenting (78 hrs)
Step 6 : Draining
Step 7 : Dehydrating (80°C for 8 hrs)
Step 8 : Dried chips

Step 9 : Size reduction
Step 10 : Milling
Step 11 : Sifting
Step 12 : Cassava fermented sour flour
Step 13 : Allowing to settle in excess water (24 hrs)
Step 14 : Decanting
Step 15 : Drying
Step 16 : Fermented sweet flour

2.5.3. Processing of noodles

Step 1 : Cassava
Step 2 : Cassava flour + refined wheat flour
Step 3 : Sieving (twice a time)
Step 4 : Blanching (for 15 min)
Step 5 : Sieving to remove dots
Step 6 : Mixing for 15 min. in extruder
Step 7 : Extrusion (using dye)
Step 8 : Tempering (4 hours at room temperature)
Step 9 : Drying (60°C for 2 hrs)
Step 10 : Noodles
Step 11 : Packaging
Step 12 : Storing

2.6. Colocasia

2.6.1. Processing of colocasia flour incorporated noodles

Ingredients

Flour	50 g
Water	50 ml
Salt	2 g

Step 1 : Colocasia flour + refined wheat flour
Step 2 : Sieving (twice a time)
Step 3 : Blanching (for 15 min)
Step 4 : Sieving to remove dots
Step 5 : Mixing for 15 min in extruder
Step 6 : Extrusion (using die)

Step 7 : Tempering (4 hrs at room temp)

Step 8 : Drying (60ºC for 2 hrs)

Step 9 : Noodles

Step 10 : Packaging

Step 11 : Storing

2.6.2. Processing of colocasia flour incorporated papad

Ingredients	
Flour	50 g
Water	50 ml
Salt	7 g
$NaHCO_3$	1 g
Cumin	1 g

Processing of papads

Step 1 : Colocasia flour + black gram flour

Step 2 : Addition of papad khar (solution)

Step 3 : Mixing

Step 4 : Portioning into small balls

Step 5 : Pressing the papads

Step 6 : Stocking

Step 7 : Application of weights

Step 8 : Sun drying for 30 min.

Step 9 : Drying

Step 10 : Packaging

Step 11 : Sealing

Step 12 : Storing

2.7. Sweet potato

2.7.1. Processing of sweet potato flour

Step 1 : Selection of tubers

Step 2 : Washing

Step 3 : Peeling

Step 4 : Slicing (0.5 cm thickness)

Step 5 : Soaking (1% KMS for 10 minutes)

Step 6 : Drying (40ºC for 6 hrs)

Step 7 : Dried chips
Step 8 : Milling
Step 9 : Sieving
Step 10 : Packing

2.7.2. Processing of sweet potato and pulse flour fortified extruded products

Step 1 : Refined wheat flour + sweet potato flour + pulse flour (soya flour/ green gram flour)
Step 2 : Mixing
Step 3 : Steaming
Step 4 : Cooling
Step 5 : Drying
Step 6 : Powdering
Step 7 : Addition of salt + water
Step 8 : Kneading (30 minutes)
Step 9 : Extrusion (noodles, macaroni and vermicelli)
Step 10 : Steaming (5 minutes)
Step 11 : Drying (60°C for 2 hrs)
Step 12 : Packaging and sealing
Step 13 : Storage

2.8. Onion

2.8.1. Onion pickle

- Onion 1 kg, vinegar 1 litre, salt 250 g, red chilli powder 10 g, cardamom (large), black pepper, cumin (powdered) each 10 g, clove (headless) 5 numbers.

Processing of onion pickle

Step 1 : Onion (small)
Step 2 : Peeling
Step 3 : Blanching for 5 minutes
Step 4 : Filling blanched onions in jar
Step 5 : Addition of salt
Step 6 : Standing for 24-48 hrs
Step 7 : Draining of water
Step 8 : Addition of vinegar and spices
Step 9 : Storage

2.8.2. Onion vadagam

Preparation of onion vadagam

Ingredients

Onion	1 kg
Black gram dhal	200g
Garlic	100g
Cumin	20g
Mustard	20g
Chillies	100g
Salt	To taste

Method

1. Soak black gram dhal in water for 2 hrs and grind into paste.
2. Peel and wash onions, make a slit in the onion.
3. Make red chillies into powder.
4. Mix onion, mustard, salt, chillies and cumin. Make into small balls without pressure.
5. In a tray, put the polythene covers and place the vadagam and sun dried for 2-3 days.

3

Value Addition in Spices

- Aromatic food substances, which enhance flavour are classified into spices.
- Spices are usually dried roots, barks and seeds used whole or crushed, powdered.
- Spices add a glorious touch to food with its flavour and fragrance.
- Spices add flavour, colour to food and make the food palatable.
- Stimulates salivation and acid secretion of digestive enzymes like amylase.
- Has anti-inflammatory, antibacterial and antioxidant properties.
- Reduces cholesterol levels useful in preventing heart diseases.
- Kerala is considered as the land of spices.

3.1. Different types of spices and their use

- **All spice or Pimento**
  - A small berry, size of the pea dried to a dark brown colour.
  - Flavouring component is a volatile oil present at 3-4% levels containing eugenol and other closely related phenols.
  - Used whole in pickling and cooking meats and fish.
  - Used ground in cakes, puddings and preserves.
- **Ani seed (Somfu/owa)**
  - Chief compound is anethole.
  - Also known as sweet cumin and has the flavour of liquorice.
  - Chewed after meals and used as mouth freshener.
  - Used in cakes, breads, cookies and candles; vadacurry, kurma, non-vegetarian dishes, briyani, thandai, kachori and in pickles.
  - An infusion of fennel is used to counteract flatulence.
  - Mildly carminative and used in treating colic pain.
- **Asafoetida**
  - Oleo gum resin exuded from the rhizome or root of *Ferula asafoetida*.
  - Odour component consists of a ferulic ester and sulphur containing volatile oil.

  - Used as an antimicrobial agent.
  - Also used in treating chronic bronchitis and whooping cough.
  - Used in counteracting intestinal flatulence.
  - Increases the levels of detoxification enzymes in the body.
- **Bay leaves (Birinj leaves)**
  - Dried aromatic leaves of laurel tree.
  - Contain 1-3% of highly aromatic volatile oil.
  - Oil of bay leaves is used in the preparation of pickling spice and in the flavouring of vinegar.
- **Caraway seeds**
  - Contains about 5% volatile oil.
  - Chief flavouring principle of which is D-carvone and D-limonene.
  - Seeds are used to flavour cakes, biscuits, cheese, apple sauce and cookies.
  - Oil from caraway seed is used to flavour meats, canned foods, sausages, soups, cakes, certain bread rolls, cheese, confectionery and also flavouring liquor.
- **Cardamom**
  - Fruit contains brownish black seeds, which have about 2-10% volatile oil with characteristic pleasant odour.
  - Active principle present in oil is cineole, terpinyl acetate, pinene, sabinene and porneol.
- **Chillies**
  - Red colouring matter of chillies is due to a carotenoid pigment.
  - Ground chilli is used in most of the Indian gravies and vegetable dishes.
  - Used in making chutneys, pickles, dehydrated chillies, vattal and in day to day preparations.
  - Dry chillies are used for seasoning.
  - Contain a substance called capsaicin, which increases gastric secretion and it causes destruction of mucosal cells.
- **Cinnamon**
  - Thin inner bark of a cinnamon tree.
  - Bark contains about 1% essential oil.
  - Active principles are eugenol, cineole and cinnamaldehyde.
  - Used in stick form in fruit preserves.

    - Ground and used for cakes, cookies and puddings.
    - Used in spicing sauces, pickles, pulav, biriyani.
- **Clove**
    - Dried flower bud of the clove tree.
    - Contains about 15% essential oil.
    - Chief active principle is eugenol which has anti mutagenic effect.
    - Cloves are used whole in meats, pickling and fish.
    - Used in cakes, cookies and puddings.
    - Used as flavouring agent in pulav, rice puttu and fruit cake.
    - Used for tooth pain.
- **Coriander seed**
    - Unmatched for its fresh delicate spring like aroma.
    - Seeds contain 0.5-1% essential oil which has an active principle coriander an isomer of geraniol.
    - Roasted powdered seed are used as an ingredient of curry powders.
    - Used as flavouring and thickening agent in cookery.
    - In the form of powder, it is used in rasam, all curries and vegetables, and chutney powders.
    - Also used as whole in kamandhokla, samosas and kachori.
    - Used for sprinkling on prepared food, flavouring sambar and rasam, pounded or mixced as a base for sauces or chutneys.
    - An infusion of coriander seeds is used in flatulence, vomiting and intestinal disorders.
- **Cumin seeds**
    - Pungent, sharp and astringent.
    - Contains 2-4% essential oil.
    - Active principle is an aldehyde cumino.
    - Used as stimulant and carminative agent.
- **Fenugreek seeds**
    - Hard lentil seed.
    - Colour is dark brown and has astringent aroma
    - Contain 5% bitter fixed oil.
    - Oil has strong celery like odour.
    - Aid in maintaining the blood glucose levels in non-insulin dependent diabetes.

- Fibre present in the seed may be responsible for this.
- Also used with the butter milk in the treatment of dysentery.

- **Garlic**
  - Contains an antibiotic principle "allin" (inactive form) which is converted to allicin (active form) by the enzyme allinase.
  - Allicin further breaks down to allyl disulphide, which is responsible for characteristic flavour.
  - Used in recipes like rasam, pickle, chutney, pulav, sauces.
  - Used to treat various digestive disorders.
  - Extracts of garlic can lower tissue cholesterol levels and prevent heart diseases.
  - Also an anti fungal agent and helps to reduce the fat content in blood (helps in blood thinning) and thus, reduce blood pressure.
- **Ginger**
  - Root of the plant *Zingiber officinale.*
  - Volatile oil present is gingerol.
  - Flavouring compound has sharp burning sensory stimulation.
  - Reported to reduce inflammation and pain in joints.
  - Also has potential prophylactic use in treating migraine headaches.
  - May also be effective in alleviating nausea.
- **Kokam**
  - Fruit dried and used as sour agent in cookery.
  - Since anthocyanin is present, it is also used in making sherbet.
- **Mango powder**
  - During the early stages of growth, the tree may have a heavy fruit drop.
- **Mint leaves**
  - Essentially an aromatic culinary herb.
  - Leaves yield an essential oil (pepper mint oil) which is used for flavouring gum, confectionery, tooth paste, perfumes and pharmaceutical preparations.
  - Principle components of the pepper mint oil are menthol, menthyl acetate, menthyl isovalerate and menthone.
  - Used fresh for beverages, salad dressing and for garnishing.
  - Also used in making many dishes like raitha, pulav, chutney, vada and pani puri water.

- **Mustard**
  - Small reddish black seeds of annual herbs.
  - Leaves of this plant are consumed as vegetable.
  - Mustard seeds have a pungent flavour.
  - Characteristic flavour of mustard seed is due to an allyl isothiocyanate.
  - Mustard paste is used to flavour hotdogs, sandwiches, cheese, eggs, meat and salad dressings.
  - Dry mustard is used in meat, sauces, gravies and mayonnaise.
  - Used in the powdered form in pickles.
  - Used in the vegetable preparations and raitha.
  - Used mainly for the seasoning.
  - Mustard is used without the skin for pickling (chilli pickles).
  - Mustard oil is also used in cookery.
  - Mustard seeds are rich in sulphur containing compounds namely the dithiol thiones, which protect against the toxic effects of aflatoxin.
  - Dithiothione is also used as antischistosomal drug.
  - Consumption of Brassica vegetables is associated with negative cancer risks.
- **Nutmeg**
  - Dry hard wrinkled seed or pit of the nutmeg fruit.
  - Mace is the orange red fleshy covering of the nutmeg kernel.
  - Contains 7-14% essential oil.
  - Essential oil contains highly toxic compound known as myristicin, used in small amounts to flavour puddings and fruit
  - Myristicin present in nutmeg seeds could lead to delirium and deep stupor.
  - Smaller amounts of nutmeg may cause vomiting and colic.
  - Nutmeg has antimicrobial property.
- **Mace**
  - More delicate and used for flavouring fish sauces, certain meat and fish dishes, pickles and preserves.
  - Used in cakes, cookies, pies and chocolate dishes.
  - Used in garam masala.
- **Onion**
  - Used as a flavouring agent in food preparations.
  - Contains an essential oil, the active principle allyl propyl disulphide which is responsible for the characteristic cooked flavour.

- Used in cookery to improve flavour or to mask undesirable flavours and to increase thickness of the gravies.
- Consumption of raw or cooked onion aids in maintenance of normal glucose levels.
- Onion has antibacterial properties.
- Sulphur containing compounds strongly act against both gram positive and gram negative bacteria.
- Extracts of onion to inhibit growth of many pathogenic fungi belonging to *Aspergillus* and *Candida.*
- Can lower blood cholesterol and lipid levels and useful in preventing heart diseases.

- **Omum**
  - Allied to the liquorice family and a powerful spice.
  - Used in ompodi, rusks, namkin and biscuits.
  - Omum water is given to children for digestion.
- **Poppy seeds**
  - Tiny dark cream colour seed of poppy plant, which is used for topping of bread, cake, rolls or fillings for buns.
  - Oil extract is used for salads.
  - Used as thickening agent in the preparation of gravies.
  - Also used in kurma, non-vegetarian dishes and sweets.
- **Pepper**
  - Dried small round berry of a tropical vine with small white flower.
  - White pepper is mature berry with the black coat removed.
  - Pepper corns are used whole or crushed.
  - Ground pepper is used for seasoning many dishes and is also used as a condiment at the table.
  - Pepper corns are generally used whole in pickling, meats and stews.
  - Used for general seasoning of meat, fish, poultry, vegetables and salads.
  - White pepper is used in dishes that require a less pungent flavour.
  - Used as substitute for chilli powder.
  - Also used in marinating the non-vegetarian foods.
  - Flavour can be improved by the addition of pepper powder to omelets, sandwich, salads, papads, soups and chips.
  - Used along with hot milk for throat infections.

- **Saffron**
  - Name given to the fragrant stigma found in the Crocus flower.
  - Regal spice of matchless aroma and the most costly in the world.
  - 75,000 flowers are needed to make one ounce of pure saffron.
  - Each filament can colour 700 times its own weight in water.
  - Used mainly for its yellow colour.
  - Have a pleasant aroma and an essential oil crocin and the colouring principle crocetin.
  - Used in soups, sauces especially rice dishes to give them bright yellow colour and distinctive flavour.
  - Used in sweets, like sandesh, rasmalai, thandai, kesar milk, ice cream, halwa and srikand.
  - May be adulterated with styles, anthers and parts of corolla of saffron and various materials coloured with coal tar dyes.
  - Used as sedative and also used for eye infections.
- **Tamarind**
  - Pulp of the tamarind fruit after the outer shell and seeds are removed.
  - Extract is used in making rasam and sambar.
  - Used as souring agent in chutneys, chat, pickle, pani puri and in tamarind rice.
  - Also used as thickening agent in gravies.
  - Tamarind puree is available in the market.
- **Turmeric**
  - Ground dried aromatic root.
  - Contains 5% essential oil.
  - Colouring substance present is known as curcumin.
  - Natural aroma is most appetizing and on its own without the use of other spices, can produce delicious food.
  - Help in detoxifying harmful drugs or chemicals that are converted to toxic metabolites.
  - Also helps in increasing mucin content of gastric juice and reduces irritation in stomach.
  - Also used to relieve sore throat, cough, cold and against flatulence.
  - National Institute of Nutrition, Hyderabad suggests that turmeric can be a potent anticancer agent.

  - Isolated from turmeric showed a potent antioxidant and anticarcinogenic properties.
  - Curcumin is known for its inhibitory action on bacteria and arrests growth of fungi.
  - Helps to reduce the levels of cholesterol in experimental animals given high cholesterol containing diet.
- **Vanilla**
  - Vanilla beans are being cultivated in Kerala.
  - Active principle is vanillin.
  - Synthetic vanilla is much cheaper than that of the natural product.
  - Imparts the flavour and it is used in preparing ice creams, custards, puddings and cakes.

4

Value Addition in Plantation Crops

4.1. Coconut

- Coconut water is a refreshing drinks which also hygienic and nutritive.
- White flesh is a calorie rich extensively used for cookery in South India.
- From grated wet coconut kernel, good quality oil as well as protein and carbohydrate can be extracted to prepare coconut honey.
- Undesirable flavour changes in desiccated coconut are due to oxidative rancidity of the oil.
- Residual cake after the oil extraction is used as cattle food.
- Dried coconut meat is known as copra and is the source of coconut oil, which is used in enormous quantities for making fats for baking and confectionery.
- An important product used in the chocolate and confectionery industry is desiccated coconut.
- Nut is removed from hard shell and thin brown rind pared off.
- Pared nuts are then washed free of milk and skin residues and put through shredding machines, which give a milled product of varying degree of fineness.
- After shredding, ground meat is subjected to a steam or hot water treatment to destroy *Salmonella* and lipases and to reduce bacteria counts to low levels.
- Pasteurized meat is then dried on wire mesh belts by means of hot air, which reduces moisture content to less than 4%.
- Critical moisture level of desiccated coconut is 5% and product should be well below this to ensure good shelf life.
- After drying, desiccated coconut is put through a multiple sieving operation to produce the four main grades *viz.,* extra fine, macaroon (fine), medium and coarse.
- A variety of other coconut products is available to chocolate and confectionery trade such as, moist white sweetened coconut, toasted coconut, creamed coconut and tenderized coconut.

- Top quality desiccated coconut should be white in colour, free from foreign matter, and uniform in granulation against standard and expressed oil should have a free fatty acid content of less than 0.1%.

4.1.1. Extraction of coconut oil

- A whole coconut consists of 50% husk, 15% shell, 25% meat and 10% water.
- The coconut shells contain 34% cellulose, 36% lignin, 29% pentosans and 1% ash.

Composition of fresh coconut kernel and copra

Product	Moisture %	Fat %	Protein %	Carbohydrate %	Mineral %	Fibre %
Kernel (wet)	40-50	35	4	8-20	1	2
Copra	6-7	62-65	5-7	15	2	3-4

Traditional method

- Coconut oil is traditionally extracted from copra.
- Copra can be obtained by separating the coconut meat from the whole coconut shell followed by sun drying and smoke drying for about 6 to 8 days.
- Copra is then, chopped to small sizes and cooked by heating small pieces of copra in a cooker for about 30 mins.
- Oil is squeezed out from cooked copra by an expeller and then, filtered by a filter press.

Mechanical extraction of coconut oil from the fresh coconut meat

- Coconut shells are cracked and coconut meat is separated.
- Fresh coconut meat is chopped and sliced into thin flakes of the desired size.
- Thin flakes of coconut meat are cooked in a cooker by heating them at an elevated temperature for 90 mins.
- Oil is expelled from heat treated cooked coconut meat (at a temperature of about 70°C) in an oil expeller.
- Oil is stored in an air tight container in a dark cool and dry place for a long and safe storage.
- Deoiled cakes are suitable for human consumption.

4.1.2. Value added coconut products

- Used as staple part of the diets of almost all Polynesian and many Asian people for centuries.
- Used as food, as flavouring and made into beverages.
- Important products of coconut are whole coconut (tender and mature), copra, toddy, neera, gur, coconut palm candy, sugar, vinegar etc. and few novel recipes like coconut boli, banana coconut cake.

1. Coconut honey

- Product from coconut water containing many growth promoting trace elements besides glucose, fructose and laevulose.
- Coconut water was filtered, evaporated and blended with a little golden syrup to produce coconut honey, a palatable nutty flavored breakfast food, soft drinks additive and a sweetener.

2. Coco sauce

- A tangy sauce prepared from coconut water with red chilli, onion powder and little vinegar.

3. Coconut lemonade

- Prepared by boiling coconut water, sugar and lemon juice.
- Traditionally a popular drink in Kerala.

4. Nata-de-coco

- Nata from fresh liquid endosperm of the matured coconut is collected and filtered by using cheese cloth.
- Pasteurized after adding 8 per cent sucrose, 0.5 per cent ammonium sulphate and adjusting the pH to 4.5 by adding acetic acid.
- Medium is developed by inoculating *Acetobacter aceti* at 10 per cent and incubated at room temperature for about two weeks.
- After two weeks, a gel like mass developed with a film like layer on the top.
- Film is then, removed and mass is cleaned and cut into cubes.
- Cubes (nata) are further processed to improve colour, flavour and taste by soaking it in sugar syrup.
- Used to decorate desserts, ice cream, puddings and fruit salad.

5. Coconut kernel or white meat

- Preservation and pickling of coconut is done by soaking coconut kernels in brine solution which maintained their colour, flavour, texture and taste and resembled like fresh coconut for 90 days.

- Salt and acid penetration in pickle is directly proportional to the concentration in soak solution.
- Steeped coconut kernel is kept for 90 days under study, in which there is a remarkable reduction in reducing sugar, total sugar and free fatty acid content.
- Moisture content after storage ranged between 38.05 and 38.61 per cent.
- Fresh coconut kernel could be preserved as long as for 3 to 6 months in the form of pieces and scrapings by steeping in a solution containing 4 per cent salt, acetic acid, sulphur dioxide and antioxidant.
- Product required washing in water prior to use.

6. Desiccated coconut

- Edible dried-out shredded coconut meat was prepared from fresh kernel of fully matured coconut and it is available in coarse, medium and fine grades and also in special grades such as threads, strips, granules etc.
- Crisp, snow white in colour with a sweet, pleasant and fresh taste of coconut kernel.
- Commercial product was manufactured from white part of the meat after removing the brown parings.
- Meat was shredded or disintegrated and dried in hot air drier at 140-170°F to 2 per cent moisture content and used in manufacture of cakes, pastries and chocolates.
- Disintegrated, white kernel of coconut is processed under strict hygienic conditions and dried to a moisture content of below 3.0 per cent.

7. Products from desiccated coconut

- Matured coconut kernels are steam blanched and soaked in sugar syrup of 30° B for a period of 48 hrs.
- Drained pieces are sulphited for 20 min and dried in cabinet drier at a temperature of 60°C for 8 hrs.
- Sugar acts as an osmotic agent for the preservation of coconut.
- Osmotically, dehydrated coconut is well suited for preparation of products in homes as well in commercial units.
- Desiccated coconut powder (2%) or soy protein concentrate (SPC) (4.5%) is used in the preparation of mango bar, which increased the percentage of protein.
- Total soluble solids of pulp is raised to 30° B with powdered cane sugar adding 0.6% citric acid and drying in an air cabinet drier at 63±2°C for 14 hrs.

- Coconut powder is used in the preparation of ragi based mixes *viz.*, sweetened and spiced mix.
- Other ingredients for sweetened mix are puffed ragi flour, sugar, beaten rice flakes, cardamom, coconut powder and puffed ragi flour, sugar, dehydrated coriander leaves, hydrogenated fat, coconut powder and groundnut for spiced mix.
- In sweetened mix, addition of beaten rice improved the colour, but increased the rate of lipid oxidation. While incorporation of coconut powder lowered the lipid oxidation.
- Value added products from coconut, such as dehydrated coconut (both savoury and sweet), ready-to-use coconut chutney mix, coconut kuruma mix, coconut rice mix and coconut burfi mix are developed.
- Acidified salt solution and sugar syrup are used as osmotic agents for the preparation of dehydrated coconut for savoury and sweet dishes.

8. Coconut chips

- Thinly sliced crispy coconut meat which may be sweetened or salted and may come in handy as a ready-to-eat snack food.
- Prepared by slicing coconut meat of eleven to twelve month old nuts, thinly into strands, soaked in syrup, drained and dried in a dryer or oven.

9. Coconut crisps

- Prepared from young coconut endosperm of nine to ten month old nuts.
- White in colour, has pleasant coconut flavour and does not leave any fibrous feeling after taste.
- Processed by slicing the meat into 0.6-0.7 mm thickness, blanched in boiling water, cooked in light syrup and then, dried which is considered as a high energy food and of a good quality product.

10. Roasted young coconut

- Prepared by a process in order to sweeten its water and tender meat as well as to enhance their flavour.
- Nuts are boiled in a solution (2% sodium meta bisulphite) for 20 min, dried and burnt for finishing touch and it is exposed to the fire for a minute or until the shell itself showed signs of burning.
- Fruit is kept at room temperature for 3 days or in the refrigerator for larger storage.

11. Honey roasted coconut

- A sweet, thinly sliced crispy coconut meat is eaten as a snack food.

- Sliced matured coconut meats is mixed with honey, sugar, margarine, sweeten condensed skim milk, molasses, salt and dried in oven for half an hour.
- Then, agitated frequently for 15 min until it become crispy, cooled and packed.

12. Dehydrated sweet coconut

- Young coconut meat is used for preparation of dehydrated sweet coconut.
- Coconut meat is washed with water and then, soaked in coconut water, again washed with clean water.
- Meat is cut into pieces, mixed with refined sugar and water in the ratio of 6:3:1 and boiled for 1 hr or until it is dried, cooled and packed.

13. Dehydrated coconut chutney

- Dehydrated coconut chutney using simple hot air drying technique was developed.
- Reconstitutes well in cold water and had all characteristics of fresh chutney.
- Product had a shelf life of 3 months at 37°C and 6 months at ambient temperature when packed in flexible pouches.

14. Coconut milk

- Refers to the milky fluid, freshly extracted from the coconut kernel with or without added water and coconut cream to the high fat cream like material obtained from the coconut milk by either gravitational separation or centrifugation.
- Prepared by blending skim milk powder with coconut milk obtained from freshly grated coconut and pasteurized at 70-72°C for 10 min.
- Contains 6 per cent skim milk powder and 9.65 per cent total solids.
- Coconut milk, a generic term for the aqueous extract of the solid coconut endosperm, plays an important role in the cuisines of South East Asia as well as in other parts of the world.
- Fermented beverage from coconut skim milk using different combinations of coconut skim milk and non-fat dry milk and the culture used is *Lactobacillus bulgaricus*.
- Based on sensory evaluation, 50 per cent of coconut skim milk: 50 per cent of non-fat dry milk ratio is the most acceptable formulation.
- Ready-to-drink fermented beverage concentrate (diluted at 1:3) contains protein (1%), fat (0.74%), sugar (18.70%) and water (79.25%).
- Fermented beverage concentrate was found to be stable even after 2 months of storage and coconut skim milk can be utilized successfully for the processing of cultured milk.

- Coconut cream, the concentrated milk extracted from fresh matured coconuts can either be used directly or diluted with water to make various curry preparation, sweets, desserts, puddings etc.
- Processed and packed coconut cream has a shelf life of six months and once opened, it should be stored in refrigerator for subsequent use.
- Coconut cream, a concentrated form of coconut milk, which is a convenient product prepared from mature and fresh coconut.

15. Coconut syrup

- A translucent, free flowing liquid is prepared by cooking coconut milk with an equal amount of refined sugar and di-sodium phosphate equivalent to 0.25 per cent of the volume of the milk, until mixture attained a TSS content of 68-70 per cent.
- Hot mixture is poured in sterile containers and sealed hermetically.
- Used as a topping for bakery products or as a mixer in alcoholic drinks or may be diluted in water and used in cooking rice cakes and other delicacies.

16. Coconut honey

- Viscous, free flowing syrup, similar to coconut syrup, but less creamy and less nutty in flavour is used as topping for pancakes and waffles.
- One part of skim milk is mixed with ½ part of refined sugar and ½ part of glucose, and then blended with sodium alginate at 0.5 per cent as stabilizer.
- Coconut cream may be added to improve flavour of the product.
- Mixture is heated for 15 minutes, homogenized and cooked with constant stirring in steam jacketed kettle to a TSS of 75 per cent.
- Poured hot into sterile containers and then, sealed hermetically.

17. Sweetened condensed coconut milk

- Coconut skim milk is used in the preparation of sweetened condensed coconut milk.
- Powdered dairy skim milk is added for protein fortification and other ingredients are corn oil, coconut cream and sugar.
- Skim milk is first pasteurized for 30 mins at 80-90°C and mixed with other ingredients.
- Mixture is blended or passed through a colloid mill and heated in a steam jacketed kettle with constant stirring to a TSS content of 68%.
- Packed hot in sterile tin canes and cooled immediately in cooling tanks.

18. Coconut candy

- Prepared from grated coconut meat mixed with coconut milk.

- Grated coconut is moistened with a portion of the milk.
- Remaining milk and molasses are poured in a cooking pan and mixture is heated to boiling.
- Refined sugar is added and mixture is cooked until it gets hardened when dropped into cold water.
- Then, poured in butter-greased pans, allowed to cool slightly, cut into desired sizes and individually wrapped in cellophane sheets.

19. Coconut milk powder

- Coconut milk powder is prepared by dehydrating the milk under controlled conditions.
- Composition of the milk is adjusted with fat percentage in the range of 50-60 per cent of the total solids.
- Emulsifiers and stabilizers are also added to the formulation.
- Most crucial step is the dehydration stage for which spray drier is employed at high temperature (around 180°C).
- Instant dehydration takes place, converting each tiny droplet into micro capsules with fine droplet of the oil inside.

20. Coconut flour

- Unique product prepared from coconut residue.
- A by-product in processing of coconut milk can provide not only value-added income to the entrepreneurs, but also a nutritious and a healthy source of dietary fibre.
- Used as fillers, bulking agents and substitute for wheat flour, rice flour and potato flour at certain levels and incorporated into various food products like baked products, snack foods, steamed and extruded products.
- Sapal from coconut meal, taken after extraction of the coconut milk is made into coconut flour which contains dietary fibre.
- Offers scope for utilization of coconut flour as a dietary component for diabetes.
- Low-fat, high fibre coconut flour, a unique product from sapal is a good source of dietary fibre.

21. Tender coconut water concentrate

- Prepared using fresh coconut water collected under hygienic condition.
- Suspended solids and oil in the samples are removed by means of three way centrifuge.
- Removal of the solids and oil is necessary in order to minimize fouling of the membranes.

- Salts present in coconut water may be removed if desired, prior to concentration to produce a very sweet product.
- This is achievable by passing the centrifuged coconut water through a mixed bed ion exchange resin.
- Concentrate can be frozen or preserved in canes and used as base for the production of carbonated and non-carbonated coconut beverage.
- Crystallized fruit is prepared using candied fruit coated with sugar or sugar crystals and allowed to deposit on it.
- Crystallized candy made with tender coconut kernel can be consumed as a snack.
- Tender coconut kernel in sugar syrup can be used as a dessert or incorporated in ice cream, custard or pudding.

22. Coconut jaggery

- Sweet toddy, if carefully collected in sterile glass vessels remains unfermented for a considerable time.
- Can be strained and boiled down to the crystallizing point to obtain 12-15 per cent jaggery with nutritional and medicinal value.
- Reducing sugars and minerals made it more nutritious and health friendly.

23. Coconut vinegar

- Produced from fermenting coconut toddy and is commonly used in the preparation of fish and meat dishes.
- If toddy is fermented for more than 24 hrs, acidification is done for 10-14 weeks and the resultant is coconut vinegar.

4.1.3. Preparation of the dehydration of coconut

Step 1 : Selecting coconut

Step 2 : Breaking

Step 3 : Inspecting

Step 4 : Scraping (without testa)

Step 5 : Steam blanching (10 minutes)

Step 6 : Soaking in osmotic agent (24 hrs)

Step 7 : Separating from osmotic agent

Step 8 : Drying in 60°C for 4 to 5 hrs (cabinet dryer)

Step 9 : Cooling

Step 10: Packaging

4.1.4. Ready to use burfi mix

Ingredients

Dehydrated coconut scrapings	100 g
Powdered sugar	150 g
Cardamom powder	2g

Method

- Dehydrated coconut scrapings are powdered coarsely using mixie.
- Powdered sugar and cardamom are added to the coarsely powdered coconut and mixed thoroughly.

4.1.5. Ready-to-use coconut chutney mix

- Dehydrated coconut scrapings, roasted bengal gram, ginger, garlic, green chillies, curry leaves, tamarind and salt are ground to get uniform slurry.
- Prepared slurry is seasoned in oil with mustard and curry leaves.
- Seasoned chutney is dried at 60°C for 6 hrs in a cabinet drier.
- For rehydration, 60 ml of water is added to the ready to use coconut chutney mix and allowed to stand for 2 minutes.

4.1.6. Coconut cookies

Ingredients

Flour	60 g
Sugar	30 g
Shortening	60 g
Coconut powder	30 g
Cherry	25 g

Method

- Sieve the flour
- Place all the ingredients in a bowl except cherry and kneaded well
- Made small balls and placed it on a greased tray one inch apart.
- Placed a piece of cherry above the ball.
- Baked at 275°F for about 12-20 minutes.

4.2. Tea

- Beverage prepared by pouring boiling water over dry processed leaves.
- Different factors affect the quality of tea like soil in which plant is grown, climate, processing techniques used etc.

4.2.1. Percentage composition of tea leaves

Constituents	%	Constituents	%
Polyphenols	22.2	Ash	5.6
Protein	17.2	Reducing sugars	3.5
Caffeine	4.3	Pectin	6.5
Crude fibre	27.0	Ether extract	2.0
Starch	0.5		

4.2.2. Types of tea

- **Black tea**
  - Various processing steps involved are withering, rolling, and fermentation, drying, grading and packing.
- **Green tea**
  - Withering and fermentation are completely omitted.
  - Made by first steaming the leaves to prevent the leaves from changing colour and to inactivate the enzymes.
  - Steaming is then, followed by rolling and drying.
  - Leaves retains much of its original green colour especially the finer leaves.
- **Oolong tea**
  - Oolong tea is partially fermented.
  - Fermentation period is too short to change the colour of the leaf completely.
  - Only partially blackened.
  - Beverage is intermediate between those produced from green and black teas.

4.2.3. Characteristics of tea beverage

- High quality tea beverage is clear and bright.
- A flavour full cup of tea has a quality known as briskness, but a subtle aroma.
- Strength of beverage determines whether the tea is mild or robust.
- Green tea is a pale straw colour, whereas black tea is deep amber.
- Some astringency is inherent in tea, but green tea beverage is more astringent than black.

Processing

- Fresh tea leaves are placed in withering troughs set on wire mesh.
- Large fans blow air through the leaves to reduce the moisture content.
- Leaves must be evenly spread so that uniform withering occurs.
- Leaves must be spread by hand and all lumps or piles removed so that maximum air flow reaches all the leaves and no heat is generated in the tea.
- Moisture content is lowered by about 70% or to a predetermined level depending on the condition of the leaf.
- For proper withering to occur, air must be at the proper temperature.
- If the leaf is wet, air is heated slightly for a short time to help to remove excess moisture.
- After withering, leaves are rolled by large rolling machines to provide a twist to the leaves and to break the cell walls to expose inside to oxygen to begin the oxidation process.
- Leaves are fed down onto the machines directly from the withering troughs which are overhead on the second floor.
- Sieve separates particles by size.
- Small leaves which have been rolled successfully the first time fall to the bottom and then, are moved to the next process.
- Larger particles which fall off the end go back to be rolled again.
- Green teas skip the fermentation stage and go directly to the drying room.
- Leaves for black and oolong teas however, are spread evenly on special raised flat beds to allow fermentation to the desired degree.
- Although the process has been traditionally called "fermentation", it is actually an oxidation process that depends on tea enzymes reacting to oxygen.
- Temperature control is critical.
- Cooling ducts at the top of the room help to maintain the desired temperature.
- Chemical reactions taking place in the leaves directly affect the cup quality and determine the aroma and briskness of the brew.
- Fermentation must be strictly regulated to produce good tea.
- Next, tea leaves travel up a conveyor belt to the dryer.
- They pass through the dryer at controlled temperature for about 24 minutes or longer, if necessary.
- Goal is to stop the fermentation process and transform the tea into a stable product.

- After drying, tea must cool down before the sorting and grading process.
- If the sorting takes place while the tea is still warm, it can lose its character and bloom.
- First step is passing the bulk tea through machine sorters which do most of the size grading - whole leaf, broken leaf, fannings.
- Then, hand graded to further fineness.
- Tea all comes from the same plant, but the grading separates the leaves by size.
- Broken leaves and fannings work well in tea bags, while whole leaves are reserved mostly for loose tea brewing.

I. Raw material

- Two to three leaves and a bud is recommended as raw material for processing black tea in South India.
- Reason is that chemicals which contribute to the quality contained in the tender leaves are in greater proportion.
- Therefore, recommended shoot composition in South India for Quality Black Tea manufacture are:
  - Two/three leaves and a bud and single/two leaf banjis: 85%
  - Immature shoots: 5%
  - Coarse leaf: 10%

Over ten per cent of coarse leaf in the shoot composition

- Affects the chemical composition of the shoots.
- Withers very slowly, leads to uneven wither which primarily causes flake and mixed infusion.
- Complete maceration of coarse leaf is also difficult and results in improper cell disruption and poor fermentation.

II. Withering (Physical/Chemical withering)

- Physical withering is to remove the moisture from the green leaf and condition the leaf from turgid to elastic to facilitate better rolling.
- In addition to physical changes, the following biochemical changes occur to get better quality:
  - Polyphenol oxidase activity is increased which enhances the rate of fermentation.
  - Proteins are hydrolyzed to amino acid which is responsible for aroma of the tea.

- Caffeine content is optimized which influences black tea quality.
- Level of organic acids is enhanced which contribute to flavour of tea.

Quality surveillance steps during withering process

- Use ambient air when hygrometric difference is above 3°C.
- If the hygrometric difference is below 3°C, use hot air with suitable proportion of ambient air.
- Temperature of hot air should be below 35°C.
- To get even wither, the trough should be loaded with 2-3 kg/sq. ft of green leaf.
- Well ventilated lofts and access for drawing large volume of air is essential.
- Ideal qualities of air for withering are low dry bulb temperature with high hygrometric difference.
- Use reversible air flow system to get better withering and to avoid more handling of leaves manually.
- Tapering cross section in the plenum chamber helps to maintain uniform pressure throughout the length of the trough.

III. Rolling

- The primary objective of rolling is to crush the leaf into small sizes.
- In addition, it presses out the juice and a thin film of juice is coated on the surface of the leaf particle to enhance chemical changes.

Quality assurance in the rolling process

- Use green leaf sifter to avoid jamming of machine.
- Use leaf shredder to avoid over heating of dhool in the rotorvane.
- A ball breaker is essential after the rotorvane, which will cool the tea before feeding into the CTC.
- Temperature of the rotorvane tea should not reach above 35°C which reduces the enzyme activity.

CTC machine

- To maintain zero defect in quality, the following should be checked regularly
  - Temperature after each cut every two hours.
  - Size and style of the leaf from all the machines every two hours.
  - Moisture content after shredding (70-72%) and after fifth cut (60%).
  - Magnet may be fixed on all the conveyers
  - Ensure daily cleaning of CTC machine.

IV. Fermentation

- Fermentation is to bring about the chemical changes necessary to make tea liquor palatable.
- Oxidation process by which the rolled leaf loses its green colour and acquires copper colour.
- Also results in the formation of aroma of the fermented tea.

Error prevention process during fermentation

- Co-current air flow arrangement is essential to supply more oxygen during the initial stage of fermentation.
- Air should be fresh, cool and saturated with moisture.
- Optimum temperature of dry bulb should range between 25°C and 27°C and the relative humidity of the air should be around 95%.
- A post fermentation ball breaker is essential.
- Fermenting drum must be cleaned with hot water or steam to avoid bacterial contamination.

V. Drying

- Drying process arrests fermentation and reduces the moisture content in the fermented tea from 55-60% to 2-3%.

Key points for value-adding to quality during drying

- Maintaining exact inlet and exhaust temperature is important to get uniform firing.
- Proper loading of drier is also essential to get uniform firing.
  - Over loading causes back pressure and results in uneven drying.
- Maintaining drying time and temperature is very important.
  - Longer the drying time - detrimental to quality and flavour.
  - Shorter the time - results in thin liquor.
  - Higher the temperature - results in bitter tea.
- In the ECP drier, optimum inlet temperature for CTC process is 100±5°C, while exhaust temperature is maintained at 55°C.
- In the FBD drier, desirable inlet temperature ranges from 120-125°C, while the exhaust temperature is maintained at 60-65°C.
- Magnet can be fixed in the drier mouth, which will help in removing iron particles from made tea.

VI. Sorting and grading

- Sorting is to remove fibre from the bulk tea and separate the tea particles into various shapes and sizes relevant to trade requirements.
- Sorting enhances the appearances and quality of liquor.
- Flakes of coarse leaf in the primary grade can be separated to avoid harshness.
- Use of stainless steel meshes is desirable for good sorting.
- Meshes should always be tight; loose meshes cause loss of bloom and uneven grading.

VII. Storage and packing

- Black tea, being a highly hygroscopic material picks up moisture from atmosphere, if packing is not proper.
    - Three per cent (3%) moisture level is advisable when packed; if the moisture content of tea reaches 6%, the keeping quality is impaired.
    - Should not be stored in open condition.
    - Use of bins for storage of the finished product is very essential.
    - Should be cooled to ambient condition, before it is taken to bins.
    - Can be passed under powerful magnet grills to ensure iron free tea while packing.

4.2.4. Preparation of herbal tea

Step 1 : Collection of herbs

Step 2 : Fresh and dried herbs in various forms (Sepals, petals, barks, leaves etc).

Step 3 : Herbs taken in 1, 2, 3, 5 and 7 percentages.

Step 4 : Boiled in 500 ml of water for 20 minutes in low flame.

Step 5 : Test dust (3 g) boiled in 300 ml of water for 10 minutes.

Step 6 : Tea decoction and decoction of the herbs are mixed in 4:1 ratio.

Step 7 : Herbal tea.

4.3. Coffee

- Coffee is an important beverage used all over the world.

4.3.1. Processing procedure

- Coffee processing consists of removing the skin, pulp, parchment and silver screen.
- Quality of the final product depends upon the manner of processing.

- Curing process that prepares coffee beans for market.
- Two methods are employed for processing - the dry and wet.
    - In the dry method, beans are sun dried.
    - In the wet method, known as the washed coffee process, pectin enzymes are used on selectively
- Picked cherries to replace spontaneous fermentation.

4.3.2. Coffee curing process

Step 1 : Coffee berries

Step 2 : Dry method and wet method

Step 3 : Sun dried and pulp is removed by pulper

Step 4 : Outer skin and parchment dry up. Mucilage is removed by spontaneous fermentation or added enzyme and seeds are dried.

Step 5 : Pounded, hulled and seeds are dried

Step 6 : Polishing by rubbing the beans against each other and removing the outer silvery skin

- Cured coffee is graded according to size and shape. Different plantation grades are
    - Pea berry (oval shaped beans) or A (first size in flats - bold, heavy and well formed)
    - B (slightly smaller than O or A)
    - C (slightly smaller than B)
    - Triage (pale, discoloured, black spotted beans including bits)
- Manufacture of coffee powder involves roasting, grinding, blending and packing.

Roasting

- During roasting, many physical and chemical changes occur which develop pleasant aroma, flavour, brown colour and good taste.

Grinding

- Roasted beans are ground to three sizes namely, fine, medium and coarse.
- Coarse ground powder retains aroma and flavour better and longer than fine ground powder.
- Coarse ground powder is more suitable for preparing coffee decoction by percolation.

Blending

- Two types of coffee powder are marketed namely pure coffee prepared from coffee beans only.

- French coffee containing chicory.
- Proportion of chicory should not exceed 50%.
- Strength, flavour, aroma and acidity are the chief criteria in judging the quality of coffee and judicious blending of different grades bring out these qualities at the best advantage.

Coffee beverage

- Coffee has no food value.
- Constituents that are of chief significance in the making of the beverage are caffeine, flavour substances and bitter substances.
- Several factors like water, temperature, material from which pot is made, agitation affects the quality of coffee.

4.3.3. Caffeine

- Alkaloid substance producing the stimulating property.
- The longer the brewing time of coffee, the more is caffeine extracted.
- Can be removed chemically from the bean to produce decaffeinated coffee.

4.3.4. Flavour

- Sulphur compounds that are the main contributors to the flavour.
- Flavour substances are volatile.

4.3.5. Bitter substances

- Polyphenol substances or tannins are hot water soluble.
- Hence, longer the coffee is brewed, the greater will be the tannin content and the bitterness becomes pronounced.

4.3.6. Methods of coffee

Method	Flavour	Tannins	Limitations
Filtration	Retains well	Extracts less	Decoction may not be hot
Vacuum coffee maker	Retains well	Extraction is more due to the contact with the grounds at a high temperature for few minutes.	More bitter
Percolator	Loss of flavour due to constant aeration of the brew as the liquid is forced up and sprayed the grounds	Extraction is more due to recirculation of hot water through the coffee grounds.	More bitter
Steeping	Retains best	Extracts less	Extraction of flavour may not complete

4.3.7. Coffee products

1. Soluble coffee

- Dried, powdered and water soluble solid made from very strong coffee brew.
- Marketed as instant and is freeze dried coffee.

2. Iced coffee

- A sparkling iced beverage that possesses that maximum amount of flavour is made by pouring a freshly made strong coffee infusion over crushed ice in a glass.

3. Coffee tablets

- Marketed in the form of tablets.
- Ground coffee is mixed with chicory powder and mixture pressed in a machine after the addition of a binding material such as glucose.

4. Decaffeinated coffee

- By a chemical process, most of the caffeine can be removed from the beans to give decaffeinated coffee which has good flavor.

4.4. Nuts

- Seeds or fruits consisting of an edible fat containing kernel and surrounded by a hard or a brittle shell.
- Have rich flavour and have a great commercial importance.
- Important ingredient in the chocolate and confectionery industry.
- Compatible in dark or milk chocolate because of their low moisture content.
- Flavour, particularly, if they are roasted, also combines well with chocolate and reduces sweetness.
- Nuts ground to pastes or to small particles are used in such confections as marzipan, persipan, noisette, and praline pastes.

4.4.1. Nutritive value

- Oil seed nuts are rich in protein and in addition they contain a high level of fat.
- Concentrated source of energy.
- Do not contain an appreciable amount of carbohydrate, but contain high level of B vitamins.
- In India, nuts like walnut, cashew nut, chestnut, almond are more commonly grown for edible oil purpose.

4.4.2. Varieties of nuts

1. Almonds

- Two types of almonds - sweet and bitter.
- Sweet almonds are used by the confectionery industry and for desserts
- Bitter almonds are used in small amounts for flavouring almond flours and pastes.
- Almonds may be eaten raw or roasted.
- May be blanched by immersion in boiling water; brown skins, which are easily removed by squeezing the nuts between rubber rollers or bands.
- Blanched almonds can be used for decoration after drying or light roasting, but more often are ground and passed through rollers for the manufacture of almond paste or marzipan.
- Blanching process is also essential for the destruction of enzymes and microorganisms when the nuts are subsequently processed to confectionery pastes.
- Ingredient of toffee brittles when they are roasted in high boiled syrups.
- Brittles may be ground to a paste called praline nougat and becomes an ingredient of other confections such as truffles and noisette.

2. Brazil nuts

- Tropical nut greatly favoured for dessert purposes and very popular as an ingredient in chocolate bars and blocks, both as a whole nut and chopped.
- Smaller Brazil nuts are covered in chocolate by enrobing or panning for use in chocolate assortments (Native of Brazil).

3. Cashew nuts

- Shell of the nut is really a tough skin and contains a juice that causes skin blisters.
- Roasting destroys the blistering compound and nuts can then, be opened by hand cracking.
- Roasting improves the flavour.
- Used to be carried out by the pickers who heated a few pounds at a time in a pan over an open fire, but mechanical roasters are now established in the larger factories.
- Have a soft, mealy texture and have gained greatly in popularity as salted nuts, but their use in confectionery has so far been confined to nut pastes and pralines.

4. Chestnuts

- Very popular in some countries
- Found very little application in the confectionery industry except for marron glaces.
- Made by heating the chestnuts in syrups of gradually increasing concentration until they are thoroughly impregnated when they can be packed attractively and will keep as well as any other sweet.

5. Pecans

- Resembles the walnut in appearance. Large quantities of the nut are consumed in the United States of America.
- Popular with vegetarians and also used in confectionery mainly for topping chocolate and cakes.

6. Pistachio nuts

- Probably better known for their appearance than for their eating qualities.
- Pale green throughout and are attractive as toppings or may be included in nougat with cherries and citrus peel, they make a colourful candy.
- Also an ingredient of the original Turkish delight.
- In India, nut is regarded as a delicacy and many dishes are prepared from it.

7. Walnuts

- Gathered nuts are washed, cured and dried (artificial drying: 38 - 43°C).
- Fresh kernel has a bitter skin, but this bitterness disappears with keeping.
- Mechanical methods of cracking and the cleaning and sorting of the kernels are complicated because of their intricate shape and 3 basic grades of walnut come on to the market such as halves, quarters and pieces.
- Halves are used for decoration of the tops of assortment chocolates, walnut whirls and cakes
- Quarters can be used in confectionery centers and cakes
- Pieces are used for inclusion in chocolate blocks, cakes and caramels.

4.4.3. Forms of nuts used in cookery

- For kitchen use, the following types of prepared nuts are available
  - Ground - Almonds
  - Nibbed - Almonds, peanuts for decorating iced dishes, pastry goods
  - Flaked - Almonds for decoration of cakes and pastries halwa.
  - Halved - Almond, walnuts for salads, in cakes, decorating cakes.

- Whole - Almonds, hazelnuts for decorating cakes and pastries.
- Desiccated - Coconut for flavouring cakes, decorating cakes, curries.
- Essence - Almond for flavouring cakes and fillings.

4.4.4. Role of nuts in cookery

- Used as fresh, raw, roasted or boiled or salted forms and also fried forms.
- Used as thickening agents.
- Coconut, poppy seeds and cashewnut are used as thickening agents in the preparation of gravy.
- Chutneys can be made, e.g. groundnut and coconut.
- Sweets are made from nuts, e.g. chikki, burfi, kozhukattai, cashewnut cake.
- Oil is used as cooking media for frying and seasoning.
- Oil is also used as preservative in pickles.
- Powders made out of nuts like groundnut and coconuts are used as chutneys and salad dressing.
- Also used in ice creams, cakes, pastries, payasam and chocolate.
- Also used in beverages, e.g. badam kheer.
- Used as garnishing material as raw, roasted, salted or boiled forms.

4.5. Cocoa

- Chocolates are made by grinding the seeds of the pods of the cocoa tree (*Theobroma* cocoa)
- Cocoa beans are the primary raw material for confectioneries, beverages, chocolates and other edible products.
- Majority of the processed cocoa products are consumed within India.

4.5.1. Steps in processing of cocoa

1. Fermentation

- Fermentation is essential to remove the mucilaginous pulp to develop flavour and aroma precursors, reduce bitterness and kill the germ of the seed and to loosen the testa.
- Heap, box tray and basket methods are considered as the standard methods.

Heap method

- Involves keeping a mass of not less than 50 kg of wet beans over a layer of banana leaves.
- Banana leaves are spread over a few sticks to keep them a little raised over the ground level to facilitate the flow of sweating.

- Leaves are folded and kept over a heap of beans and a few wooden pieces kept over it to keep leaves in position.
- Heaps are dismantled and beans mixed on the third and fifth days.
- Needs about six days for completion of fermentation and beans can be taken out for drying on the seventh day.
- Minimum quantity of beans required for effective fermentation is 50 kg; a further increase in quantity of beans in a heap will be beneficial.
- Heaps of more than about 500 kg may be difficult to handle.

Tray method

- Wooden trays of size 90 x 60 x 13 cm with battens or reapers fixed at bottom with gaps in between are filled with beans.
- Each tray can contain about 45 kg wet beans.
- Six such trays are stacked one over the other
- Empty tray is kept at bottom to allow for drainage of sweating.
- After stacking, beans of the top most trays are kept covered with banana leaves.
- After 24 hours of setting, stack of trays is kept covered with gunny sacking to conserve heat that develops.
- There is no need for mixing the beans and fermentation will be completed in four days.
- On the fifth day, beans can be taken out for drying.
- Minimum number of trays required to be stacked is about six, but as many as 12 trays can be used simultaneously.

Box method

- Wooden boxes of 1.2 x 0.95 x 0.75 m with holes at the bottom and sides of the box are filled with wet beans.
- Boxes can hold one metric ton of wet beans.
- Beans are to be mixed on alternate days.
- As the quantity of beans is high, best done by changing the beans from one box to another at the time of mixing.
- This would necessitate having a minimum of three boxes.
- Wet beans taken for fermentation should be sufficiently ripe so as to separate the beans from the polacuta and husk easily.
- Minimum quantity of wet beans for a normal fermentation is about 100 kg.

- Duration of fermentation is commonly for 3-5 days *i.e.,* 72-120 hrs.
- Fermentation over 120 hrs will cause loss of chocolate flavour and development of off flavour.

2. Drying and storage

- Fermented beans can be dried either in the sun or by artificial means.
- Sun drying can be done in thin layers of 2-3 cm depth and stirring from time to time.
- When the beans are dried properly, they produce a characteristic cracking sound on compressing a fistful of beans in the palm.
- More scientific method is to use moisture meter.
- Dried beans after cooling maintaining 6-8% moisture should be cleared before storage.
- Fruits broken, shrivelled and other extraneous materials are removed.
- Cleared bags are kept on a raised platform of wooden planks.

3. Roasting

- Roasting develops characteristic flavour and colour.
- Also causes change in the chemical structure of polyphenols producing less astringency compounds.
- Roasting also changes the colour to dark brown.

4. Grinding and defatting

- Roasted beans are cooled and gently crushed soon after, to fracture husk and break down the kernels into their natural angular fragment or nibs.
- Roasted nibs are ground using stone mills to a fine paste or liquor.
- Heat produced during grinding causes the cacao fat to melt.

4.5.2. Percentage nutritive value of cocoa and chocolate

Constituents	Cocoa	Chocolate
Water	4.6	5.9
Protein	21.6	12.9
Fat	28.6	48.7
Carbohydrate	37.7	30.0

4.5.3. Steps for cocoa and chocolate manufacturing

1. Cocoa

Step 1 : Fermented and dried beans

Step 2 : Cleaning

Step 3 : Roasting

Step 4 : Breaking and winnowing

Step 5 : Nib-shell mixtures

Step 6 : Germ separation

Step 7 : Germ-free nib

Step 8 : Milling

Step 9 : Cocoa-mass (Chocolate liquor)

Step 10 : Cocoa manufacture (alkalization)

Step 11 : Removal of excess moisture

Step 12 : Fat process

Step 13 : Press cake

Step 14 : Breaking

Step 15 : Grinding

Step 16 : Sitting

Step 17 : Cocoa powder

2. Chocolate

Step 1 : Fermented and dried beans

Step 2 : Cleaning

Step 3 : Roasting

Step 4 : Breaking and winnowing

Step 5 : Nib-shell

Step 6 : Germ separation

Step 7 : Germ-free nib

Step 8 : Milling

Step 9 : Cocoa-mass (Chocolate liquor)

Step 10 : Addition of sugar, flavor, milk etc. and cocoa butter

Step 11 : Mixing

Step 12 : Refining

Step 13 : Conching

Step 14 : Tempering

Step 15 : Moulding and enrobing

Step 16 : Plain or milk chocolate and chocolate coated goods

3. Fudge

- Generally prepared from brown sugar.
- As brown sugar contains a higher percentage of invert sugar than white sugar, sucrose crystallizes less readily.
- Fudge ripens during storage and becomes soft and velvetty after 24 hrs storage.

Ingredients

Sugar	200 g
Milk	120 g
Butter	14 g
Chocolate	21 g

Procedure

- Add chocolate and butter to cooking pan and heats it on a steam bath till the chocolate and butter have melted and add sugar.
- Mix chocolate and butter well with sugar.
- Then, add milk and heat till the sugar dissolves completely.
- Cook the syrup to a soft ball stage (112°C).
- Allow it to cool to about 70°C and transfer to a greased moulding pan.
- Cut thc candy when it is cool and wrap in butter or foil and store in an air-tight container.

5

Value Addition in Medicinal Plants

5.1. Aonla

- Fruit of this deciduous tree found mainly in India.
- Aonla is rich in natural vitamin C.
- Aonla has cooling, diuretic and laxative properties.
- It helps in cleansing the mouth, and strengthens teeth and bones.
- It increases the red blood cell count and helps to promote good health.
- It also has an antioxidant properties.
- The dried aonla fruit is useful in the treatment of hemorrhage, diarrhoea, and dysentery.
- Aonla has laxative properties and therefore, useful in the treatment of constipation and piles.
- It also has antibacterial properties and helps in preventing infections and healing ulcers.
- It also helps in preventing skin infections.
- Aonla has cell rejuvenating properties and therefore, used in maintaining good health of skin and hair.
- Aonla helps to keep the hair glossy and shining.
- It also helps to prevent dandruff.
- Aonla is also known to have anti aging properties.
- Apart from vitamin C, aonla also provides many other essential minerals to the body.
- These minerals help in maintaining proper functioning of the metabolic activities of the body.
- Aonla contains minerals such as chromium, zinc, and copper.
- The chromium present in it helps in anti diabetic effect.
- Aonla is also used in the treatment of gastritis.
- It is found to be very effective in the treatment of heart burn due to increased hydrochloric acid secretion of the stomach.

5.1.1. Aonla preserve

- A mature fruit / vegetable or its pieces impregnated with heavy sugar syrup till it becomes tender and transparent is known as preserve.
- FPO specifications for preserve are TSS - 68°B and fruit pulp - 55%.

Ingredients

Aonla	1 kg
Sugar	1 kg
Alum	20 g
Salt	20 g
Citric acid	1.5 g
Water	1 lit

- Prick with fork, needle or goose berry pricker.
- Steep in 2% salt solution for 24 hrs to remove astringency.
- Wash and dip in 2% alum solution for 24 hours and then, wash thoroughly.
- Blanch until soft, but segments do not break or crack.

General considerations

- Cooking of aonla directly in syrup causes shrinking of fruit and reduces absorption of sugar.
- Fruit should be blanched first to make it soft enough to absorb water, before steeping in syrup.
- Fruits may be cooked in syrup by two processes as given below:

1. Rapid process

- Aonla are cooked in low sugar syrup.
- Boiling is continued with gentle heating until the syrup becomes sufficiently thick.
- Rapid boiling should, however be avoided as it makes the fruit tough, especially when heating is done in a large shallow pan with only a small quantity of syrup.
- Final concentration of sugar should not be less than 68% which corresponds to a boiling point of 106°C.
- Simple and cheap process, but the flavour and colour of the product are lost considerably during boiling.

2. Slow process

- Aonla is blanched until it becomes soft.

- Sugar, equal to the weight of fruit, is then, added to the fruit in alternate layers and the mixture is allowed to stand for 24 hrs.
- During this period, aonla gives out water and sugar goes into solution, resulting in a syrup containing 37-38% TSS.
- Next day, syrup is boiled after removal of aonla to raise its strength to about 60% TSS.
- A small quantity of citric acid (1 to 1.5 g/ kg sugar) is also added to invert a portion of cane sugar and thus, prevent crystallization.
- Whole mass is then, boiled for 4-5 min. and kept for 24 hrs.
- On the third day, strength of syrup is raised to about 65% TSS by boiling.
- Fruit is then, left in the syrup for a day.
- Finally, strength of the syrup is raised to 70% TSS and aonla are left in it for a week.
- Preserve is now ready and is packed in containers.

5.1.2. Aonla candy

Ingredients

Aonla	1 kg
Sugar	1.120 kg
Water	500 ml
Citric acid	6.4 g
KMS	1.2 g

Method

- Preparation of sugar syrup (addition of 765 g of sugar in 500ml of water).
- Addition of citric acid and KMS.
- Soaking the fruit for 24 hrs.
- Boiling of sugar syrup to 60° B.
- Addition of remaining sugar.
- Soaking for 24 hr and increased the brix to 70°B for 7 days.
- Aonla pieces and sugar should be 1:1.5 ratio.
- Packed the aonla preserve in glass jar.
- Dried in shade to get aonla candy.

5.1.3. Aonla spiced candy

Ingredients

Aonla	1 kg
Sugar	300 g
Water	700 ml
Chilli powder	20 g
Salt	20 g
Asafoetida	5 g

Method

- Select well matured and sound aonla fruits
- Wash and soak in 2 per cent brine for 24 hrs
- Steam aonla fruits for 10 minutes and cut into segments.
- Mix aonla with the other ingredients and boil for 30 minutes.
- Separate the boiled aonla candy and dry at 60°C for 5 hours in a mechanical drier.

5.1.4. Aonla jam

Ingredients

Aonla, guava	1 kg
Water	150 ml
Sugar	750 g
KMS	40 ppm
Pectin	1g

Step 1 : Selection of sound fruits

Step 2 : Cutting into pieces

Step 3 : Pulping

Step 4 : Addition of sugar, pectin (add water, if necessary)

Step 5 : Boiling

Step 6 : Judging of end point (68° B/ temperature105°C)

Step 7 : Addition of KMS

Step 8 : Filling into sterilized bottles

Step 9 : Cooling

Step 10 : Storage

5.1.5. Aonla juice

Aonla	1 kg
Water	1 lit
KMS	6.5 g

Step 1 : Fruits
Step 2 : Washing
Step 3 : Seed removal and separation into segments
Step 4 : Addition of water (1:1)
Step 5 : Passing through pulper
Step 6 : Filtering
Step 7 : Boiling at temperature 78°C
Step 8 : Addition of KMS
Step 9 : Filling in sterilized bottles
Step 10 : Pasteurization for 20 minutes
Step 11 : Cooling

5.1.6. Aonla RTS

Aonla juice	1 lit
Lemon juice	200 ml
Ginger juice	100 ml
Sugar	1.6 kg
Citric acid	22 g
Water	7.7 lit

Step 1 : Aonla juice
Step 2 : Addition of sugar, citric acid, water
Step 3 : Mixing of juice with lemo and ginger juices
Step 4 : Filling in sterilized bottles
Step 5 : Pasteurization for 20 minutes
Step 6 : Cooling

5.1.7. Aonla squash

Step 1 : Fruits
Step 2 : Washing
Step 3 : Steam blanching (5 to 10 min)
Step 4 : Cooling and separating the seeds

Step 5 : Extracting
Step 6 : Boiling (Sugar, water and citric acid)
Step 7 : Filtering
Step 8 : Cooling (Sugar syrup)
Step 9 : Mixing with juice
Step 10 : Adding colour and preservatives
Step 11 : Filling
Step 12 : Labeling
Step 13 : Marketing

5.1.8. Aonla sauce

Ingredients

Aonla pulp	1 kg
Sugar	75 g
Salt	10 g
Onion	50 g
Garlic	5 g
Chilli powder	5 g
Cinnamon, cardamom, aniseed, cumin	10 g each

Step 1 : Aonla juice
Step 2 : Addition of sugar, citric acid, water
Step 3 : Mixing of juice with other ingredients
Step 4 : Filling in sterilized bottles
Step 5 : Pasteurization for 20 minutes
Step 6 : Cooling

5.1.9. Aonla pickle

Aonla	1 kg
Salt	150 g
Turmeric powder	10 g
Red chilli powder	10 g
Fenugreek	30 g
Clove	5 nos.
Oil	350 ml

Step 1 : Aonla
Step 2 : Washing
Step 3 : Boiling for 15 minutes to soften segments

Step 4 : Separation of segments and removal of seeds

Step 5 : Frying spices in oil

Step 6 : Mixing segments with spices

Step 7 : Frying for 5 minutes

Step 8 : Mixing salt

Step 9 : Filling in jar

Step 10 : Storage

5.1.10. Aonla based mouth freshener

- Select matured, firm and fibreless aonla fruits and washed in cold water.
- Fruits are deseeded using aonla deseeder and sliced into small pieces with vegetable slicer.
- Desired spice combinations are weighed and mixed with aonla pieces.
- Soaked for 48 hrs, mixed once or twice during curing.
- Aonla + lime + salt (aonla 100 g, lime juice 15%, salt 5%)
- Aonla + ginger + salt (aonla 100 g, ginger 5%, salt 5%)
- Aonla + pepper + salt (aonla 100 g, pepper 5%, salt 5%)
- Aonla+ lime + ginger + salt (aonla 100g, lime juice 15%, ginger 5%, salt 5%)
- Aonla + lime + pepper +salt (aonla 100g, lime juice 15%, pepper 5%, salt 5%)
- Aonla + pepper + ginger + salt (aonla 100g, pepper and ginger 5%, salt 5%)
- Aonla + pepper + ginger + lime + salt (aonla 100 g, pepper 3%, ginger 3%, lime juice 15, salt 5%)
- Dried in a cabinet drier at 60 to 80°C for 4 to 5 hrs, packed and stored in an airtight container.

5.1.11. Aonla sweet candy

Step 1 : Aonla fruit

Step 2 : Dipping alternatively in 2% salt and alum solution (24 hrs)

Step 3 : Soaking in cold water (1h)

Step 4 : Blanching for 15 minutes

Step 5 : Removal of seeds

Step 6 : Segments

Step 7 : Taking equal parts of sugar

Step 8 : Separating two equal parts of sugar

Step 9 : Heating (aonla segments + 1 part of the sugar)

Step 10 : Steeping for 24 hrs

Step 11 : Remaining sugar is separated in 3 equal parts

Step 12 : One part of sugar is added to the above solution

Step 13 : Steeping for 24 hrs

Step 14 : Repeat the process for remaining two parts of sugar

Step 15 : Sugar syrup concentration increase to 75° B

Step 16 : Aonla segments washed with cold water

Step 17 : Drying in a cabinet drier (60°C for 5 hrs)

Step 18 : Packed in polyethylene bag

Step 19 : Stored at room temperature

5.1.12. Preparation of aonla-tomato blended sauce

Step 1 : Aonla fruit + tomato pulp

Step 2 : Mixing

Step 3 : Adding one third of sugar

Step 4 : Heating

Step 5 : Adding spice bag containing chopped onions, garlic, ginger, chilles and powdered cinnamon, cardamom, cumin, pepper

Step 6 : Cooling on low flame with continuous stirring till the volume reduces by about 1/3

Step 7 : Adding remaining sugar and vinegar

Step 8 : Cooking

Step 9 : Adding salt

Step 10 : Bottling

Step 11 : Capping

Step 12 : Processing in boiling water

Step 13 : Labelling

5.1.13. Preparation of aonla - payaya fruit bar

Step 1 : Aonla pulp and papaya pulp

Step 2 : Mixing

Step 3 : Adding sugar

Step 4 : Heating along with continuous stirring (80-85°C)

Step 5 : Judging the end point with refractometer

Step 6 : Spreading on trays lined with butter paper

Step 7 : Drying in cabinet drier at 60°C for 7 hrs

Step 8 : Cutting into rectangular shapes

Step 9 : Packaging

5.2. Bael

- Bael has numerous seeds which are densely covered with fibrous hair and are embedded in a thick aromatic pulp.
- Flesh is eaten fresh or dried.
- The bael fruit consists of moisture 61.5 per cent, protein 1.8 per cent, fat 0.3 per cent, minerals 1.7 per cent, fiber 2.9 per cent and carbohydrates 31.8 per cent per 100 grams of edible portion.
- Its mineral and vitamin contents include calcium, phosphorus, iron, carotene, riboflavin, thiamin, niacin and vitamin C.

5.2.1. Food uses

- Ripe bael fruit is regarded as the best natural laxative.
- For best results, it should be taken in the form of sherbet, which is prepared from the pulp of the ripe fruit.
- After breaking the shell, the seeds are removed, with the contents spooned out and sieved.
- Milk and sugar are added to make it more palatable.
- The pulp of the ripe fruit can also be taken without the addition of milk or sugar.
- The unripe or half-ripe fruit is very effective in treating chronic diarrhoea and dysentery where there is no fever.
- Best results are obtained by the use of dried bael or its powder.
- The bael fruit, when it is still green, is sliced and dried in the sun.
- The dried bael slices are powdered and preserved in airtight bottles.
- The unripe bael can also be baked and used with jaggery or brown sugar.
- The pulp is often processed as nectar or squash (diluted nectar).
- A popular drink (called sherbet in India) is made by beating the seeded pulp together with milk and sugar.
- A beverage is also made by combining bael fruit pulp with that of tamarind. These drinks are consumed perhaps less as food or refreshment than for their medicinal effects.

- Mature, but still unripe fruits are made into jam, with the addition of citric acid. The pulp is also converted into marmalade or syrup for both food and therapeutic use and the marmalade being eaten at breakfast by those convalescing from diarrhoea and dysentery.
- Bael pulp is steeped in water, strained, preserved with 350 ppm $S0_2$, blended with 30% sugar, then dehydrated for 15 hrs at 120° F (48.89°C) and pulverized. The powder is enriched with 66 mg ascorbic acid per 100 g and can be stored for 3 months for use in making cold drinks (squashes).
- A confection, bael fruit toffee, is prepared by combining the pulp with sugar, glucose, skim milk powder and hydrogenated fat. Indian food technologists view the prospects for expanded bael fruit processing as highly promising.

5.2.2. Therapeutic bael fruit beverage

- Important indigenous fruit of India.
- One of the most therapeutic and nutritious fruits.
- Has been known in India from pre-historic times.

Ingredients

Bael fruit pulp	1.0 kg
Jaggery	880 g
Water	4100 ml
Citric acid	18 g
Potassium meta bisulphite	2.8 g

Method

- Fully matured ripe fruits are washed, broken; pulp along with seeds and fibres is scooped.
- Equal quantity of water is mixed with bael fruit pulp to extract the pulp.
- Extracted pulp is heated at 80°C for one minute passed through hand pulper.
- Required amount of water and citric acid are dissolved in water by gentle heating, strained through muslin cloth and mixed thoroughly with bael fruit pulp.
- Beverage is filled in sterilized bottles, capped, processed in boiling water at 80°C for 20 minutes.
- Beverage is cooled quickly to room temperature, wiped, dry and stored in a cool dry environment.

5.2.3. Bael toffee

Step 1 : Bael fruit pulp (500 g) + water (100 ml)

Step 2 : Cooking (20 minutes)

Step 3 : Adding sugar (700 g)
Step 4 : Pouring ghee (50 g) in a thick bottomed pan
Step 5 : Adding pulp
Step 6 : Dissolving 25 g corn flour, 200 g milk powder, 2.5 g citric acid in water and adding to the pulp
Step 7 : Cooking (20 minutes)
Step 8 : Testing consistency (soft ball stage)
Step 9 : Pouring in greased tray
Step 10 : Shaping (small balls)
Step 11 : Toffee

5.2.4. Processing of cereal bar

Step 1 : Bael fruit pulp (500 g) + water (100 ml)
Step 2 : Cooking (20 minutes)
Step 3 : Adding sugar (700 g)
Step 4 : Pouring ghee (50 g) in a thick bottomed pan
Step 5 : Adding pulp
Step 6 : Dissolving 25 g corn flour, 200 g milk powder, 2.5 g citric acid in water and adding to the pulp
Step 7 : Cooking (20 minutes)
Step 8 : Testing consistency (soft ball stage)
Step 9 : Pouring in greased tray
Step 10 : Shaping (small balls)
Step 11 : Toffee

5.3. Stevia

- Calorie free sweetener and flavour enhancer; it contains a variety of constituents besides the steviosides and rebaudiosides.
- Contains sterols, triterpenes, flavonoids, tannins, and an extremely rich volatile oil comprises rich proportions of aromatics, aldehyde, monoterpenes and sesqui terpenes.
- Beneficial in regulating blood sugar levels, bringing them into normal range.
- Also used as a digestive aid.
- Used to clear blemishes, tighten skin to remove wrinkles, to heal mouth sores and to treat a variety of wounds.
- Used to treat eczema, seborrhea and dermatitis.

1. **Hypoglycemic action**

- Helpful for hypoglycemia and diabetes, because it nourishes the pancreas and thereby, helps to restore normal pancreatic function.

2. **Cardiovascular action**

- Cardiotonic action, that is, would produce, a mild strengthening of the heart and vascular system.

3. **Antimicrobial action**

- Ability of stevia to inhibit the growth and reproduction of bacteria and other infectious organisms is important.
- Lower the incidence of dental caries.

4. **Digestive tonic action**

- Improves digestion, and that it improved overall gastrointestinal function.
- Stevia tea, made from either hot or cold water, is used as a low calorie, sweet tasting tea, as an appetite stimulant, as a digestive aid and as an aid to weight management.

5.3.1. Preparation of stevia tea

Step 1 : Milk + water

Step 2 : Boiling

Step 3 : Addition of tea + stevia powder

Step 4 : Boiling (2 minutes)

Step 5 : Straining

Step 6 : Serving

5.3.2. Preparation of grape squash using stevia

Step 1 : Selection of ripe grapes

Step 2 : Washing and cleaning

Step 3 : Pulping with hand pulper

Step 4 : Filter through muslin cloth

Step 5 : Stevia solution (stevia + citric acid + water)

Step 6 : Boiling (2 minutes)

Step 7 : Filtering through muslin cloth

Step 8 : Mixing syrup with juice

Step 9 : Adding essence

Step 10 : Filling in sterilized bottles (head space ¼")

Step 11 : Crowning the bottles

Step 12 : Labelling

Step 13 : Storing (at room temperature)

5.3.3. Preparation of grape RTS beverage using stevia

Step 1 : Selection of ripe grapes

Step 2 : Washing and cleaning

Step 3 : Pulping with hand pulper

Step 4 : Filter through muslin cloth

Step 5 : Adding stevia and citric acid (0.3%)

Step 6 : Boiling at 80°C (5 min)

Step 7 : Adding essence

Step 8 : Filling in sterilized bottles (head space ½")

Step 9 : Crown/Corking

Step 10 : Pasteurizing in boiling water (30 minutes)

Step 11 : Cooling

Step 12 : Labelling

Step 13 : Storing (at room temperature)

5.3.4. Preparation of gulab jamun in stevia syrup

Step 1 : Gulab jamun mix

Step 2 : Mixing with dalda

Step 3 : Adding milk

Step 4 : Making dough

Step 5 : Portioning into small balls

Step 6 : Frying at 18°C (5-7 minutes)

Step 7 : Sweet solution (stevia + water)

Step 8 : Dipping jamun in hot solution for 1 hour.

Step 9 : Cooling

Step 10 : Serving

5.3.5. Preparation of kesari with stevia

Step 1 : Melting ghee in a pan

Step 2 : Frying till light brown

Step 3 : Adding water + stevia

Step 4 : Boiling till rawa is cooked

Step 5 : Adding cardamom, cashew, raisins and colour

Step 6 : Serving hot

5.3.6. Preparation of sweet biscuits with stevia

Step 1 : Creaming shortening + stevia powder

Step 2 : Shifting maida and baking powder

Step 3 : Sieve twice

Step 4 : Mixing of cream + maida + baking powder + salt

Step 5 : Adding water and essence

Step 6 : Mixing of ingredients of dough

Step 7 : Rolling the dough to a sheet of 0.3 cm thick

Step 8 : Sizing of dough into biscuit using biscuit cutter

Step 9 : Arranging on greased and dusted tray

Step 10 : Baking at 160°C for 25-30 minutes.

Step 11 : Cooling

Step 12 : Packing

Step 13 : Storing (at room temperature)

5.3.7. Preparation of cake with stevia

Step 1 : Maida + baking powder

Step 2 : Sieve twice

Step 3 : Creaming of fat and sugar

Step 4 : Beating egg with essence

Step 5 : Mixing egg foam with cream

Step 6 : Mixing maida with cream

Step 7 : Adding milk

Step 8 : Baking in greased pans at 120°C (25-30 minutes)

Step 9 : Cooling

Step 10 : Serving